Grassland Invertebrates

Grassland Invertebrates

Ecology, influence on soil fertility and effects on plant growth

J.P. Curry

Department of Environmental Resource Management
University College Dublin
Ireland

CHAPMAN & HALL

London · Glasgow · New York · Tokyo · Melbourne · Madras

Published by Chapman & Hall, 2–6 Boundary Row, London SE1 8HN, UK

Chapman & Hall, 2–6 Boundary Row, London SE1 8HN, UK

Blackie Academic & Professional, Wester Cleddens Road, Bishopbriggs, Glasgow G64 2NZ, UK

Chapman & Hall Inc., One Penn Plaza, 41st Floor, New York NY10119, USA

Chapman & Hall Japan, Thomson Publishing Japan, Hirakawacho Nemoto Building, 6F, 1-7-11 Hirakawa-cho, Chiyoda-ku, Tokyo 102, Japan

Chapman & Hall Australia, Thomas Nelson Australia, 102 Dodds Street, South Melbourne, Victoria 3205, Australia

Chapman & Hall India, R. Seshadri, 32 Second Main Road, CIT East, Madras 600 035, India

First edition 1994

© 1994 J.P. Curry

Typeset in 10/12 Palatino by Acorn Bookwork, Salisbury, Wiltshire
Printed in Great Britain at The University Press, Cambridge

ISBN 0 412 16520 1

A catalogue record for this book is available from the British Library

Library of Congress Cataloging-in-Publication data available

Contents

Preface

There is now a good deal of information on many aspects of the ecology of grassland invertebrates. Three major themes feature prominently in recent research on these animals. The first relates to economically important pest species, with an increasing volume of population studies carried out in response to the need to devise more ecologically sound strategies for pest management and pest control. The second is concerned with the influence of invertebrates on grassland ecosystem processes, especially the roles of invertebrates in the maintenance of soil fertility through the modification of soil properties and through involvement in the decomposition of organic matter and mineralization. The third major area of interest relates to the structure and properties of invertebrate communities – how these are organized, and how they change in response to changes in management and other factors. The objective in writing this book was to draw together material from those three major areas in an attempt to present an overview of the ecology of grassland invertebrates, with particular emphasis on their impact on soil fertility and sward productivity.

The book starts with a review of the major grassland types and grassland features of greatest importance in relation to the fauna. Next, the main faunal groups occurring in grassland are considered in turn. The nature of grassland invertebrate communities and the factors that influence their constituent populations are then considered, with particular emphasis on how populations are affected by management practices. Finally, the influence of invertebrate activities on grassland through feeding on litter and organic matter, through soil working, and through interacting in various ways with plants is dealt with and prospects for the application of pest-management principles in the control of troublesome pest species are considered. Inevitably, the coverage is heavily biased in favour of temperate grasslands, reflecting the much greater volume of information that exists for temperate than for tropical or for other habitats.

I am indebted to Dr Mark Hassell, University of East Anglia, UK, and to Dr Bob Clements of IGER, North Wyke, UK for their valuable comments on earlier drafts; to Bernard Kaye and Gordon Purvis of the department of Environmental Resource Management, UCD for their help with preparing the artwork, and to Michele Keogh and Aoife Carne for typing the various verdions of the manuscript. Finally, I wish to express my appreciation to my family and friends for their support, encouragement and patience throughout the development of this project.

J.P. Curry
Dublin

1

The grassland habitat

Estimates of the total extent of the world's grasslands depend on the definition used. Moore (1964) defined natural grassland as a plant community in which the dominant species are perennial grasses, there are few or no shrubs and trees are absent. He included in his definition any community, whether natural or developed by humans, in which grasses provide a substantial proportion of the feed for domestic stock, thus encompassing plant communities in which grasses form a substantial proportion of the ground flora and which are utilized for grazing (savannah, woodland, shrubland, etc.). Spedding (1971) takes a similarly broad view and extends the definition to include agricultural swards where legumes are a major constituent, in some cases the main constituent. Thus, broadly defined, grasslands encompass a wide range of habitat and floral types, ranging from tropical savannah to tundra, and including those arid and semiarid lands of desert steppe and prairie as well as cool temperate meadow and moorland that sustain grazing. Shantz (1954) estimated that grassland is the potential natural vegetation of 33 million km^2, occurring mainly in areas of low rainfall, in the rain shadow of major mountain ranges or between desert and forest. Whittaker and Likens (1975) cite comparable figures: savannah 15.0, temperate grassland 9, tundra and alpine meadows 8 million km^2, totalling 32 million km^2, or 21.5% of the earth's total land area.

Several different methods of classification have been applied to grassland; some of the more widely used approaches are briefly discussed below and the distribution of the main grassland types is considered. Habitat features that have a major influence on the distribution and abundance of grassland invertebrates are then reviewed.

1.1 CLASSIFICATION OF GRASSLANDS

There are many different approaches to plant classification, none of which has gained universal acceptance (Whittaker, 1978). Plant communities are usually not sharply delineated; the boundaries between adjacent communities are often arbitrary and different groupings may be arrived at depending on the criteria used in classifying them.

An early approach to formal classification that has come into wide use utilizes plant structure or physiogomy as the main criterion. Community types on a given continent defined by dominant growth forms and major environmental factors are described as formations and a grouping of similar formations found in similar climates in different continents is a formation-type. The corresponding units when the fauna are included are biome and biome-types. Temperate grassland is the characteristic grassland biome-type, comprising the prairies and desert grasslands of North America, the steppes of Eurasia, the pampas and llanos of South America, and South African veldt (Whittaker, 1975). Other biome-types with significant grassland include tropical savannah, arctic tundra and alpine habitats, temperate woodlands and shrublands, and tropical and temperate forests that have been cleared for agriculture. The physiognomic approach has been criticized in that it lumps together vegetation types that are vastly different in their ecological relations, thus resulting in an artificial classification. Its main merit is in defining major plant groupings appropriate for broadly based international studies.

By contrast, the floristic approach classifies communities on the basis of their botanical composition. This approach is agglomerative, starting with small categories of vegetation that may be combined into successively larger and more heterogenous units. A widely applied and effectively standardized example of this approach is the Braun-Blanquet system. Beard (1978) considers the floristic approach to be most suitable for studies of local areas, but considers the higher-level units of the Braun-Blanquet school to be of less and less practical value as you go up the hierarchy; he concludes that meaningful world systems of vegetation classification can be formulated only on a physiognomic basis.

Many authors combine floristic criteria for lower-level classification with formations for higher levels. An example is the Tansley (1939) system. This recognizes stable and climax communities in a given area as formations, which are divided into broadly defined 'associations' characterized by their dominant species.

The approach taken in the International Biological Programme (IBP) was a simple and practical one, essentially based on climatic and utilization criteria (Coupland, 1979). Two major climatic subdivisions were used, temperate and tropical. Temperate sites were subdivided into natural grasslands, where climate was considered to be the prime controlling factor, and seminatural pastures and meadows, which were considered to be essentially deforested areas in regions of forest climate. Arable grassland was a utilization subdivision, which could be either temperate or tropical. The seminatural grasslands are essentially anthropogenic plagioclimaxes that are maintained by human activities; when abandoned, they quickly revert to scrub and undergo succession to their climatic climax. Natural grasslands, on the other hand, are believed to be climatic climaxes, occurring in areas of low rainfall where soil moisture is insufficient to support forest growth but adequate to prevent the development of desert. For present purposes, four major grassland types are considered – tropical grasslands, prairie and steppe, semi-natural temperate grasslands and tundra – but these four categories are not mutually exclusive.

1.2 DISTRIBUTION OF MAJOR GRASSLAND TYPES

No single factor is solely responsible for the development of grassland, but climate undoubtedly has a major influence. Since climate is also a significant factor in soil development, it is not surprising that some general correlations can be seen between the distributions of climates,

Table 1.1 Major grassland types and the climatic and soil types in which they are most prominent

Grassland type	Soil class (FAO–UNESCO, 1974)	Climate (Strahler, 1970)
Tropical grasslands	Arenosols (= sands)	Tropical wet–dry, tropical desert and steppe
	Ferralsols (= latosols)	Trade-wind littoral, tropical wet–dry
Prairie/Steppe	Chernozems	Humid continental, middle-latitude desert and steppe
	Phaenozems (= brunizems, degraded chernozems)	Humid continential
	Kastanozems (chestnut)	Middle-latitude steppe
	Vertisols	Tropical and middle-latitude desert and steppe
Seminatural, temperate grasslands	Cambisols (brown earths)	Marine west coast, humid continental
	Podzoluvisols (grey-brown podzolic)	Cool moist continental
	Rendzina	Various
Tundra	Regosols (inceptisols), folisols, podzoluvisols, etc.	Arctic; continental subarctic; marine subarctic

soils and grasslands. Table 1.1 indicates the main climatic and soil types associated with the major categories of grassland; their geographical distribution is shown in Figure 1.1. Some aspects of the interrelationships between grassland, climate and soil are considered below; more detailed treatment of grassland distribution can be found in Moore (1964), Whittaker (1978), Coupland (1979) and Anderson (1982) among other sources.

1.2.1 Tropical grasslands

Tropical and subtropical grasslands occur between 30° N and 30° S latitude and are found mainly in three low-latitude climates: tropical wet–dry, trade-wind littoral and tropical desert and steppe. All three climates experience relatively high mean annual temperatures (21–27 °C), with greatest seasonal variation occurring under tropical desert and steppe conditions, where the warmest month has mean temperatures in the range of 35–38 °C and the coldest month 12–16 °C. Annual precipitation is highest in trade-wind littoral climates (1500–3000 mm per year), intermediate in tropical wet–dry (1000–1700 mm) and lowest in tropical desert and steppe (100–500 mm; desert, 10–100 mm). Precipita-

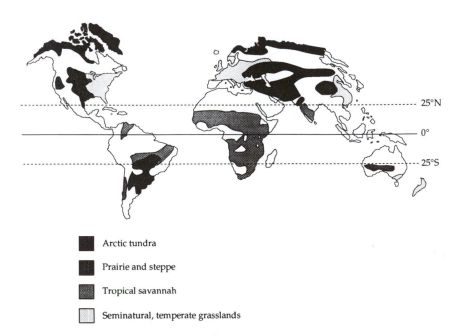

Figure 1.1   Distribution of major grassland types. (After Cox, Healey and Moore, 1976.)

tion is strongly seasonal, with a summer maximum and a pronounced dry season. In the low-rainfall areas of tropical desert and steppe, precipitation is often in the form of heavy showers. Trade-wind littoral regions experience strong winds and a fairly cool, winter dry season. The natural vegetation ranges from tropical rain forest through deciduous and semideciduous tropical forest, thorn woodland, savannah, steppe and desert depending largely on precipitation and utilization. Tropical grasslands are mainly seral, maintained at various successional stages by grazing and especially by recurrent fire. Savannah is a common formation, with a continuous grass stratum interrupted by trees and shrubs. Differences in soil, climate, age and mode of origin, intensity of grazing, periodicity of burning, etc. produce a very diverse array of grassland communities in savannah.

Tropical grasslands typically occur on arenosols and ferralsols. The arenosols are sandy soils, derived from unconsolidated deposits of aeolian, colluvial or alluvial origin that occur principally throughout east and south-east Africa, southern Sahara, central and western Australia and various parts of equatorial South America. Ferralsols are typically derived from consolidated rocks, but also form on unconsolidated deposits such as Tertiary sediments and old volcanic ash. They generally have a hard laterite horizon, contain little organic matter and are low in nutrients. They occupy large areas on both sides of the equator in Africa and South America, in northern Australia and in India.

1.2.2 Prairie and steppe grasslands

These occur mainly under tropical and middle-latitude desert and steppe, and humid continental climatic conditions. Middle-latitude desert and steppe climates occur between 35° and 50° N and S latitude and differ from their tropical counterparts mainly in temperature: the mean annual temperature is 4–16 °C and there is marked seasonal variation, from 18–27 °C for the warmest month to 1–4 °C for the coldest, and frost occurs on 180–220 days. Humid continental climates occur in the latitude range 35–60° N. Precipitation is in the range 400–700 mm yr^{-1}, with a weak summer maximum. The mean annual temperature is 2–7 °C; the warmest monthly mean is 18–24 °C and that of the coldest month is −4 to −18 °C; there are 120–140 frost-free days. Cool moist summers, heavy winter snowfalls and a wide temperature range are characteristic of this climate.

Prairie and steppe grasslands occur mainly on chernozems, phaenozems, kastanozems and (to some extent) vertisols. A characteristic feature of their development is the process of melanization or darkening of the soil by the addition of organic matter, resulting in the extension of the dark surface horizon down into the profile (Buol, Hole and McCracken,

1973). Factors involved in this process include the extension of roots down into the soil profile and the development of relatively stable dark compounds, resulting from the partial decay of organic matter, and the reworking of soil and organic matter by earthworms and other invertebrates, notably cicada nymphs. The role of the fauna in the development of prairie soils is considerable and Curtis (1959) concluded that the upper 60 cm is turned over by ants, worms and rodents once each century.

Chernozems are derived mainly from loess; they are highly fertile soils, rich in organic matter with a typical black coloration. Phaenozems likewise have high amount of organic matter and high fertility; they are derived from glacial drift, loess and alluvium. They are more widely distributed than chernozems, occurring in large areas of central North America, where they are the principal prairie soils, and in Eurasia and South Africa. Kastanozems are widespread in the semiarid regions of the world, particularly in midwestern North America and southern Russia. They have a high inherent fertility but inadequate moisture. Vertisols are clay soils with low organic matter, occurring typically in tropical arid and semiarid areas. The natural vegetation of these soils depends very much on precipitation: where soil moisture is adequate oak forest may be found on chernozems and phaenozems; in drier regions tall grasses predominate. Tall grass and short grass occur on kastanozems, depending on the degree of aridity. Under increasingly arid conditions vegetation becomes sparser, with isolated areas of bunch grass typical of more arid prairies. The predominant plant communities on vertisols are tall grass and acacia thorn forest.

According to Buol, Hole and McCracken (1973), the prairie and steppe grasslands of the middle latitudes occupy about 13 million km^2, comprising 5.5 million km^2 tall grass and 7.5 short grass. Short grass is 13–30 cm high, with patches of taller grass in unusually wet years. Feather-grass and chee grass (*Stipa*) are the main grasses in the western Russian steppe; blue grama grass (*Bouteloua gracilis*) is common on the drier brown soils of North American prairie and buffalo grass (*Buchloe dactyloides*) and western wheatgrass (*Agropyron smithii*) grow on chestnut and reddish chestnut soils. Sage brush (*Artemisia* spp.) is a major component of the flora of drier prairie. Tall grasses grow luxuriously to a height of 1–3.3 m at maturity; these are typical of undisturbed Argentine pampas. Big and little bluestem grasses (*Andropogon gerardi* and *A. scoparius*) are common in American tall-grass prairies. A third category (mid-grasses) is also recognized, comprising grasses 30–100 cm high.

1.2.3 Seminatural, temperate grasslands

Large areas of highly productive grasslands are found in middle-latitude climatic areas of Europe, North America, Australia, New Zealand and parts of Asia, where growth is not constrained by prolonged drought and extremes of heat and cold and where the original vegetation was predominantly deciduous forest. The main climates involved are marine west coast and humid and moist continental. Marine west coast climate has annual precipitation ranging from 500 to 2500 mm yr^{-1} with mean annual temperature of 7–13 °C. Temperature extremes are less than in continental climates; the warmest month has a mean of 16–18 °C and the coldest 2–7 °C. This climate occurs at 40–60° N and S latitude and is typified by dull, drizzly weather with cool, wet summers and mild, wet winters. The main soils within those climatic areas that support grassland are cambisols, podzoluvisols and rendzinas. Cambisols and podzoluvisols are widespread in areas of Europe, North America, Australia and New Zealand. The natural vegetation is deciduous forest, much of which has been cleared for agriculture. Cambisols have greater inherent natural fertility and a wider range of land-use capabilities than rendzinas and podzoluvisols. Rendzinas are not restricted to any climatic or geographic area; they are shallow soils the characteristics of which are determined primarily by the high carbonate content of their parent materials, which are usually consolidated rocks such as limestone and chalk.

Duffey *et al.* (1974) classified the seminatural grasslands of Britain into three main groups based on soil type: calcareous, neutral and acidic. Calcareous grasslands occur on soils derived from a variety of calcareous parent materials, often cambisols or rendzinas, with high pH and organic matter content. Neutral grasslands are widespread in Europe, mostly occurring on clays or loams. Their botanical composition is determined mainly by water regime, management, soil and geographical location. Acidic grasslands are the most widespread type of seminatural grasslands in the British Isles. They occur on a great variety of soil types, which include free-draining acidic sandy heaths and podzoluvisols developed on alluvial deposits of sand, silt or clay and upland moorland mineral soils of various types as well as raw humus peats in waterlogged lowland areas and montane areas.

1.2.4 Tundra

Tundra is usually defined on the basis of climate to include areas where the mean annual temperature is less than 0 °C. It includes a range of polar and subpolar habitats – arctic, subarctic and subantarctic – as well as alpine habitats at high altitudes in temperate latitudes. Relatively warm temperate moorlands were included in this category for IBP purposes, but such areas can equally well be classified as temperate habitats.

Tundra grasslands occur on a wide range of soil types, including immature regosols, folisols with raw humus layers on the surface, lithosols on slopes and podzoluvisols. Some of the main features, and the variability, may be illustrated with reference to the IBP tundra site on the arctic coastal plain at Barrow, Alaska (Bunnell, MacLean and Brown, 1975). The air temperature is below freezing for most of the year and mean temperature rises above 0 °C on only 87 days. Wet mineral soils thaw to a mean depth of about 40 cm, but this can vary from 20 cm to 80 cm, depending on the content of organic matter and moisture. Vegetational and soil characteristics vary along a moisture-dominated gradient through wet meadow and marsh to open water ponds and lakes. The soils of the arctic coastal plain near the shore are marine deposits of Pleistocene origin. A high proportion have organic surface horizons over gleyed mineral soil with variable amounts of enmixed organic matter. Few have sufficient thickness to be classified as organic soils (histosols); most belong to the regosol group. The vascular vegetation of the Alaskan Coastal Plain is meadow-like, with sedges, grasses, forbs and a few dwarf shrub species.

1.3 THE FAUNA OF DIFFERENT GRASSLAND TYPES

In many cases there is as much variation within as there is between major grassland types in terms of their invertebrate faunas, so that only very general comparisons between grassland types can be made.

Base-rich, temperate grasslands with abundant earthworms tend to support the highest faunal biomass, which may often exceed 100 g fresh mass m^{-2}. At the other extreme, faunal biomass in arid and semiarid steppe and desert soils may be little more than 1 g m^{-2}, mainly comprising microfauna such as protozoans and nematodes. Tropical grasslands and tundra tend to be somewhere in-between, with humid tropical and some subarctic tundra sites where earthworms are abundant being comparable with some temperate grasslands in terms of biomass. Termites and, to a lesser extent, ants, tend to be the dominant groups in drier tropical soils, while mesofauna (e.g. mites, collembolans and enchytraeid worms) predominate in the acidic raw-humus horizons of tundra and temperate moorland soils.

Some marked taxonomic and ecological differences are apparent between the invertebrate faunas of tropical and mesic temperate habitats. For example, the lumbricid earthworms of temperate latitudes are replaced by taxa that are able to survive the dry season in the tropics: these include the large Microchaetidae of South Africa and Megascolicidae of Australia, which escape the adverse effects of drought by living more or less permanently deep in the soil. Termites are tropical/ subtropical in distribution while some surface dwelling predatory arachnids such as scorpions and solifugids are restricted to warm, arid soils. At the other climatic extreme, under arctic and antarctic conditions, biological diversity is restricted and cold tolerance severely limits the range of invertebrates present. However, most invertebrate groups in the middle latitudes have a wide range of distribution and the size and composition of the community will be determined by local habitat factors, such as the botanical composition and structure of the sward, weather and climate, soil physical and chemical properties, and topographical features. Some of the ways in which these factors may influence the fauna are discussed in the remainder of this chapter.

1.4 BOTANICAL COMPOSITION AND GROWTH PATTERNS

Natural and seminatural grasslands support a rich assemblage of plant species, comprising grasses (Gramineae), grass-like plants such as sedges (Cyperaceae) and a variety of dicotyledons, especially Compositae. Forbs (Leguminosae principally) may be abundant in successional, old field grasslands in previously arable abandoned land; a variety of shrubs may be present in grasslands undergoing succession towards climax deciduous forest and scattered trees are a prominent feature of the savannah landscape. Floristic diversity is conditioned by many factors that include soil type, water regime, geographical location, grazing pressure and human management. Species richness is typically greatest where human influence is least and where local conditions provide a mosaic of microhabitats. Thus, relatively uniform semiarid, natural grasslands may support about 50 vascular plant species while subhumid undulating tropical grassland may have over 200. Despite this great floristic diversity, graminoids typically provide over 90% of the biomass in the canopy and two or three species frequently provide 60% or more of the total biomass (Coupland, 1979).

1.4.1 Role of grazing in maintaining grassland

It is a moot point whether natural grasslands owe their existence to grazing as some authors suggest, but, together with fire and climate, grazing is undoubtedly a major factor in determining grassland distribution, and the evolution of grasses and ungulate mammals (Bovidae) are closely interdependent. Grasses have been known from the Tertiary and probably emerged as a distinct group during the late Cretaceous or even earlier (Van Dyne *et al.*, 1980). They are uniquely adapted to grazing, because leaf formation can continue during and after each defoliation. Whereas dicotyledons grow from terminal meristems that are highly vulnerable to destruction by large ungulates, during the vegetative phase of growth the meristematic zones of grasses are located close to the soil surface and are thus protected from grazing animals. According to McNaughton (1979), meristematic protection by physical isolation and physiological evolution leading to compensatory growth have been major features in grass evolution. Mechanisms of compensation for grazing include enhanced photosynthetic capacity, more efficient light use due to reduction in mutual leaf shading, hormonal redistribution promoting tillering, division and expansion of leaf cells, and reduced rate of leaf senescence. A feature of long-established grasslands is the high degree of selection for grazing resistance and the long vegetative persistence of grazing-resistant genotypes. Thus, Harberd (1962) estimated that some clones in British grasslands are over 1000 years old.

The importance of grazing in maintaining grassland in areas of climatic forest climax is evident from the many studies that have documented the floristic changes in pasture following cessation of grazing. While large herbivores feed mainly on grass, they also consume significant amounts of forbs and shrubs. Thus, the diet of 28 large herbivore species in the USA that have been adequately studied consists of grass (60%), forbs (20%) and herbs (20%) while some species of pronghorns prefer forbs and shrubs to grass (Van Dyne *et al.*, 1980). Many species such as sheep show strong seasonal preference for forbs. When grazers are removed, grasses lose their competitive advantage and forbs and shrubs quickly become established. Wells (1971) described how chalk grasslands in England grazed by rabbits and sheep maintained a species-rich sward. Following the reduction in the rabbit population caused by a myxomatosis epizootic in 1954, colonization by coarse grasses and scrub occurred, with a fall in floristic diversity.

1.4.2 Fire and grassland maintenance

The basal location of the meristem also affords grasses a high degree of protection against fire when compared with trees and shrubs where the buds are exposed, and fire is a major factor in maintaining grasslands in many parts of the world. Annual burning maintains an open community in savannah and there is a longstanding practice of deliberate burning of vegetation in tropical and subtropical Africa to control the spread of bush (Spence and Angus, 1971) and to improve the quality of the sward for grazing animals. Fire, of course, drastically alters the habitat for grass-land-dwelling invertebrates and has considerable effects on their populations (Chapter 5).

1.4.3 Botanical composition and the fauna

The significance of botanical composition *per se* for the invertebrate fauna varies from group to group. In general, the fauna of a particular grassland is recognizably different from that of neighbouring habitats such as woodland, but it is often difficult to say whether this is due to botanical composition as such or to differences in structural or microclimatic features. Floristic composition is undoubtedly important for many species of monophagous or oligophagous herbivores that are associated with particular plant species or closely related groups of species. Close association with particular graminaceous hosts is found, for instance, among the Cecidomyidae or gall midges (Barnes, 1946) and among some grass-feeding plant bugs (Southwood and Leston, 1959; Gibson, 1976). Strong preferences for particular grass species have been demonstrated among stem-boring shoot flies (Nye, 1959; Mowat, 1974), and many species of graminivorous grasshoppers often thought of as omnivores exhibit distinct food preferences (Mulkern, 1972). Many dicotyledonous species in grassland such as *Urtica* spp. (Davis, 1973) and *Centaurea nigra* and *Campanula rotundifolia* in chalk grassland (Morris, 1967, 1971a) have more or less specialized feeders associated with them. However, many of the invertebrate herbivores that are common in grassland, particularly root feeders such as scarab beetle larvae (Scarabaeidae), wireworms (Elateridae) and tipulid larvae (Tipulidae), are polyphagous. Most of the non-herbivorous invertebrates (decomposers, predators, parasites) show no direct relationship with floral composition *per se* although the flora undoubtedly exerts strong indirect effects on them in various ways.

1.5 SWARD STRUCTURE

Structurally, grasslands vary enormously, ranging from the multilayered vegetation of many natural and seminatural grasslands that may comprise tall grasses and forbs, short grasses and variously shaped dicotyledons, and lichens and mosses on the soil surface, to highly uniform, sown leys. Structural complexity often reflects management intensity, with the most intensively managed grasslands displaying the highest degree of uniformity. There are many examples that suggest that sward architecture exerts a profound influence on the invertebrate fauna. Thus, Andrzejewska (1965) reported that different species of leafhoppers, planthoppers and froghoppers (Auchenorrhyncha, Homoptera) tend to occupy definite horizontal layers in the meadow habitat. The type of habitat and the formation of the different layers determine the number and abundance of species as well as the development rate of the population, the number of generations and the dominance structure of the community. In meadows with a well-developed litter layer, species such as *Delphacodes venosus*, *Kelisia vittipennis*, *K. pallidula* and *Athysanus quadrum* that remain in the litter layer throughout their life are abundant; these are absent from cultivated meadow with no litter layer. Morris (1971a) carried out an extensive survey of various chalk grassland sites in Britain and found significant correlations between mean vegetation height and numbers of individuals and species diversity of Auchenorrhyncha. In general, this was also true for Heteroptera (bugs) and Curculionidae (weevils).

The influence of vegetation structure on the distribution of spiders has been demonstrated for limestone grassland in Britain by Duffey (1962) while Kajak (1965) found a linear relationship between the diversity of vegetation and the population of web-spinning spiders in a Polish meadow. The nature of the vegetation determined the number of microhabitats in the environment for webs, cocoons and also retreats for living in. Likewise, Cherrett (1964), investigating the distribution of spiders in montane moorland in Britain, found that coarse, ungrazed vegetation supports large numbers of individuals and species and he also detected stratification of species in a *Calluna/Eriophorum* stand.

Plant–invertebrate interactions involve many facets in addition to sward architecture, and this topic will be discussed more fully in Chapter 3.

1.6 INFLUENCE OF WEATHER AND CLIMATE ON INVERTEBRATES

1.6.1 Effects on distribution and population density

Climate affects the distribution and abundance of invertebrates directly by influencing their biology and life processes, and indirectly by determining the nature of the habitats they occupy and their food supply. As previously mentioned invertebrate groups restricted by climate include the predominantly tropical and subtropical termites and the lumbricid earthworms of temperate regions, while only the most cold-hardy invertebrates such as some species of mites (Acari) and collembolans are found in Antarctica. Behan and Hill (1980) attributed an overall sevenfold decrease in mite diversity along a climatic gradient from subarctic soils to polar deserts to increasing severity of abiotic factors. Breymeyer (1978), in a comparison of the trophic structure of tropical savannah, temperate meadows and pastures, and semiarid prairie, concluded that invertebrate decomposers in the tropical and prairie habitats are reduced by the drastic abiotic soil conditions as compared with the temperate habitat. Zlotin and Khodashova (1974) likewise concluded that invertebrate saprophages in steppe grassland are limited by adverse climatic factors.

Severe constraints may be imposed on the life history and phenology of temperate species living at the edge of their climatic range. Whittaker (1965) reported that *Neophilaenus lineatus*, a common hemipteran species in lowland areas of Britain, underwent marked fluctuations in numbers in an upland moorland site at Moor House, Yorkshire. It became locally extinct at this location when climatic conditions prevented its life cycle from being completed; in contrast, its populations in lowland sites remained stable.

A spectacular instance of the influence of severe weather on the soil fauna is provided by Bro-Larsen (1949), who recorded a drastic reduction in numbers of tipulid and bibionid flies, elaterid beetles and earthworms after three abnormally cold winters and two very dry springs. He noted that species active in late autumn and early spring suffered particularly whereas species that have a short breeding period in the summer and vegetate for the rest of the year were less affected.

Weather has been identified as a major factor in determining the population densities of a wide range of grassland invertebrates, including grasshoppers in Britain (Richards and Waloff, 1954), isopods in California (Paris, 1963), tropical ants (Brown, 1973), tipulids in upland Britain (Coulson, 1962), enchytraeids in Danish soils (Nielsen, 1955b), the grass bug *Leptoterna dolabrata* in British grassland (McNeill, 1973), grass grub *Costelytra zealandica* larvae in New Zealand pastures (Kain, 1975), prairie grasshoppers in the USA (Rodell, 1978) and overwintering frit fly (*Oscinella frit*) larvae in Britain (Southwood and Jepson, 1962).

1.6.2 Adaptations to unfavourable climatic conditions

(a) Life cycle extension

Extension of the life cycle appears to be a common adaptation to cold in a range of different taxa. Kontkanen (1954) compared the life cycles of leafhopper species in eastern Finland with those of an area of less rigorous climate and longer growing season in western Germany. Of the 10 species that the two sites had in common, three had one generation a year each in both Finland and Germany, and seven had only one generation annually in Finland but two in Germany. The tipulid species *Tipula excisa* has two generations a year under mid-alpine conditions in southern Norway compared with a 1-year life cycle in less rigorous climates (Hofsvang, 1972). In the even harsher climate of Barrow, Alaska, Bunnell, MacLean and Brown (1975) found that the dominant tipulids required at least 4 years to complete larval development. Life-cycle extension is also a feature of the biology of mites in Antarctica (Booth and Usher, 1986).

(b) Quiescence and diapause

The capacity to enter a resting stage is an adaptation commonly found in a wide range of invertebrates living in areas with strongly seasonal climates. Two categories of resting behaviour are generally recognized – quiescence and diapause (Tauber, Tauber and Masaki, 1986). Quiescence is a reversible state of suppressed metabolism in response to adverse conditions, with an immediate resumption of development when favourable conditions return, while diapause is a neurohormonally mediated state of low metabolic activity that occurs during a genetically determined stage in the life cycle.

Diapause enables the organism to survive unfavourable conditions and permits a high degree of synchronization of life cycle with weather, ensuring that the active stage of the life cycle is present when food is abundant and when weather favours rapid development. Diapause may occur in the egg, larval, pupal, nymphal or adult stage of insects, depending on the species. Egg diapause is a feature that allows many species of aphids to overwinter in northern European climates (Vickerman and Wratten, 1979) and that allows the collembolan *Sminthurus viridis* to survive summer drought in Australia (Davidson, 1932, 1933). Diapause in earthworms occurs in response to drought (Olive and Clark, 1978); it is considered to be obligatory in some species but facultative in others. Before entering diapause earthworms cease feeding, empty their guts and construct spherical chambers lined with mucus in the soil, in which they coil into a tight ball, one earthworm per chamber. Earthworms lose weight during diapause but do not suffer tissue dehydration. Earthworms may enter a quiescent stage in response to drought or low soil temperature. This is a state of torpor that does not involve specially excavated resting chambers, and severe tissue dehydration may occur during the quiescent stage.

(c) Vertical migration

Vertical migrations have been reported for many invertebrate groups in grassland; these may involve diurnal movements of foliage-dwelling invertebrates into the litter layer and topsoil during unfavourably hot or cold periods of the day or movement by litter and topsoil-dwelling invertebrates into deeper layers of soil during cold winter and hot summer weather. Diurnal movements probably reflect short-term changes in humidity (Joosse and Groen, 1970), while seasonal changes in vertical distribution have usually been interpreted as a response to weather-induced changes in conditions in the upper soil layers. Very pronounced downward shifts in soil microarthropods in tropical soils coinciding with the dry season have been observed by Strickland (1947) and Belfield (1956); likewise, a great increase in the density of fauna in the top 5 cm of soil accompanied the onset of wetter weather. Changes in vertical distribution of the fauna in the soil profile have also been reported from temperate grasslands (Dhillon and Gibson, 1962; Curry, 1971; Whelan, 1985), but these changes often do not follow any definite seasonal pattern and are less easily explained in terms of seasonal weather changes.

Despite behavioural and physiological adaptations, however, weather-related mortality is very significant for most grassland invertebrates. Pronounced seasonal changes in numbers of many soil invertebrates have often been reported, with minimum populations coinciding with unfavourable winter and summer weather. The two factors that together largely determine weather patterns are temperature and precipitation.

1.6.3 Temperature

Since their body temperatures vary with external conditions, all invertebrate functions are profoundly influenced by temperature. In general, the temperatures within which animals thrive range between 0 and 50 °C but the range tolerated by any given species is much shorter and the preferred temperature range is usually quite narrow. Surface-dwelling invertebrates are often faced with marked variations in diurnal and seasonal temperature: differences between day and night temperatures frequently exceed 20 °C in temperate areas and may be as much as 40 °C in arid areas, while seasonal temperatures commonly vary from winter minima of −30 °C to summer maxima of 35 °C in regions with continental climates.

Temperature is determined by the amount of radiation reaching the earth, which in turn depends on many factors, such as time of the day, season, latitude, aspect, altitude, cloud cover, vegetation cover, soil colour, etc. The temperature close to the soil surface is strongly influenced by the soil. During the day, short-wave solar radiation absorbed by the soil and radiated back as long waves heats a thin layer of air above the soil, which is then considerably warmer than the soil below or the air above, while on clear nights after a heavy dew a thin layer of chilled air lies over the soil surface. Biel (1961) found that on a typical summer afternoon when the air temperature at 2 m is 28 °C the temperature 5 cm above the ground may be 6 degrees higher, while at sunrise the 5-cm temperature may be 3 degrees lower than that at 2 m.

Vegetation profoundly influences the local microclimate. The temperature of sun leaves on still sunny days may be 20 °C greater than that of the air (Edwards and Wratten, 1980), thereby posing considerable problems for leaf feeding herbivores. Unfavourable temperatures in such exposed situations, however, can be considerably ameliorated by wind: convection cooling can permit insects to feed in areas that would be too hot in the absence of wind (Waterhouse, 1955). Waterhouse studied the microclimatological profiles in grass cover and found that seasonal and diurnal air temperatures and humidity profiles are related to height, density, mode of growth of the crop and solar altitude (Figure 1.2). In tall, dense grass the active surface of interception of solar radiation is raised

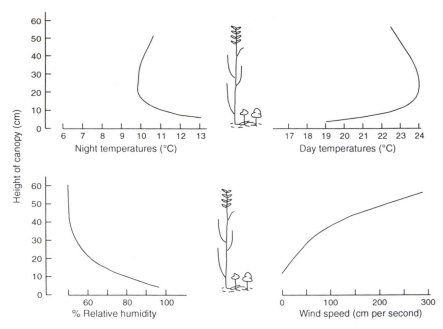

Figure 1.2 Typical temperatures, relative humidity, and wind-speed profiles within a grass–clover sward under marine west coast climatic conditions. (After Waterhouse, 1955.)

above the ground. Maximum absorbtion of solar radiation occurs just below the upper surface of the vegetation; a zone of elevated temperature occurs above this and temperatures at the soil surface are lower: by contrast, in short grass, maximum temperatures are at the ground level. At night, dense vegetation prevents chilled air from settling to the ground and minimum temperatures occur at some distance above the ground. The height of the vegetation has an important influence on the temperature in the sward. Norman, Kemp and Tayler (1957) compared temperatures throughout the winters of 1953–54 and 1954–55 in adjoining grass swards, one of which had been cut to 2–3 cm while the other had reached a height of 30–45 cm. The number of nights on which the temperature at 2.5 cm dropped below 0 °C was only half as much in the long grass as in the short. Vegetation also strongly influences air movements: in dense grass, complete calm exists at ground level and this creates stable temperature and humidity conditions that are favourable for many invertebrates. Likewise, the presence of a surface mat of dead vegetation and broad-leaved plants such as clover promotes favourable cool humid conditions next to the ground. The buffering effect of a surface mat layer on soil temperature in a Dutch polder grassland was demonstrated by aerial infrared photography, which showed that areas with a mat were warmer by night and cooler by day than areas from which the mat had been removed by earthworm activity (Hoogerkamp, Rogäar and Eijsackers, 1983). In high-altitude areas, the presence of a surface organic horizon and snow cover has an important influence in moderating winter soil temperatures.

Soil temperatures undergo fairly regular seasonal and diurnal variations. The amplitude of seasonal variation is greatest close to the soil surface and is progressively dampened with depth (Figure 1.3). Diurnal temperature variations are also greatest close to the surface (Figure 1.4); this example also illustrates the important role of plant cover in dampening fluctuations in soil temperature.

The microclimates of north-facing and south-facing slopes are significantly different, since south-facing slopes receive the most solar energy and north facing slopes receive the least. According to Smith (1966), at latitude 41° N midday insolation on a 20° slope is on average 40% greater on south slopes than on north during all seasons. This markedly affects moisture and heat budgets and tends to produce warm, xeric microclimates with wide extremes on the south slope in contrast to

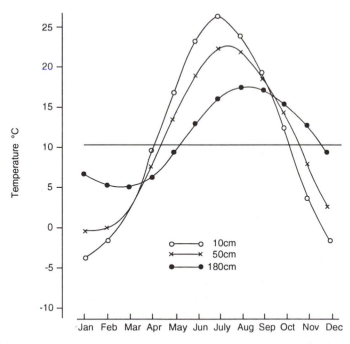

Figure 1.3 Seasonal variation in soil temperatures at various depths in Ames, Iowa, USA; 10 cm (○); 50 cm (×); 180 cm (●). (From Smith, Newhall and Robinson, 1964.)

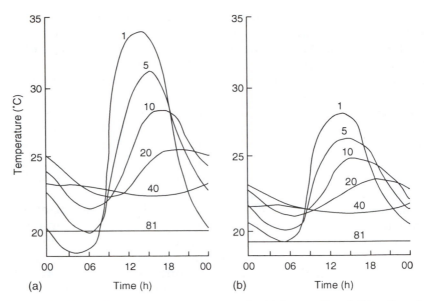

Figure 1.4 Diurnal variation in temperature in (a) bare soil and (b) soil under grass at St Paul, Minnesota, USA, in July. The soil depth is shown in centimetres. (After Baker, 1965.)

cooler but less variable conditions on the north slope. In Waterford, California, the mean annual temperature of a grassland soil on a 20-30° south-facing slope was 6.4 °C higher than that on a similar north-facing slope (Smith, Newhall and Robinson, 1964).

Valleys and depressions experience greater extremes of temperature than exposed areas because of the absence of the mixing influence of wind. Temperatures tend to be lower at night and higher during the day than in exposed areas.

It is often difficult to separate the effects of temperature and moisture on invertebrate distribution and abundance as the two factors interact. One example in which the key role of temperature has been fairly thoroughly documented is in the case of the black beetle, *Heteronychus arator* (East, King and Watson, 1981). This is a sporadic pest of New Zealand pasture, confined to areas with annual screen temperature greater than or equal to 12.8 °C, with outbreaks being associated with above-average spring temperatures. Flight, oviposition, feeding, egg and larval development cease when air temperature falls below 15–17 °C.

Temperature has been invoked as the main factor responsible for inducing vertical migration of invertebrates in the soil in some situations: Dowdy (1944) found that downward movements of soil animals in autumn and winter and upward movements in spring coincide with the periods of temperature overturns, and concluded that in areas greatly disturbed by people temperature is the main stimulus behind vertical movements. Athias (1976) likewise concluded that vertical migrations of microarthropods mainly occur in response to changes in soil temperature in savannah at Lamto, Ivory Coast.

1.6.4 Precipitation

Since most invertebrates must maintain the moisture content of their bodies within fairly narrow limits, the seasonal distribution of rainfall is more important than average annual precipitation. Where periods of drought are a predictable feature of the climate it is common to find drought-resistant stages in the life cycle that enable the organism to survive. For example, *Sminthurus viridis* is unable to survive during the nymphal and adult stages for more than a few hours except when the atmosphere is nearly saturated with water vapour. However, this collembolan species is widespread and abundant in South Australia in regions where 5 months or more of continuous summer drought occur. Eggs laid as summer approaches complete only about half their development and then aestivate in an extremely drought hardy state (Davidson, 1932, 1933).

1.7 SOIL MOISTURE

The actual amount of total annual precipitation entering soil depends on the intensity and duration of rainfall, vegetational cover, slope, speed of snow melt, soil organic matter, infiltration capacity and soil permeability. After heavy precipitation water is lost by drainage under the influence of gravity before stabilizing at field capacity, a value that is fairly consistent for any given soil. Water is rapidly lost through plant evapotranspiration until the roots can no longer extract water and wilting occurs. The forces holding water in the soil are represented by the suction pressure or tension measured in centimetres of water. This may be expressed on a logarithmic (pF) scale, or on the equivalent osmotic pressure scale in atmospheres, bars, or kilopascals (kPa). The tension needed to support a column of water 1000 cm high is about 1 bar or 100 kPa. The moisture content of a given soil can be related to suction pressure by the moisture-characteristic curve, the shape of which varies according to soil texture. Since pore space can be defined as the volume of water withdrawn between complete saturation and the point when pores are empty, the moisture-content scale can be converted into a percentage pore-space scale, and, since moisture tension is a function of aperture size, the suction scale can be converted to a neck-diameter scale. The shape of the moisture-characteristic curve thus reflects pore-size distribution; in a

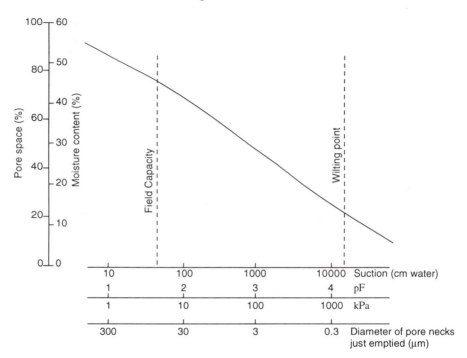

Figure 1.5 Moisture-characteristic curve for a medium-textured soil.

coarse soil with large pores the shape of the curve will be sigmoid with a distinct point of inflection occurring at relatively low suction whereas in a fine soil with small pores the curve will approach a straight line. These relationships are illustrated for a hypothetical medium-textured soil in Figure 1.5. The relative humidity of the soil atmosphere below the soil surface usually approaches 100% and only falls below 98% at pF 4.45. The tension range between field capacity and wilting point corresponds to the plant-available range; this varies with soil type but is normally between 5 and 1500 kPa. It should be noted that percentage moisture *per se* gives little idea of its availability; thus, in a coarse sandy soil the percentage water held at 50 kPa was 3% of dry mass and the corresponding figure was 52% for an organic soil (Nielsen, 1967).

Moisture tension is of direct relevance to microfauna such as protozoans and nematodes living within the water film, and these organisms can be relatively abundant even in semiarid soils (Smolik and Lewis, 1982; Elliott and Coleman, 1977), whereas larger forms occupying air-filled pores are more influenced by the actual water content and the relative humidity of the soil pores. These are particularly susceptible to desiccation and many groups undergo severe mortality during dry weather. Nielsen (1955b) studied the water relationships of enchytraeid worms in Denmark and concluded that desiccation was the main factor causing summer mortality. High populations of tipulid larvae are associated with wet soils; in dry years, their distribution and abundance are reduced. Coulson (1962) recorded high mean densities of *Tipula subnodicornis* in upland moorland sites in Britain, but during a dry period in 1955 this species died out in many areas or was severely reduced in numbers. Only in restricted areas that remained wet was density unaffected. Earthworms experience high seasonal mortality associated with unfavourable soil conditions. Gerard (1967) found that most *Allolobophora* juveniles that had hatched in May, June and July had died by August in permanent pasture in southern England. Other developmental stages avoided desiccation by moving deeper in the soil profile and sometimes by becoming inactive in the soil, usually during hot dry periods in summer. These species also moved downwards in the soil during periods in winter when the surface soil was frozen. The activity of soil protozoans is also strongly influenced by soil moisture levels. Dash and Guru (1980) reported a strong correlation between soil moisture content and numbers of active testate amoebae Rhizopoda in Indian grassland ($r = + 0.89$). Summer drought coincided with minimum density and prolonged sustained increase in soil moisture levels was accompanied by an increase in the number of active forms and a decrease in the number of encysted, drought-resistant forms in the soil.

1.8 MICROCLIMATE AND SHELTER

Local features such as the nature and density of the sward, the presence of a litter layer, aspect, and the availability of refuges from adverse conditions can greatly influence habitat suitability for surface-dwelling invertebrates in grassland. Denno (1977) attributed the diverse sap-feeding herbivorous fauna on the salt-marsh grass *Spartina patens* along the northeastern coast of the USA to the presence of a dense, persistent thatch composed of dead plants. By contrast, *Spartina alterniflora* without a thatch layer supported a much poorer fauna. When the thatch layer of *S. patens* was removed the species diversity and evenness of sap feeders associated with this species declined.

Plants with a tussock form of growth have an important influence on the distribution and abundance of grassland invertebrates. Luff (1966) found a greater density of beetles in tussocks of *Dacytlis glomerata* (cocksfoot) and *Deschampsia caespitosa* (tufted hair-grass) in English grassland than occurred between tussocks; the difference was particularly marked in winter. These trends reflect the more favourable temperature and humidity conditions within the tussock during dry, warm summer weather and during cold winter weather (Figure 1.6). The significance of tussocks as refuges for invertebrates was also demonstrated by Bossen-

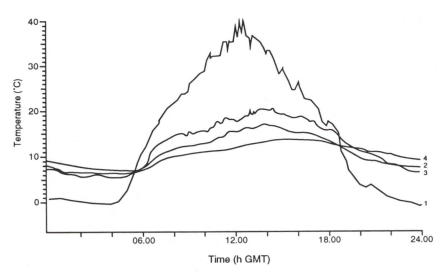

Figure 1.6 Daily temperature fluctuations in *Dactylis* tussocks compared with those in non-tussock (*Holcus*) vegetation in fine summer weather. Curve 1, 7.5 cm level in *Holcus*; 2, soil level in *Holcus*; 3, 7.5 cm level in *Dactylis*; 4, soil level in *Dactylis*. (After Luff, 1966.)

brock *et al.* (1977), who reported that grass tussocks of *Holcus lanatus, Poa trivialis* and *Deschampsia caespitosa* have a greater capacity to regulate temperature and provide shelter for invertebrates in winter than species with rosette and loose-structured growth forms. The importance of tussocks is greater in areas of low plant cover and in open areas; in such areas under adverse weather conditions animals tend to seek shelter in tussocks more than in areas with good plant cover. Tussocks of *Lomandra fibrata* provide summer refuges for the millipede *Ommatoiulus moreletii* in South Australian grassland: Baker (1978b) found aggregations of this species beneath these tussocks in summer in preference to surrounding pasture grasses. The soil beneath the tussocks was moister than under grass and the temperature was more stable: whereas the temperature at the soil–litter interface varied from 10 °C to 43 °C beneath pasture grasses, it remained in the range 14–26 °C beneath the tussocks. Laboratory experiments indicate that this species had a preference for high relative humidity and a temperature of 20–25 °C. However, Baker (1980) made the interesting observation that not all the population aggregated in tussocks during the summer; some remained in the litter under pasture grasses. These had developed the capacity to survive desiccation and high temperature during summer but had lost this by autumn. Grassy field edges of intensively grazed pastures can have a role analogous to that of tussocks under more extensive management: Desender *et al.* (1981) reported that untrampled field edges can provide favourable microclimatic conditions for hibernating beetles in Belgian pasture.

Refuges are particularly important in regions with a pronounced dry season. Janzen (1973) found a strong movement of foliage insects into moist refuges during the dry season in Costa Rica and, at night, he recorded a dramatic increase in the numbers and species in nearby dry sites. In dry thornbush savannah in northern Senegal, Gillon and Gillon (1973) found that many arthropods seek shelter during the dry season in small heaps of dead wood or in shallow depressions where water has accumulated for some time during the rains. The latter are selected as breeding grounds by many phytophagous species.

Logs and stones provide stable microclimates with high humidity and support a fauna that seems strikingly similar in all regions. Cole (1946) studied the fauna under boards in an Illinois woodland and concluded that typical members are refugees from physical conditions or biotic factors such as predation and competition in the outside environment. The fauna included many permanent residents, comprising groups that are highly susceptible to desiccation, such as isopods, slugs and snails, and also many members that spend part of their life cycle elsewhere, such as insect larvae and hibernating adults. Hågvar and Östbye (1972) considered that stones are important for the fauna of high mountain sites because of the favourable, relatively constant, microclimate they provide. They found a rich fauna under stones in alpine sites at Finse, southern Norway, representing to a certain extent the same orders and families as in the surrounding litter and vegetation. Stones 100–400 cm^2 in area appear to provide optimum shelter (Schonborn, 1961).

Interest in the possible importance of hedgerows as reservoirs for crop pests and predators has stimulated several studies. Lewis (1969) found that the insect community of a hedge in Britain was more diverse than that in adjacent crop and pasture fields. The diversity of the aerial population decreased with increasing distance from the hedge and the resulting pattern of diversity resembled the pattern of shelter produced by the hedge. The presence of the hedge enriched the aerial population for a distance of three to ten times its height to leeward and two to three times to windward. In a follow-up study of air movements near windbreaks (Figure 1.7), Lewis and Dibley (1970) concluded that accumulations of small airborne insects are probably caused when the insects are diffused or convected into recirculating air behind barriers.

The significance of hedgerows as refuges for members of the open-field fauna appears to be limited. Many authors consider that the hedge fauna is really an impoverished woodland fauna that interacts very little with that of neighbouring fields (Pollard, 1968; Thiele, 1977; Desender, 1982),

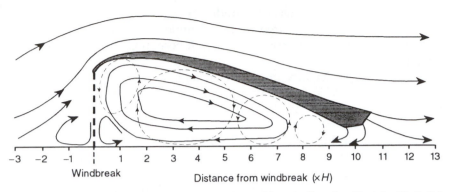

Figure 1.7 Wind-flow patterns near a permeable windbreak of height *H*. Solid lines show time-average motion; light dotted lines indicate erratic movements within the general pattern. (After Lewis and Dibley, 1970.)

although Pollard (1968) did report that removal of the bottom flora of a hawthorn hedgerow reduced the numbers of two carabid beetle species (*Bembidion guttula* and *Agonum dorsale*) and one hemipteran (*Anthocoris nemorum*) that overwinter in hedgerows and are active in crop fields in summer.

Although cow pats attract a specialized coprophagous fauna initially (Chapter 3), they can be viewed as a temporary extension of the soil-litter habitat, acting as refuges for some faunal groups in pasture grazed by cattle. Davidson (1981) demonstrated that collembolans and mites respond to gradients of moisture, organic matter and microbial activity at cattle droppings, showing increased population density at the droppings.

1.9 SOIL PROPERTIES

Soils result from the weathering of organic and inorganic parent materials by climatic and biotic agencies over time. The inorganic parent materials are predominantly superficial deposits of glacial drift, alluvium and aeolian deposits, resulting from the widespread climatic changes and surface disturbances of the Pleistocene glaciations. In tropical and subtropical regions, large areas of soil have developed from the decomposition *in situ* of mainly acidic rocks – granite, gneiss, schist, slate, sandstone and shale. Outside these areas, shallow soils developed from the underlying rock occur in mountain ranges. Soils developed from limestone and other basic rock material are of more limited distribution, but locally widespread. The chemical and mineralogical composition of parent materials are largely responsible for the physical and chemical composition of the soil.

The physical and chemical characteristics of soil indirectly affect invertebrates above ground through their influence on vegetation and plant growth; they affect soil-dwelling invertebrates in a more intimate and direct way. Soil attributes of greatest significance include temperature and moisture content (which have already been considered), soil organic matter, texture, structure and porosity, soil pH and base status. It is convenient to consider these factors individually, but it is important to remember that they do not operate independently but are strongly interconnected.

1.9.1 Soil organic matter

Soil organic matter consists of the remains and decomposition products of vegetation and to a lesser extent, microbial and animal tissues. It ranges from freshly fallen leaf litter on the surface through various stages where cellular structure is still visible to amorphous soil humus. The distribution of organic matter in the soil profile is related to local site characteristics and decomposition rates; under conditions where the litter is resistant to decomposition and where soil aeration and/or pH do not favour rapid decomposition a discrete surface organic horizon develops. Under conditions that favour rapid decomposition (neutral to alkaline, well-aerated soils; readily decomposable litter) no surface organic horizon is visible and finely dispersed organic material is mixed intimately with the soil. Such contrasting organic profiles, named, respectively, mor and mull by Müller (1878, 1884), are seen in many forest soils, mor typically occurring in acidic sandy soils under conifers and mull in neutral to alkaline forest soils under deciduous trees with a relatively high calcium content. Many intermediate conditions between mull and mor occur. These contrasting organic profiles are associated with very different faunal communities. In mull soils, earthworms are abundant and earthworm activity is important in the incorporation of surface litter into the soil, while under mor conditions earthworms are scarce or absent and the fauna tends to be dominated by microarthropods (mainly Acari and Collembola) and enchytraeid worms living in the superficial organic horizons (Macfadyen, 1963; Satchell, 1967; Standen, 1982). An extreme form of mor development is seen under anaerobic, acidic conditions when undecomposed organic matter accumulates as peat.

Undisturbed natural grasslands are characterized by relatively higher organic matter content and a more even distribution with depth in the profile (Figure 1.8) when compared with adjacent deciduous woodland on comparable soil. This is due to the more rapid turnover of the relatively short-lived grasses, and the greater proportion of roots to soil organic matter in grassland. Soil organic matter influences many soil properties, including increased water-holding capacity and enhanced

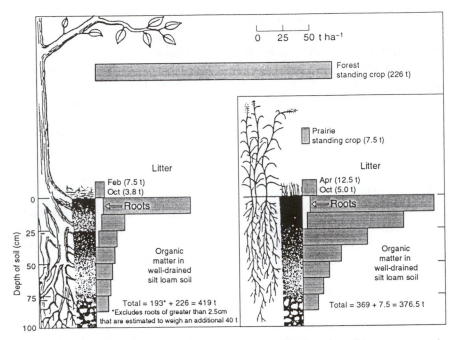

Figure 1.8 Distribution of organic matter in oak forest and prairie ecosystems in Wisconsin, USA. (After Nielsen and Hole, 1963 in Foth, 1978.)

cation exchange capacity (*c.* 3000 mEq kg^{-1} for humus), which influence the nature of the soil habitat, and it provides the energy and nutrient base for the community of soil decomposers. The suitability of this food source is determined in large measure by the properties of the litter input.

1.9.2 The nature of the litter

Among the many chemical and physical factors that determine the palatability of plant litter to soil animals, the ratio of carbon to nitrogen is usually considered to be of prime importance. The C:N ratio of grassland litter is usually low (< 20:1) and this is reflected in rapid decomposition. By contrast, many coniferous litters have C:N ratios exceeding 60:1 and their decomposition may take several years. Factors known to influence palatability of woody leaf litter include physical toughness, tannin and polyphenol content (Swift, Heal and Anderson, 1979), although less is known about the role of these factors in determining the palatability of herbaceous litter. Litter from plants in acidic grassland is likely to be poor in nutrients and to decay slowly, leaving acidic raw humus that is unpalatable to earthworms. In addition, many acidic grassland species such as *Nardus stricta* are unpalatable to grazing animals and there is, therefore, a high return of litter from them to the soil, which enhances the build up of acid raw humus. Factors influencing the food quality of grassland litter for decomposer invertebrates are considered further in Chapter 4.

1.9.3 Soil texture, structure and porosity

Soil texture refers to the 'feel' of moist soil and reflects the size distribution of particles or the mechanical composition. Soils may be divided into textural classes on the basis of sand, silt and clay content. Soil structure is determined by the degree of aggregation of soil particles and by the nature and distribution of pores and pore spaces. Well-developed aggregates are found in horizons of medium-textured soils whereas coarse sand has a loose structure without well-formed aggregates. Clay soils tend to have a coherent or massive structure, with few large aggregates. The degree and type of aggregation influences aeration, permeability, infiltration capacity and water movements in the soil; it has a major role in determining pore space and the volume of the soil atmosphere. Pores are occupied by water or by soil atmosphere; pore space is discrete in clay soils and continuous in well-aggregated, medium-textured soils and sandy soils. The soil atmosphere is typically saturated with moisture in humid climates; it consists of about 80% N, 15–20% O_2 and 0.25–5% CO_2. The concentration of CO_2 thus considerably exceeds that of the above-ground atmosphere (0.03%). Gas diffusion in the soil is related to the volume of gas-filled pore space and not to pore size: poor soil aeration is likely to reflect excessive water content rather than the amount and size of pores. Clay soils are particularly susceptible to low aeration when wet because most of the pore space becomes filled with water and continuous avenues for gas diffusion are absent. In a typically well-aggregated, medium-textured soil, 50% of the total pore space is occupied by water at field capacity and about 25% at wilting point.

Many factors are involved in the development of soil structure, including the soil fauna. FitzPatrick (1980, 1984) describes 21 types of soil structure, three of which are of faunal origin (Figure 1.9). The labyrinthine type, with abundant, intertwining faunal passages produced by vigorous arthropod activity is well represented in the upper and middle horizons of tropical and subtropical soils where termites are abundant. The vermicular type is of earthworm origin, with vermiform, wormlike aggregates and contains various amounts of faecal material (vermiforms). This type of structure is well represented in the chernozems of Europe and ferralsols of Australia where earthworms are active. The channel type, also of earthworm origin but lacking aggregates, is common in the upper and middle horizons of cultivated and natural grasslands. Intertwining

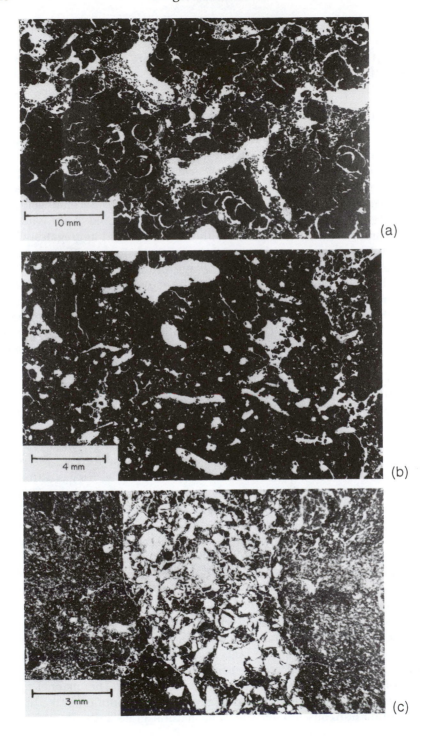

channels are abundant and some may be completely filled with vermiforms. Faunal activity is also a major influence in the formation of composite-type soil structure, which is common in cultivated soils. Indeed, in any soil with abundant soil animals the upper horizons become a tangled mass of passages filled to a greater or lesser extent with invertebrate faeces. In some soils, certain horizons consist almost entirely of faecal material; a good example is seen in shallow alpine rendzina soils where coprogenic horizons occur (Kubiena, 1955).

The significance of soil structure, porosity and aeration for soil animals is considerable. The animals are restricted to aerobic horizons and are confined to the surface layers of soils prone to waterlogging. Tischler (1955) concluded that mites and collembolans penetrate to a greater depth in sandy soils than in loam as a result of better ventilation. Invertebrates particularly sensitive to desiccation tend to be more abundant in heavy clay soils that are less subject to drying out than in light sandy soils. The size of cavities in the soil can impose restrictions on the vertical distribution of non-burrowing invertebrates in the profile and many larger invertebrates such as isopods, diplopods and molluscs that are confined to the surface of undisturbed soils may be found deeper in loose textured, cultivated soil. Weis-Fogh (1948) found a good correspondence between the vertical distribution of microarthropods in a Danish soil and the distribution of pore spaces. He concluded that it is not total pore volume as such but the size of microcaverns that restricts the fauna. These results were supported by Haarløv (1960), who found a correlation between pore size and depth distribution of microarthropods.

1.9.4 Soil pH

Soil pH, which is of central importance for many soil processes, is largely determined by the balance between H^+ and, to some extent, Al^{3+} ions and basic cations such as Ca^{2+}, K^+ and Mg^{2+}. Organic matter has an important influence on pH since the products of decomposition are acidic and tend to depress pH unless counteracted by a high concentration of basic cations.

Organic soils such as peat are typically acid (pH 3–4) but they may be nearly neutral in the case of fens of high base status. High rainfall can accelerate leaching of cations and can lower pH. Soil biological activity

---

Figure 1.9 Types of soil structure created by faunal activity.
(a) Labyrinthine structure showing abundant passages formed by termites; (b) channel structure showing channels produced by earthworms; (c) A single earthworm vermiform containing coarser soil material translocated from a different horizon. (From FitzPatrick, 1984.)

tends to be greater in neutral or slightly alkaline soils. In general, bacteria have a relatively narrow pH tolerance, in the alkaline range, while fungi have a broad range of tolerance but are more active in acid soils.

Soil animals in general are not very sensitive to pH and are active in soils over a wide range of pH. Some groups such as snails, millipedes and woodlice with high Ca requirements are restricted to base-rich, alkaline soils. Considerable differences in tolerance for low pH are found among earthworm species but few are found in acid soils with pH < 4–4.5 and most reach highest abundance in soils of neutral reaction.

1.10 SUMMARY

Broadly defined to include all communities in which grasses are the dominant component of the ground flora, grasslands comprise over 20% of the earth's total land area.

Climate has a major influence on the development of grasslands and soils, and general correlations exist, therefore, between climates, soils and grassland types. Tropical grasslands occur mainly on arenosols and ferralsols under tropical wet–dry and tropical desert and steppe climatic regimes in areas with relatively high mean temperatures throughout the year. Prairie and steppe grasslands occur mainly under tropical, middle-latitude desert and steppe, and humid continental climatic conditions. These climates are strongly seasonal, with cold winters and summers ranging from hot to cool. Precipitation ranges from moderate to low. Prairie and steppe occur mainly on chernozems, phaenozems, kastanozems and to some extent vertisols.

Large areas of seminatural grassland are found in the middle latitudes under marine west coast and humid and moist continental climatic regimes where the natural vegetation is deciduous forest, much of which has been cleared for agriculture. Tundra grasslands occur at high latitudes, and at high altitude in the middle latitudes, in areas with a mean annual temperature less than 0 °C.

Features of the grassland habitat that have a major influence on the invertebrate fauna include botanical composition and sward structure, weather and climate, and soil properties. The degree to which botanical composition directly influences grassland invertebrates is variable. Some groups, including the Cecidomyidae (gall midges) and grass-feeding Hemiptera, have close associations with particular species, while others such as stem-boring shoot flies and some grasshoppers have a wide range of graminaceous hosts but can have strong preferences for particular grass species. Many of the dicotyledonous species in grassland have more or less specialized feeders associated with them. Many root-feeding invertebrates – for example, Scarabacidae (scarab beetle larvae), Elateridae (wireworms) and Tipulid larvae – are polyphagous, while non-herbivorous invertebrates are usually not strongly influenced by floral composition *per se*. Sward structure has an important influence on the surface fauna, with architecturally varied swards tending to support more varied and dense invertebrate communities. Climate can limit the geographical distribution and/or impose severe seasonal restrictions on the activities of many invertebrate groups. Adaptations to unfavourable climatic conditions may include life-cycle extension, diapause or periods of quiescence, and diurnal and seasonal vertical migrations in the soil.

Temperature and precipitation are the two components of weather that most influence the fauna. Air temperatures are often unfavourable, but a dense sward and surface litter can have important ameliorating effects, while factors such as aspect and slope can greatly influence microclimatic conditions. Soil moisture is often limiting and high summer mortality can occur in earthworms and other groups susceptible to desiccation. Plant tussocks, grassy field edges, hedgerows, moist depressions, logs, stones and other microsites with favourable microclimates can provide important refuges from adverse conditions for surface fauna.

Soil properties that influence the fauna include temperature and moisture, organic matter, texture, structure and porosity, soil pH and base status, while the fauna in turn can modify soil properties considerably.

2

The composition of the invertebrate fauna

There is no single grassland site for which a complete invertebrate species list is available. Data for those groups that have received most attention indicate a very wide range in faunal diversity among sites, reflecting differences in such factors as climate, floristic composition and management. Thus, Purvis and Curry (1981) recorded fewer than 200 arthropod species from the herbage layer of a managed grass ley pasture in Ireland while Evans and Murdoch (1968) recorded close to 1600 insect species from an unmanaged old field in Michigan, USA.

The invertebrate groups of greatest significance in grassland are the Protozoa, Nematoda, Annelida, Mollusca, Acari, Araneae, Insecta, Apterygota, Isopoda and Myriapoda. In this chapter, their distribution, abundance and biomass are considered and their feeding habits, biological activity and influence on grassland processes are briefly reviewed. A wide range of sizes is represented among these groups, ranging from microscopic protozoans to earthworms several centimetres long. A traditional classification of soil fauna based on size has been into microfauna, comprising Protozoa, Nematoda and similarly sized animals; mesofauna, comprising Enchytraeidae, Acari, Apterygota, smaller Myriapoda and other invertebrates less than *c*. 0.5 cm long; and macrofauna, comprising the larger invertebrates. This is a somewhat arbitrary but convenient classification when considering physical interactions between the fauna and their habitat. Alternative classifications based on feeding habits are more appropriate for studies of food chains and ecosystem processes.

In most cases, no specific information is available on the ways in which different invertebrate groups influence ecosystem processes in grassland; their relative 'importance' is often inferred from the quantities of food energy they utilize. Representative estimates of energy consumed and/or respired by different groups are given in this chapter, while the contribution of invertebrates to energy flow and related ecosystem processes is considered more generally in Chapter 6.

2.1 PROTOZOA

2.1.1 Distribution

These unicellular animals, ranging from 2 to 1000 μm in size, are extremely abundant and widely distributed, occurring in a wide range of climatic and soil types from hot desert to tundra (Stout and Heal, 1967). In grassland, they are associated with every stage of the plant cycle – on the surface of stems and leaves of growing plants, on senescent and dead tissue, on living and dead roots, in the litter layer and in the soil profile. Ciliates and flagellates occurring on the growing vegetation typically complete their life cycles rapidly and form strongly resistant cysts that enable them to survive desiccation (Stout, 1974). They are active on leaf surfaces when moisture conditions are favourable, such as after dew formation. In tussock vegetation, where microclimatic conditions are more stable and where there is a greater variety of microhabitats in the form of green and senescent vegetation and a lower stratum of decaying vegetation, a wider range of species occurs. Likewise, grassland soil offers more favourable conditions than vegetation and harbours a wider range of amoebae, ciliates, flagellates and the slow-growing testate amoebae. High populations are associated with old decaying roots and root rhizospheres, presumably stimulated by microbial activity associated with the metabolism of root exudates. Stout and Heal (1967) identify the characteristics favouring the existence of protozoans in soil as their small size and simple structure, their capacity for rapid multiplication under favourable conditions, encystment and excystment mechanisms adapted to fluctuations in soil moisture, salinity, aeration and food supply, tolerance for a wide range of pH and temperatures and their ability to absorb nutrients in dissolved or in particulate form.

Protozoans are aquatic and therefore limited in distribution and activity by moisture. Dash and Guru (1980) reported strongly seasonal activity of testate amoebae in Indian grassland soils, with an increase in active forms and a decrease in numbers of drought-resistant encysted forms accompanying an increase in soil moisture levels. Elliott and Coleman (1977) likewise found that numbers of soil protozoans increased markedly following irrigation of semiarid short-grass prairie: numbers of active forms ranged from $c.$ 20 000 g^{-1} dry soil in control plots to over 100 000 g^{-1} in irrigated and fertilized plots. According to Stout and Heal (1967), protozoan metabolism and reproduction are favoured by increasing temperatures up to about 30 °C but higher temperatures are unfavourable with the death point at about 35–40 °C: cysts can survive much higher temperatures. Flagellates, amoebae and ciliates grow over a wide range of pH, but testate amoebae appear to have a more limited range of tolerance. In general, amoebae, flagellates and ciliates favour conditions of high base status with mull humus and high mineralization rates whereas testate amoebae favour the slow organic turnover characteristic of raw humus (Stout, 1968).

2.1.2 Abundance and biomass

Soil protozoan populations are frequently estimated by dilution culture techniques, which are considered to give satisfactory results for flagellates and amoebae in mineral soils but are unsuitable for the testate amoebae because of their slow growth; these are more satisfactorily estimated by direct counting. A micromethod for estimating most probable numbers of protozoans developed by Darbyshire *et al.* (1974) has been used in several recent studies.

Flagellates and amoebae are the most numerous protozoans under most soil conditions, with amoebae being most abundant in the plant rhizosphere where microbial activity is greatest (Darbyshire and Greaves, 1967). Amoebae counts of over 10^6 g^{-1} of wet soil have been recorded from temperate soils, although the more usual range appears to be 10^3– 10^5 g^{-1} for flagellates and amoebae combined (Elliott and Coleman, 1977; Schnürer, Clarholm and Rosswall, 1986). Because of uncertain assumptions about vertical distribution in the profile, results are not usually expressed in terms of populations per square metre but Elliott and Coleman (1977) calculated a population density of 7.2 $\times$ 10^8 m^{-2} to a depth of 5 cm in a short-grass prairie soil in Colorado, USA. Ciliate population estimates are somewhat unreliable because of their delicate structure, but they generally appear to number less than 10^3 g^{-1} wet mass in temperate grassland soils (Stout and Heal, 1967).

Testate amoebae are usually fairly scarce in temperate grassland soils while up to 2500 g^{-1} wet soil have been recorded in tropical grassland (Dash, 1975) and Heal (1965) recorded 890 $\pm$ 150 $\times$ 10^6 m^{-2} from subantarctic *Deschampsia* grassland.

Protozoan biomass estimates range from 0.3 g m^{-2} in semiarid short-grass prairie (Elliott and Coleman, 1977) to about 20 g m^{-2} in woodland and old grassland with accumulated organic matter where conditions favour the development of the larger amoebae and testate amoebae (Stout and Heal, 1967). Mean biomass in productive temperate grassland and arable land may be in the order of 5–10 g dry mass m^{-2} (Clarholm, 1984; Andrén *et al.*, 1990).

2.1.3 Feeding habits

Most soil protozoans feed on bacteria with yeasts, algae, fungal mycelia and (rarely) fungal spores being consumed to a lesser extent (Stout and Heal, 1967; Wright, Redhead and Maudsley, 1981; Coûteaux and Pussard,

1982). Some predatory forms can feed on smaller protozoans, nematodes and other microfauna (Sayre, 1973). Since testate amoebae thrive in raw humus where microbial activity is restricted it has been suggested that they can digest cellulose (Stout and Heal, 1967).

2.1.4 Ecological significance

Although protozoan biomass in the soil may be low, their rapid production and turnover rates suggest that they could have a considerable role in the transformation of soil organic matter. Elliott and Coleman (1977) estimated protozoan production of 1.3 g m^{-2} yr^{-1} in short-grass prairie, giving a secondary production:standing crop ratio of 4:1 in control, semiarid plots but where soil was subjected to cycles of wetting and drying this ratio became 51:1. Stout and Heal (1967) estimated that for small amoebae and flagellates in temperate soils the amount of protoplasm produced and recycled during the year was 50–300 times the standing crop.

Considerable attention has been focused on the significance of microbial predation by protozoans for soil fertility. While some authors in the past have considered that the effects of predation on beneficial organisms could be detrimental, the modern view is that protozoan predation promotes the turnover of readily available nutrients and enhances biochemical activity in the soil (Coleman *et al.*, 1977; Clarholm, 1981, 1985; Woods *et al.*, 1982). Elliott and Coleman (1977) estimated that 55 g bacteria m^{-2} yr^{-1} were consumed by protozoans in short-grass prairie, equivalent to 3–4 times the mean microbial standing crop, while Stout and Heal (1967) calculated that up to 85 times the standing crop of readily available (zymogenous) bacteria may be consumed in arable land that has been fertilized by farmyard manure. This is equivalent to a consumption rate of 150–900 g bacteria m^{-2} yr^{-1} (Clarholm, 1984). Since the end product of N metabolism in protozoans appears to be mainly ammonia, protozoan predation on microorganisms is probably of considerable significance for mineralization of organic N and other plant nutrients.

2.2 NEMATODA

2.2.1 Distribution

Nematodes are a widely distributed and successful group of invertebrates. They are found in all sorts of habitats – marine, freshwater and terrestrial – where there is sufficient moisture and organic food. In soil and vegetation, there is a wide range of freeliving forms as well as many plant and animal parasitic forms.

They constitute a significant component of the fauna of grassland soils in a range of climatic and soil types, including temperate mineral soils (Nielsen, 1949; Wasilewska, 1976), tropical soils (Lamotte *et al.*, 1979), semiarid prairie and steppe (Zlotin and Khodashova, 1974; Smolik and Lewis, 1982), hot desert (Freckman and Mankau, 1977) and tundra (Bunnell, MacLean and Brown, 1975). They are usually classified with the soil microfauna, being typically 0.5–5 mm long and 20–100 µm wide. Like protozoans they are hydrophilous and depend on water for their activity. Their cuticle is permeable to water and they are restricted to habitats with a saturated atmosphere. They appear to be inhibited by high moisture content (Wallace, 1963) and to be most active under conditions where the bulk of the pore space is occupied by air except for a thin film of water covering the soil particles and where larger quantities of water are retained at the necks between soil crumbs. They are unable to burrow and are, therefore, restricted by their size to macropores with necks greater than 20 µm diameter; that is, they cannot penetrate the spaces between soil particles but are confined to those between aggregates. Such pores empty of water at a suction pressure of around 15 kPa, about field capacity, and Nielsen (1967) suggests that nematode activity is confined to suction pressures between this point and a lower moisture limit somewhat short of permanent wilting point. There are thus many situations where nematode activity is likely to be limited by either excessive or insufficient moisture content, even in temperate grassland. For example, the upper few centimetres of sandy Danish soils dry out to pF 3–4 several times during most summers, probably resulting in dehydration and mortality of many nematodes that occur abundantly in the upper 5 cm of the soil layer (Nielsen, 1967). Some surface-dwelling species such as *Ditylenchus dipsaci*, the stem nematode, can withstand desiccation.

While nematodes as a group are found in virtually all soil types considerable differences occur among species in their habitat tolerance. In his study of nematode distribution in Danish soils, Nielsen (1949) recognized a eurytopic group of species in sand, clay, raw humus and peaty soils, and a stenotopic group with a more restricted distribution in sandy soils and raw humus. Yeates (1974) studied nematode populations at six sites representing a climatic sequence in New Zealand tussock grassland: *Paratylenchus* was dominant at the two driest sites, *Macroposthonia* occurred at all the wetter sites while *Radopholus* occurred at the wettest site.

Soil nematodes are usually concentrated around root systems and are, therefore, most abundant in the top 10 cm of most grasslands where there is the greatest concentration of plant roots and biological activity associated with root exudates and decaying organic matter. Nielsen (1949) found virtually no nematodes below 20 cm in Danish pasture. In prairie soils with a greater depth of rooting, the vertical distribution of nema-

The composition of the invertebrate fauna

todes may be similarly extended and Smolik (1974) reported only about 70% of nematodes in the top 20 cm in mixed prairie at the US/IBP Cottonwood grassland biome site in South Dakota, USA. Although most nematodes live in soil, many species are also found on the aerial parts of plants. These include free-living stages of animal parasitic forms that ascend vegetation in search of vertebrate hosts, saprophytic forms that feed on phylloplane microflora, and plant parasitic species (notably *Aphelenchoides* and *Ditylenchus*) that invade leaves, stems and flowers of suitable host plants. Wallace (1963) suggests that a characteristic of such aerial forms is their inherently high activity, which enables them to swim upwards in water films on plant surfaces.

Table 2.1 Population density and biomass of nematodes in various grasslands

Author	Site	Numbers $\times 10^6$ (m^{-2})	Biomass $(g\ m^{-2})$*
Wasilewska (1974)	Mountain sheep pasture, Poland	3.2–3.7	1.4–3.0(F)
Wasilewska (1974)	Mountain meadows, Poland	1–8	
Wasilewska (1976)	Meadow, Poland	2.2–2.5	0.84(F)
Nielsen (1949)	Range of sites, Denmark	4–20	6.0–17.8(F)
Törmälä (1979)	Meadow, Finland	6.8; max 12	
Banage (1963)	*Juncus* moorland, England	1.9–3.1	0.5–0.75(F)
Yeates (1979)	Nine New Zealand pastures	0.8–5	
Ricou (1979)	Pasture, France	15–55	0.4–1.5(D)
Zlotin and Khodashova (1974)	Meadow steppe, Russia		5.6(F)
Smolik and Lewis (1982)	Mixed prairie, USA	2–6	0.2–0.8(D)
Sohlenius, Boström and Sandor (1987, 1988)	Grass leys, Sweden	7.8–9.3	0.33(D)
	Lucerne leys, Sweden	4.3–12.8	0.32(D)
Freckman, Duncan and Larson (1979)	Ungrazed grassland, California	2.7–7.3	0.1–0.9(D)
Freckman and Mankau (1977)	Warm desert, USA	0.4	
Banage and Visser (1967)	Tropical savannah, Uganda	3.8	
Bunnell, MacLean and Brown (1975)	Tundra meadow, Alaska	0.65	0.065(D)
Chernova *et al.* (1975)	Tundra herb-grassland, Russia	10.0	7.7(F)

*F = fresh mass; D = dry mass.

2.2.2 Abundance and biomass

Nematodes vary enormously in abundance both within and between habitats; Table 2.1 gives estimated abundances in a range of grasslands. The range represents a hundred-fold variation, ranging from about 0.5 million m^{-2} in climatic extremes, such as warm desert and tundra, to a winter maximum of 55 million m^{-2} under marine west-coast climatic conditions in northern France. Most values fall within the range 2–10 million m^{-2}. Sohlenius (1980) calculated a mean of 9 million (range 2.4– 30 million) for 20 temperate grassland sites for which he had reliable published and unpublished data. The mean biomass was 3800 mg fresh mass m^{-2} (range 650–17800). However, Yeates (1979) calculated much lower means of 1.9 million individuals and 217 mg dry mass m^{-2} based on sites sampled intensively enough to allow annual means to be calculated. There appears to be a good general relationship between nematode abundance and plant cover. Nielsen (1949) noted that the size of the population increased as the density of vegetation increased in mineral soils in Denmark. Yeates (1979) confirmed the relationship between nematode abundance and plant growth when he related herbage production in nine New Zealand grazed pastures over a wide range of soil types and conditions with nematode populations and found a significant linear regression at eight sites (Figure 2.1); at the aberrant ninth site the exceptionally high mean monthly value reflected favourable soil moisture and temperature conditions throughout the year. Wooded and non-

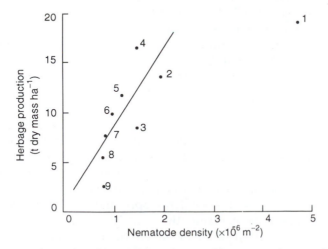

Figure 2.1 Relationship between mean monthly nematode populations and herbage production at nine New Zealand pasture sites; $Y = 0.5 + 8X$, $P < 0.001\%$ if outlier is excluded. (After Yeates, 1979.)

wooded ecosystems appear to have similar nematode population ranges and populations tend to be lower in raw humus and peat soils than in mineral soils (Yeates, 1979). Available data indicate that populations in arable soils tend to be of the same order of magnitude as those in grassland (Sohlenius, Boström and Sandor, 1987; Wasilewska and Paplińska, 1975).

2.2.3 Feeding habits

Nematodes as a group derive their nutrition from a wide range of sources including living plants, bacteria, fungi, actinomycetes, protozoans and other animals. They require living protoplasm for food and are unable to digest cell wall material and dead organic matter, although this may provide a rich source of microbial food. Bacterial feeders constitute a significant proportion of the fauna in most temperate grasslands and arable soils: they have been recorded as forming 30–50% of the nematode fauna in a range of European soils, including Danish pastures (Nielsen, 1949), Polish meadow (Wasilewska, 1976), and Swedish grass and lucerne leys (Sohlenius, Boström and Sandor, 1987). Fungal feeders accounted for 3–65% of nematodes in Polish meadows (Wasilewska, 1976), and for 18–31% in Swedish grass and lucerne leys (Sohlenius *et al.*, 1987). Yeates (1979) found bacterial feeders and omnivores to predominate in a grazed New Zealand pasture. The proportion of plant feeders in the population varies widely under mesic climatic conditions. They constituted 7–66% of the population in Polish meadows (Wasilewska, 1976), 28–41% in Swedish grass and lucerne leys (Sohlenius, Boström and Sandor, 1987), and 4–16% in Danish soils (Nielsen, 1949). Under more arid conditions Yeates (1979) found plant feeders to predominate and Smolik (1974) reported this group as constituting 66% of the nematode fauna in mixed prairie soil in South Dakota, USA.

Bacterial feeding is prevalent among the Rhabditida, with *Rhabditis* spp. being particularly abundant in decaying organic matter. Insect parasitic forms such as *Heterorhabditis* and *Steinernema* spp. also occur in this group. The Tylenchida contains species characterized by having a hollow protrusible mouth spear; this group contains most of the known plant parasitic species but some (e.g. *Tylenchus, Ditylenchus* and *Aphelenchus* spp.) feed on fungi, some on algae and others appear to be parasitic or predacious. The Dorylamida contains the plant-feeding *Xiphinema, Longidorus* and *Trichodórus* spp. and predatory *Dorylaimus* spp. The Mononchida contains the widely distributed carnivorous *Mononchus* spp. that feed on small invertebrates including nematodes, protozoans and small oligochaetes. The Trichosyringida contains *Mermis* spp., which parasitize a range of arthropod hosts.

Plant parasitic species of significance in grassland include ectoparasitic, migratory forms such as *Longidorus elongatus*, *Trichodorus* spp. and *Tylenchorhynchus dubius*, which are common in open textured sandy soils, and *Xiphinema diversicaudatum*, which is common in clay soils and is the vector of brome grass mosaic virus. Endoparasites include the migratory *Pratylenchus* (lesion nematodes) and *Ditylenchus dipsaci* (stem eelworm), and the non-migratory, cyst-forming *Heterodera* spp., such as *H. avenae* and *H. punctata* on grasses and *H. trifolii* and *H. galeopsidii* on clover. *Anguina* spp. form galls on flowers and leaves of many grass species.

2.2.4 Ecological significance

Total energy consumption by nematodes in a range of Polish pasture and meadow sites was estimated to be 335–761 kJ m^{-2} yr^{-1} (Wasilewska, 1974, 1976). Assuming a joulorific value of 18.5 kJ per gram dry mass, this is equivalent to 18–41 g dry plant material. In a meadow site where consumption by individual trophic groups was estimated, microbivores accounted for 109–117, plant parasites 67–125, fungivores 10–11 and omnivores 126–142 kJ m^{-2} yr^{-1}. Estimated nematode consumption in Swedish grass and lucerne leys was 32 and 47 g dry mass m^{-2} representing 7.5–10.3% of organic matter input to the soil (Sohlenius, Boström and Sandor, 1988). Plant feeders were estimated to consume 2–3% of root production. Ricou (1979) calculated that nematodes released 2–8% of the energy from underground detritus in a grassland site in northwestern France. Kajak (1974, 1975) calculated that nematode respiration accounted for almost 40% of total soil faunal respiration in a sheep-grazed Polish pasture although Sohlenius (1980) considers 10–15% to be more typical, with nematode respiration in productive sites being in the range of 84–250 kJ m^{-2} yr^{-1} and 29–42 kJ in unproductive dry or cold areas.

Losses in yield in alfalfa, clover and timothy hay in the USA due to feeding by plant parasitic nematodes have been estimated at 5–6% (Hodges, 1973) and there is evidence that losses in dry soils may be proportionately much greater. Yeates (1979) estimated from data in Smolik (1974) that nematodes consumed 22–32 g plant material m^{-2} in prairie grassland. Smolik (1974) reported 28–59% increase in herbage production following suppression of the nematode population with vydate, suggesting an effect of nematodes on yield out of all proportion to their direct consumption. Significant depression in production of newly established *Lolium* has been attributed to nematodes in the UK (Spaull, Clements and Newton, 1983), while *Ditylenchus dipsaci* and other nematodes can depress clover production in cool temperate grasslands.

Microbial grazing by nematodes may have effects on nutrient cycling that are analogous with those of protozoans. Anderson, Coleman and Cole (1981a) estimated that 19–124 kg N ha^{-1} yr^{-1} could be mineralized by bacteria-feeding nematodes. However, Nielsen (1961) concluded that nematodes were much less important in this respect than protozoans. He calculated that a population of 10 million nematodes per m^2 would consume 800 kg live bacterial biomass per hectare compared with 8000 kg by protozoans and 400 kg by enchytraeid worms under similar conditions.

Table 2.2 Population density and biomass of enchytraeid worms in various grasslands

Author	Site	Numbers $\times$ 10^3 (m^{-2})	Biomass (g m^{-2})*
Healy and Bolger (1984)	Wet soils, Ireland	55.5	
	Dry soils, Ireland	5.9	
Cragg (1961); Peachey (1963)	Moorland limestone grassland, England	25 (max.)	15(F)
Cragg (1961); Peachey (1963)	*Juncus* moorland, England	289 (max.)	53(F)
Cragg (1961); Peachey (1963)	*Nardus* moorland, England	204 (max.)	35(F)
Standen (1982)	*Molinia* peat, England	136 (max.)	
Standen (1982)	Four mineral and peat soils, England	17	
Persson and Lohm (1977)	Fen grassland, Sweden	24	0.85(D)
Törmälä (1979)	Abandoned field, Finland	50	
Nielsen (1955a)	Pasture, Denmark	10–117	0.6–7(F)
Ryl (1980)	Meadow, Poland	17	
Andrén and Lagerlöf (1983)	Permanent pasture, Sweden	25.2	
Lagerlöf, Andrén and Paustian (1989)	Grass ley, Sweden	5.5	0.34(D)
Lagerlöf, Andrén and Paustian (1989)	Lucerne ley, Sweden	9.9	0.54(D)
Möller (1969)	Three grassland sites, Germany	20–45	
King and Hutchinson (1976)	Sheep pasture, Australia	6	
Lal (1981)	Tropical grassland, India	6–18	0.48 (max.)
Lamotte (1979)	Tropical savannah, Africa	10 (max.)	
Thambi and Dash (1973)	Subhumid grassland, India	7 (max.)	
Bunnell, MacLean and Brown (1975)	Tundra meadow, Alaska	56.7	2.3(D)
Chernova *et al.* (1975)	Tundra herb-grassland, Russia	20.5	25.7(F)
Østbye *et al.* (1975)	Dry mountain meadow, Norway	14.5	0.3(D)
Østbye *et al.* (1975)	Wet mountain meadow, Norway	34.5	0.9(D)

*F = fresh mass; D = dry mass.

2.3 OLIGOCHAETA – ENCHYTRAEIDAE

2.3.1 Distribution and abundance

These small, segmented worms (potworms) are widely distributed and are especially common in moist temperate, organic soils. The factors that most influence their distribution appear to be soil moisture, pH, soil organic matter and temperature. Highest populations and biomass tend to occur in acid, organic soils such as moorland peat (Table 2.2). Species appear to vary in their preference for different soil horizons: O'Connor (1967) found *Cognettia* predominantly in the litter layer of a North Wales coniferous forest site while *Marionina* was most numerous in the humus layers and *Achaeta* occurred mainly in the mineral soil.

Enchytraeids as a group appear to be tolerant of acid conditions but Standen (1982) reported that few species (one to three) live in extremely acid soils (pH 3.7–3.8) while communities richer in species occur only where the pH approaches or exceeds 5. *Cognettia sphagnetorum* is particularly tolerant of low pH and is often the dominant species in very acid soils (Standen, 1982). Species such as *Fridericia paroniana* and *Enchytraeus bucholzi* were found to respond positively to liming of acid soil (Hågvar and Abrahamsen, 1980).

Enchytraeids are very susceptible to desiccation and soil moisture is an important factor in determining their distribution, although there does appear to be some variation in tolerance to moisture conditions. *Mesenchytraeus* and *Cognettia* prefer very wet sites and are abundant in peat whereas genera such as *Marionina, Fridericia, Enchytraeus* and *Henlea* tolerate a wider range of moisture conditions and may be found in dry sandy soils (O'Connor, 1967). Most of the common species in Irish soils have a wide range of distribution in relation to soil moisture, but fewer species and lowest population densities occur in soils subject to summer drought (Healy and Bolger, 1984).

Distinct seasonal fluctuations are commonly observed in enchytraeid populations and these often reflect changes in soil moisture level. In arid regions, mean populations are low, but where there is a distinct wet season populations increase. Lal (1981) found a strong positive correlation between seasonal fluctuations in enchytraeid numbers and moisture content in a tropical grassland soil at Varanasi, India. Likewise, Nielsen (1955b) attributed summer minima in dry sandy soil populations in Denmark to desiccation and found that population decrease is directly related to the degree and duration of the desiccation. Enchytraeids die almost instantaneously through water loss when exposed at tensions above 1000 kPa, and cocoons are killed after more prolonged exposure. However, O'Connor (1967) suggested that cocoons of certain species are drought-resistant and hypothesized that this may be related to the fact that cocoons of some species such as *Fridericia bisetosa* are covered with mineral and organic debris. Hatching of drought-resistant cocoons could explain the sudden appearance of large numbers of juveniles after the end of drought in Danish soils.

Nielsen (1955b) suggested that summer declines in populations in Danish soils where moisture conditions are not limiting could sometimes be caused by lethal soil temperatures (> 25–30 °C); this is rarely a problem in temperate soils but undoubtedly is a factor of major significance in arid regions. O'Connor (1967) considers that temperature exerts a major influence on enchytraeid populations and concluded that population fluctuations in widely different habitats can be explained in relation to moisture and temperature. He suggested that seasonal trends in density in permanently moist habitats are largely controlled by the effects of temperature on the rate of population increase. Typically in temperate regions, maximum density will occur in summer followed by a decline throughout the winter to a minimum in late winter, the actual timing of the summer peak depending on how early the spring increase in temperature occurs. Thus, the peak occurs in July in North Wales and in August in Denmark, but is delayed to September or November in the Pennine mountains, England, where spring is late. When high summer temperatures and low rainfall induce drought the spring increase in population is halted; severe mortality occurs in summer, with peak densities in spring and autumn.

2.3.2 Feeding habits and ecosystem role

O'Connor (1967) examined the gut contents of three species in a Douglas Fir coniferous plantation in North Wales and found that they consisted of a mixture of plant material, fungus and silica, the proportion varying with the species. Enchytraeids are assumed to derive their nutrition primarily from the microflora, since Nielsen (1962) did not find any evidence to indicate that they possess the enzymes necessary to digest the constituents of plant cell walls. However, Dash, Nandra and Mishra (1981) found cellulase in tropical species, suggesting that they can utilize plant polysaccharides and may have a greater role in the primary decomposition of plant material than previously thought. They concluded that microbial activity is required to break down more complex hemicellulose and other materials before enchytraeids can feed on them. Fungal biomass provides an important food source for enchytraeids in acid soils. Dash and Cragg (1972) showed that enchytraeids, and other soil fauna in the organic layer of aspen woodland in Canada, had a strong preference for *Centhospora* and *Cladosporium* spp. but were scarce on *Penicillium* and *Paecilomyces*.

The presence of ingested plant and soil material in the guts of enchytraeids suggests that these oligochaetes have a role in soil-mixing analogous to that in earthworms. The process creates a micro-sponge structure in some mineral soils, with clay–humus complexes forming water-stable aggregates (Kubiena, 1955). Nielsen (1961) estimated that enchytraeids respire 142–649 kJ m^{-2} yr^{-1} in three Danish grassland sites while Möller (1969) gives values of 167–460 kJ m^{-2} for three German grasslands. Persson and Lohm (1977) estimated 177 m^{-2} for Swedish grassland, and Hutchinson and King (1980) 131 kJ m^{-2} yr^{-1} for lightly grazed Australian sheep pasture. O'Connor (1967) calculated from the data of Peachey (1963) that enchytraeid respiration could be as high as 3105 kJ m^{-2} yr^{-1} in *Juncus* moorland in the Pennine mountains, suggesting a significant role for enchytraeids in the energy metabolism of raw humus. However, this is probably an atypical case for Standen (1973) in her detailed study of *Cognettia sphagnetorum*, which comprised 90% of the enchytraeid biomass in blanket bog; she estimated total population respiration for a mean standing crop of 12 g fresh mass to be 262 kJ m^{-2} yr^{-1}.

Taking Nielsen's (1961) estimates for respiration, and assuming that respiration represents about 15% of consumption (Lagerlöf, Andrén and Paustian, 1989), enchytraeid populations in favourable grassland could consume 950–4400 kJ m^{-2} yr^{-1}, equivalent to 50–240 g dry matter m^{-2}.

2.4 OLIGOCHAETA – EARTHWORMS

2.4.1 Distribution

Earthworms are widely distributed throughout temperate and tropical regions and, where environmental conditions allow, they dominate the invertebrate biomass. The Lumbricidae is the characteristic family of Palaearctic regions and this group has been the subject of considerable study since Müller (1878) and Darwin (1881) drew attention to its role in maintaining soil fertility.

The Lumbricidae comprise 10% of the total *c.* 3000 species of earthworms (Lee, 1985); about 19 species are common over most of Europe and some 15 have been spread by Europeans throughout temperate regions of the world. Many of these are highly adapted to agriculture and have often displaced indigenous species from cultivated areas of Australia, New Zealand and South Africa (Barley, 1959a; Martin and Charles, 1979; Reinecke and Visser, 1980). Lumbricidae are rare in tropical soils, where the dominant species are members of the families Almidae, Kynotidae, Glossoscolecidae, Megascolecidae, Eudrilidae and Oenerodrilidae (Lee, 1983b). Most studies point to the overwhelming importance of soil water in determining earthworm distribution, although temperature, pH, soil type and food can often be limiting factors.

(a) Water

Unlike the microfauna within the surface water film on soil particles, earthworms are very dependent on the presence of unbound water in the soil, even though they can tolerate considerable moisture losses – up to 63.5% of body mass in the case of *Aporrectodea caliginosa* (Grant, 1955). Their widest distribution occurs in temperate soils with adequate moisture although there is a good deal of intraspecific and interspecific variation in terms of soil-moisture preference, with some scope for adaptation to local conditions (Lee, 1985). Many species have behavioural or physiological mechanisms that enable them to overcome periods of drought. Aestivation in a state of quiescence or diapause is a commonly reported response to unfavourable moisture conditions, although there is some uncertainty as to the extent to which obligatory diapause occurs in earthworms. The evidence suggests that diapause is uncommon in temperate regions (e.g. Gerard, 1967), but may be more prevalent in species occupying soil subject to prolonged and severe seasonal moisture stress. Lamotte *et al.* (1979) reported that when soil moisture falls below a certain minimum (pF 3–4.2) in West African savannah, earthworms empty their guts, dehydrate themselves and coil up tightly. Some also make themselves a case of sand grains stuck together, which limits water loss. Activity is resumed with the onset of the wet season. Nevertheless, mortality is high during the dry period and populations decrease rapidly during prolonged drought. The severity of the dry season appears to be the main factor limiting earthworm populations in tropical grasslands, and significant populations are found only in humid regions with short dry seasons.

A downward shift in the soil profile before aestivation may occur in the cases of *Allollobophora* and *Aporrectodea* spp., which normally occupy the superficial soil layers (Gerard, 1967). This is assumed to be a quest for moister conditions, but does not appear to occur universally (Rundgren, 1975). Deep-burrowing species such as *Lumbricus terrestris* are able to remain active throughout the summer in temperate regions; presumably they can escape the effects of surface desiccation by retreating into their deep burrows. Earthworm species vary in their tolerance to desiccation. Thus, Pierce (1981) showed that surface-dwelling species such as *Lumbricus rubellus* and young *L. terrestris* remained highly mobile up to a water loss of 30% of body mass, whereas in deep-burrowing *Aporrectodea longa* and adult *L. terrestris* mobility was greatly reduced with increasing water loss. This suggests that surface-dwelling forms are more tolerant of desiccation than forms that can escape by burrowing and/or aestivation.

Earthworms are also frequently limited by excess water and are scarce for example, in wet peaty soils. In such cases it is often difficult to separate the effects of excess moisture and accompanying low oxygen tension from those of low pH and unpalatable plant litter. Some species are more tolerant of wet conditions than others and some may prefer wetter soils. Baker (1983) recorded higher numbers of the surface-dwelling *L. rubellus*, the soil-dwelling *Octolasion tyrtaeum* and the green form of *Allolobophora chlorotica* from wet areas of peat grassland in an Irish fen than from drier areas; most species, however, were more abundant in the drier areas. Perel (1977) suggested that *O. tyrtaeum* is more tolerant of wet conditions than other species because of its well-developed subcutaneous net of blood vessels and high concentration of haemoglobin, which enable it to inhabit poorly aerated soil. Surface-living, pigmented species such as *L. rubellus* are relatively independent of soil conditions and are frequently reported from wet habitats like peat. The green form of *A. chlorotica* is frequently reported as having greater tolerance for wet conditions than the pink form. The reason is not clear; possibly its green colour confers on it a cryptic advantage over the pink form in flooded situations where they are forced to live close to the soil surface (Satchell, 1967). A few species (e.g. *Eisenilla tetraedra*) are characteristic of very wet soils and can live under completely submerged conditions.

(b) Temperature

Together with soil moisture, temperature plays a major role in determining earthworm activity and biology. Satchell (1967) states that 10.5 °C is optimum for *Lumbricus terrestris* activity while Nordström (1975) reported that *Lumbricus* and *Dendrobaena* spp. are active within the range 0–20 °C in Swedish soils. *Allolobophora/Aporrectodea* spp. have a narrower range, being active between 2–4 °C and 14–16 °C. Low winter temperatures induce quiescence combined with downward movement in the profile in both Sweden and England (Gerard, 1967; Rundgren, 1975). It is unlikely that most earthworms can tolerate temperatures lower than 0 °C or greater than 30 °C, and it is, therefore, likely that temperature exerts a direct effect in limiting earthworm populations in arid regions when soil temperatures in excess of 30 °C are common during the dry season, and in arctic and subarctic regions where the soil is frozen for prolonged periods during the winter. However, species characteristic of arid soils may tolerate higher temperatures: James (1982) reported that *Diplocardia singularis* was active at 30 °C in soils that were wetted in the lab. Even in temperate regions where lethal temperature conditions are likely to affect only the surface layers of grassland soil there may be considerable mortality among juveniles with limited burrowing ability; adults can probably readily avoid unfavourable temperatures by burrowing (Satchell, 1967).

(c) pH

Maximum populations of earthworms are found in base-rich, neutral, mineral soils; populations are low in acid moorland and heathland. However, the relative importance in determining earthworm distribution of pH and factors correlated with pH, such as availability of Ca and other bases and the nature of the vegetation and litter, is not known. Studies such as those of Laverack (1961) indicate quite a wide range of pH tolerance, and the threshold values at which acid-sensitive nerves respond in 'acid-intolerant' species such as *Aporrectodea longa* (pH 4.6–4.4) and 'acid tolerant' species such as *Lumbricus rubellus* (pH 3.8) are not sufficiently different to explain the quite striking differences between the earthworm communities inhabiting acid soils with pH below about 5 and those inhabiting soils with higher pH.

(d) Soil type

The most striking differences in the composition and abundance of the earthworm fauna are seen in soils with mull and mor types of organic matter. Fertile, free-draining mull soils in temperate regions are characterized by high earthworm densities and biomass and comprise mainly the burrowing species such as *Lumbricus terrestris*, *Aporrectodea/Allolobophora* and *Octolasion* spp. By contrast, mor soils are impoverished, acidic, and often too dry (sandy heaths) or too wet (peat) for earthworms, and support vegetation such as *Calluna*, *Nardus*, *Molinia* and coniferous forest that produces unpalatable litter. Under these conditions the burrowing species are absent and earthworms are represented by sparsely distributed, surface-dwelling, litter-feeding species such as *Lumbricus rubellus*, *L. castaneus*, *Dendrobaena octaedra*, *Dendrodrilus rubidus* and *Bimastos eiseni*.

Mineral soils of medium texture provide the most suitable conditions for earthworms. Cotton and Curry (1980b) recorded very low population densities from sandy and silty coastal sites at Kilmore, Co. Wexford (maximum 73 m^{-2}) compared with nearby loam soils that were similarly managed and supported up to 516 worms m^{-2}. Guild (1948) reported that medium loams had higher populations than soils of clay or sandy texture and Nordström and Rundgren (1974) reported correlations between the abundance and occurrence of deep-burrowing species and soil factors such as clay content. However, it is likely that such relationships are secondary and reflect the effects of other factors of more direct relevance, such as moisture, base status and food.

2.4.2 Abundance and biomass

Table 2.3 lists earthworm population densities and biomass reported in numerous studies, mainly in temperate regions. As is the case with all

Table 2.3 Population density and biomass of earthworms in various grasslands

Author	Site	Numbers (m^{-2})	Biomass $(g\ m^{-2})$*
Nowak (1975)	Sheep-grazed mountain pasture, Poland	83–99	14.5–24.7 (F)
Nowak (1975)	Mountain pasture heavily fertilized with sheep dung, Poland	233–591	32.1–132.1 (F)
Nowak (1976)	Lowland meadows, Poland	61–230	5.6–26.3 (D)
Evans and Guild (1948b)	Old pasture, England	129–190	113–164 (F)
Vernon, Findlay and Lyons (1981)	Seven grassland sites, England	97–212	36–121 (F)
Reynoldson (1955)	Permanent pasture, North Wales	215–570	90–167 (F)
Curry (1976b)	Grassland, inorganically fertilized, Ireland	114–516	41–197 (F)
Cotton and Curry (1980a,b)	Grassland fertilized with animal manures, Ireland	222–651	111–324 (F)
Baker (1983)	Fen peat grassland, Ireland	< 197	
Waters (1955)	Permanent pasture, New Zealand	741–1236	146–303 (F)
Barley (1959a)	Pasture, New South Wales, Australia	260–740	39–152 (F)
Hutchinson and King (1980)	Pasture, New South Wales	27–145	29–258 (F)
Nordström and Rundgren (1973)	Permanent pasture, Sweden	109	59.4 (F)
Persson and Lohm (1977)	Fen grassland, Sweden	133	5.93 (D)
Bengtson et al. (1975)	Grass meadows, Iceland	68–312	14.5–40 (F)
Svendsen (1957a)	Moorland mineral soils, Cumbria, England	260–350	53–85 (F)
Bouché (1977)	Permanent pasture, France	288	125 (F)
Standen (1979)	Moorland mineral soils, Cumbria, England	40–100	
Zajonc (1975)	Meadows, Russia	203–251	
Zajonc (1982)	Mountain meadows, Czechoslovakia	7–221	
Törmälä (1979)	Meadow, Finland	300	
Reynolds (1973)	*Andropogon* grassland, Tennessee, USA	13–41	3–8 (F)
Lloyd et al. (1973)	Short-grass prairie, USA	1–39	< 0.3 (D)
James (1982)	Tall-grass prairie, USA	< 44	< 2 (D)
Zlotin and Khodashova (1974)	Meadow steppe, Russia		83 (F)
Chernova et al. (1975)	Herb grassland, tundra, Russia	150	60 (F)
Reinecke and Visser (1980)	Irrigated lucerne, South Africa	< 900	< 160 (F)
Lamotte (1947)	West African savannah	50	20 (F)

Table 2.3   (continued)

Author	Site	Numbers (m^{-2})	Biomass $(g\ m^{-2})$
Block and Banage (1968)	Pasture, Uganda	21.3	2 (F)
Lavelle (1973)	Savannah, Sudan	5–20	2–10 (F)
Lavelle (1974)	Humid savannah, Ivory Coast	180	44 (F)
Lavelle (1973)	Highland grassland, Guinea	150–300	50–120 (F)
Dash and Patra (1977)	Tropical grassland, India	64–800	6–60 (F)

*F = fresh mass; D = dry mass.

groups of soil fauna, the data conceal considerable variation, arising from factors such as extraction efficiency and times of sampling, but the general picture is of denser populations and greater biomass in mesic temperate grasslands than in areas with more extreme climates. It is interesting to note that sites from which highest biomass have been recorded – those studied by Evans and Guild (1948b), Waters (1955), Barley (1959a), Curry (1976b), Cotton and Curry (1980a,b), Hutchinson and King (1980), and Reinecke and Visser (1980) – are all productive, managed grasslands, supporting the conclusion of Waters (1955) that there is a direct relationship between sward productivity and worm populations.

Discrepancies between population densities and biomass in Table 2.3 often reflect the degree to which the deeper burrowing, larger species such as *Lumbricus terrestris*, which can weigh more than 5 g live mass per worm, are represented. The high biomass supported by the Australian and New Zealand sites is all the more remarkable when one considers that *L. terrestris* is absent or virtually absent and that the population largely consists of *Aporrectodea caliginosa*, which reaches a maximum live mass of about 1 g per worm.

Tropical and subtropical sites present a variable picture, largely reflecting soil-moisture status. Grasslands in humid regions with a short dry season, such as Lamto, Ivory Coast, and Orissa, India (Lavelle, 1974; Dash and Patra, 1977), support relatively large populations; highland areas with more equable climate, such as Mt Mimba in Guinea (Lavelle, 1973), support populations comparable with those in temperate grasslands (Table 2.3). Savannah with less regular rainfall has lower populations, and earthworms are totally absent from dry savannah (Lamotte *et al.*, 1979). Earthworm population densities in American prairie grasslands are relatively low, reflecting the moisture-limited nature of those habitats. However, significant earthworm populations have been recorded from some steppe sites in the former USSR under comparable climatic conditions (Ghilarov and Chernova, 1974, cited by Petersen and Luxton, 1982; Zlotin and Khodashova, 1974). Earthworms are scarce or absent in tundra soils, an exception being some habitats at Taimyr, in Russia where a dense population of one species (*Eisenia nordenskiöldi*) was recorded (Chernova *et al.*, 1975).

2.4.3 Ecological classification

Many aspects of earthworm biology and activity correlate with their size and burrowing abilities and several ecological classifications have been proposed taking those features into account. That of Bouché (1972) has been adopted by many workers; it divides the population into (1) surface active, pigmented, non-borrowing litter dwellers (epigées); (2) large, deep-burrowing forms that come to the surface at night to draw down litter (anéciques); and (3) species that live in surface mineral and organic horizons and construct branching, horizontal burrows (endogées). The epigées are best represented in soils where there is an accumulation of surface organic matter such as deciduous woodland and acidic grassland. Anéciques such as *Lumbricus terrestris* and *Aporrectodea longa* usually dominate the biomass in fertile temperate grasslands although endogées such as *Aporrectodea caliginosa*, *A. rosea* and *Allolobophora chlorotica* may be equally or more abundant. The proportions of these three groups in permanent pasture at Citeaux, France were as follows (Bouché, 1977): anéciques 49.9%, endogées 40.6% and epigées 9.5% of the numbers; 77.2%, 19.8% and 3.0% of the biomass, respectively. By contrast, earthworm communities in savannah and tropical grasslands consist predominantly of geophagous endogées. Lavelle, Sow and Schaefer (1980) reported that 97% of the earthworm biomass in grass savannah at Lamto, Ivory Coast, and almost 100% in secondary tropical pasture in Vera Cruz, Mexico, belonged to this category. The endogées derive their nutrition from organic material dispersed in the mineral soil whereas the epigées consume decaying plant residues on the soil surface. Anéciques prefer surface plant litter but can also derive their nutrition from soil organic matter when surface litter is scarce.

Life-history characteristics that appear to reflect burrowing ability include cocoon production and the phenology of juvenile emergence. Surface-dwelling epigées such as *L. rubellus* and *L. castaneus*, which are most vulnerable to desiccation and predation, produce high numbers of cocoons annually (65–106), endogées such as *A. chlorotica*, *A. caliginosa* and *A. rosea* in the topsoil produce intermediate numbers (8–27) whereas deep-burrowing endogées such as *A. longa* have the lowest numbers (3–8) (Evans and Guild, 1948a; Satchell, 1967). Rundgren (1977) studied the seasonality of emergence of young juvenile lumbricids in southern

Sweden and found that, even though small juveniles occur at all seasons, climatic influences and other factors tend to synchronize emergence to one or two periods. In general, burrowing species (endogées and anéciques) have a bimodal emergence pattern, with peaks in April/May and September/October, whereas surface-living species tend to have single and relatively prolonged emergence periods from summer through autumn. He suggested that bimodal emergence is a risk-spreading adaptation in burrowing species that produce few cocoons whereas high production and patchy distribution of cocoons in a varying surface layer may offset the disadvantage of a single emergence period for surface-dwellers.

The epigéic/anécique/endogéic classification may be less applicable to earthworm communities of tropical soils than to temperate lumbricid species. Of the seven dominant species in impoverished savannah at Lamto, Ivory Coast, two feed on litter mixed with some soil, and five are strictly geophagous (Lavelle, 1979). Species with anécique characteristics appear to be scarce in tropical soils, and Lavelle concluded that this group may only be important among the Lumbricidae.

2.4.4 Seasonal trends

Marked trends in earthworm numbers and activity occur in regions where climate is seasonal. Most studies in temperate regions indicate spring and autumn peaks (Evans and Guild, 1948b; Barley, 1959a; Gerard, 1967; Nordström 1975; Nowak, 1975; Bouché, 1976; Martin and Charles, 1979). These spring and autumn peaks coincide with favourable soil temperature and moisture conditions and augmentation of the population by flushes of emergence from cocoons. Activity declines with the onset of summer drought, during which much of the soil population is inactive and considerable mortality occurs among juveniles close to the surface. With the onset of autumn, survivors resume activity and the population is rapidly supplemented by juveniles emerging from cocoons that have survived the summer. Activity continues until the onset of winter frosts.

The duration of activity depends on the species and the climate. Barley (1959a) found earthworms (predominantly *A. caliginosa* and *A. rosea*) to be active for 24 weeks a year in New South Wales. Nordström (1975) reported comparable periods of activity for *Allolobophora* spp. (6–7 months) in pastures in southern Sweden, but found that *Lumbricus* and *Dendrobaena* spp., which did not become inactive in summer, were active for 9–10 months each year.

In more equable regions, seasonal changes in abundance cannot be explained so readily in terms of climate. Waters (1955) concluded that seasonal changes in earthworm mass and numbers are due to seasonality of food supply at Palmerston North in New Zealand where climatic conditions are very favourable for earthworms. He attributed the late summer to early winter rise in earthworm biomass to a seasonal flush of food in the form of root debris and, to a lesser extent, dead herbage, and the seasonal declines to the exhaustion of the food supply and to the adverse effects of poor aeration resulting from high soil moisture.

2.4.5 Food

Earthworms derive their nutrition mainly from dead organic material and its associated microflora although at least some species feed actively on living roots (Baylis, Cherrett and Ford, 1986). The surface-dwelling species (epigées) ingest plant litter whereas the endogées are geophagous deriving their nutrition from the organic material dispersed through ingested soil. Piearce (1978) examined the gut contents of six species and found largely undecayed plant remains in the guts of two epigées (*L. rubellus* and *L. castaneus*), well-decayed, amorphous organic material in the endogées *A. caliginosa* and *A. chlorotica*, while the anécique *A. longa* had both undecomposed and amorphous organic material in its gut. Anéciques such as *A. longa* and *L. terrestris* appear to ingest surface litter when available but when it is not they rely on soil organic matter.

Organic materials vary very much in their value as earthworm food. Barley (1959b) found that *A. caliginosa* lost weight when fed on decaying *Phalaris* leaves and roots, maintained weight on clover leaves and roots, and gained weight rapidly on dung, particularly when it was mixed with soil. Boström (1988b) and Boström and Lofs-Holmin (1986) reported marked differences in growth rates of *A. caliginosa* juveniles grown in soil amended with residues of meadow fescue, barley and lucerne. Among the factors thought to account for the differences were N content, particle size and, in the case of fresh lucerne residues, toxins that caused high mortality. Satchell and Lowe (1967) studied preferences of *L. terrestris* for leaf litter and found large differences in palatability between species. Factors positively correlated with palatability include N and soluble carbohydrate content, whereas polyphenol content (especially tannins) was negatively correlated. Palatability increased with weathering and this was attributed to microbial degradation of tannins.

Many workers have noted that fresh leaf litter is not acceptable to earthworms but must undergo a period of weathering before it is eaten. Factors believed to be involved in increasing palatability include microbial degradation and leaching of feeding inhibitors (Satchell and Lowe, 1967) and softening of hard tissues. Wright (1972) found that fresh apple leaves were not eaten by *L. terrestris* but were readily consumed after soaking in water for 48 h. However, when fresh leaves were cut in narrow strips they were readily consumed, suggesting that texture is not important if the material is small enough to be swallowed whole; otherwise, it has to be soft enough to be torn apart.

The soil microflora appears to have an important role in earthworm nutrition. Waters (1951) reported that *A. caliginosa* fed on clover litter only when it was mixed into the soil and had been attacked by bacteria and fungi. Wright (1972) noted that apple leaves drawn into burrows by *L. terrestris* that were partially eaten were heavily infested with micro-organisms. Discs of apple leaves and filter paper that had been artificially coated with bacterial cells were more readily consumed by *L. terrestris* than were uncoated leaves. Cooke and Luxton (1980) found filter paper discs contaminated with two soil fungi (*Mucor* and *Penicillium*) to be more palatable to *L. terrestris* than uncontaminated leaves, although discs contaminated with the bacterium *Pseudomonas fluorescens* were not. These results suggest that earthworms cannot derive their nutrition from normal nutrient-poor soil organic matter but depend on microfloral biomass to augment their diet.

Dung is a highly nutritious source of earthworm food and is actively sought out by surface-living species in poor-quality habitats such as moorland (Svendsen, 1957a,b). Amendment of soil with dung increases rates of reproduction and growth (Evans and Guild, 1948a; Meinhardt, 1973, 1974; Lofs-Holmin, 1983b) and increases population density (Chapter 5).

2.4.6 Ecosystem role

Earthworms have a low metabolic rate, and their direct contribution to energy utilization in grassland is small relative to their biomass. Estimates for earthworm respiration in European grasslands range from 314 kJ m^{-2} yr^{-1} in Swedish fen grassland (Persson and Lohm, 1977) to 947 kJ m^{-2} yr^{-1} in French pasture with a mean biomass of 125 g m^{-2} (Bouché, 1977). Barley (1964) calculated an annual respiratory loss of 280 kJ m^{-2} for a population of *Aporrectodea caliginosa* in an Australian temperate pasture. This represented *c.*4% of the total estimated annual litter oxidation, while Nowak (1976) estimated that earthworms metabolize 3–9% of the organic matter entering a Polish meadow soil. Rates of energy utilization in tropical soils can be considerably higher. Dash and Patra (1977) calculated that the total amount of energy assimilated by a grassland earthworm population of mean biomass 30.25 g m^{-2} in Orissa, India, was 4260 kJ m^{-2} yr^{-1}, representing 13.2% of the total energy input. Of this, respiration accounted for 1206 kJ, mucus production 2377 kJ and tissue production 678 kJ m^{-2}, a much higher rate of tissue production that those recorded from temperate soils (Nowak, 1975; Lakhani and Satchell, 1970).

Earthworm activity greatly modifies soil properties and significantly affects soil fertility and plant growth through litter incorporation and fragmentation and through burrowing and soil mixing, etc. These effects will be considered in Chapter 6.

2.5 MOLLUSCA

2.5.1 Distribution and biology

Two main groups of land molluscs, slugs and snails, are widely distributed in grassland. Newell (1967) considers that their distribution is governed primarily by moisture, calcium, shelter and food. Moisture is especially important for slugs and their eggs, which are very susceptible to desiccation. Lutman (1977) found that slug populations were lowest in drier areas under moss in montane grassland in Wales and were highest in wet, peaty soils under *Nardus*, while high populations and damage to agricultural crops are associated with heavy soils and dense vegetation (Gould, 1961). By contrast, some species of snails can withstand dryness for relatively long periods; many species enter a period of suspended activity in the summer (aestivation) and hibernate during the winter.

Soil reaction has little direct effect on mollusc distribution (Newell, 1967); the main factor is the supply of Ca ions in food plants, but since plants growing in acid soil have low Ca content molluscs are mainly associated with base-rich soils. However, Lozek's (1962) data suggest that some species inhabiting steppe and woodland soils of Central Europe can tolerate low calcium conditions.

Slugs and snails are hermaphrodite and cross-fertilization is usual. Some common grassland species such as *Deroceras reticulatum* have two generations a year with spring and autumn breeding peaks, while other species with one generation a year breed mainly in late summer in temperate regions. Hatching may be completed within 3–4 weeks under warm spring conditions while eggs laid in late autumn may not hatch until the following spring.

2.5.2 Abundance

Because of the aggregated behaviour of molluscs and sampling difficulties there are few reliable estimates of mollusc population density in grassland. The situation has now been improved considerably with the advent of the defined area trap, which provides a reliable and simple method of estimating slug populations (Ferguson, Barratt and Jones, 1989). Snails appear to be less abundant in grassland than in more sheltered deciduous woodlands where a plentiful supply of food is available in the form of leaf litter. Mason (1974) recorded up to 489 snails m^{-2} from beech woodland in England and considers populations of 1–30 m^{-2} to be typical of open grassland, although he points out that biomass may be greater in grassland because of the predominance of larger species such as *Cepaea nemoralis*.

Most interest in slugs has been in arable crops where they often cause economic damage. Factors influencing population density have been reviewed by Godan (1983); weather conditions and food appear to be of paramount importance. Populations approaching 150 m^{-2} have been reported from wheat and other crop fields (Thomas, 1944; Hunter, 1966). Estimates from temperate grasslands under favourable moisture conditions include up to 20 slugs m^{-2} for pasture in Ireland (Munnelly, 1970) and 30 m^{-2} for montane grassland in Wales (Lutman, 1977), and 68–115 *Deroceras reticulatum*, 12–102 *Arion hortensis* and 68–99 *Milax budapestensis* associated with tufts of *Dactylis glomerata* in Northumberland pasture, England (South, 1965). Molluscs are normally scarce in drier regions: for example, Lamotte (1947) recorded fewer than 1 m^{-2} from West African savannah.

2.5.3 Feeding habits

Terrestrial molluscs as a group have fairly generalized feeding habits, but some species are predominantly phytophagous while others feed mainly on decaying organic matter. The herbivores include species of the genera *Milax*, *Limax*, *Deroceras* and *Arion* and these can be important crop pests, although some species within these genera such as *Milax soworbii*, *Arion subfuscus* and *Limax* spp. appear to prefer fungi to plant material (Taylor, 1907). *Deroceras reticulatum*, the field slug, is frequently the dominant species in grassland. Pallant (1972) found that this species is primarily herbivorous, mainly ingesting grass leaves and, to a lesser extent, dicotyledon leaves and other material. Lutman (1977) came to similar conclusions about *Deroceras* but concluded that other species in montane grasslands in Wales (mainly *Arion* spp.) tend to take senescent and dead plant material. The feeding habits of *Cepaea nemoralis*, a common grassland snail, have been the subject of several studies (Richardson, 1975; Williamson and Cameron, 1976). It apparently feeds on most plant material available but has a strong preference for dead and senescent herbs, and during the year the diet changes according to the availability of preferred materials.

2.5.4 Ecosystem role

Slugs have frequently been reported as pests of various field crops and in grassland they may seriously affect the establishment of grass legume leys (Port and Port, 1986). Their effect on plant growth in established grassland is normally probably fairly small. Lutman (1977) estimated that slugs consumed 16.3 g plant material a year, representing 0.6% of primary production in montane grassland. However, he pointed out that this consumption occurred mainly in the period from mid-October to mid-April when primary production is at a standstill and slug grazing could reduce the standing crop of photosynthetic material and delay the onset of growth in the spring. Slug damage to white clover is widespread in British pasture, but the impact on productivity has not been quantified (Clements and Murray, 1991; Lewis and Thomas, 1991).

Molluscs appear to possess a large complement of enzymes, including cellulases (Nielsen, 1962; Hartenstein, 1982) and can, therefore, be assumed to have some role in the primary decomposition of plant litter. This role is probably not very important in grassland since Mason (1974) calculated that a population of 489 molluscs m^{-2} in *Fagus* woodland only consumed 0.35–0.72% of the leaf input. It has been suggested that mucoproteins in mollusc faeces and slime may have a role in the binding of soil particles and thus in the development of crumb structure in the soil (Newell, 1967).

2.6 ISOPODA

2.6.1 Distribution

Isopods are the only crustaceans that have become widespread on land; they have successfully colonized a wide variety of terrestrial habitats, ranging from humid tropics to arid desert. They are highly susceptible to desiccation and must find shelter from drought in moist microhabitats. Paris (1963) described how *Armadillidium vulgare* lives on or near the soil surface during rainy weather in Californian grassland but during drought seeks refuge from desiccation by descending into soil fissures to a depth of 45 cm. On the surface it takes refuge from desiccation in small areas of soil kept moist by precipitation of water of condensation from certain tall plants, notably *Brassica campestris*; the condensation results from low stratus that periodically envelops the California Coast Range at night.

Moisture is of central importance in determining the distribution and abundance of isopods and, in general, the distribution of the various families reflects their ability to avoid desiccation. The Trichoniscidae require very moist conditions; the Oniscidae, Porcellionidae and Armadillidiidae are progressively more adapted to drier habitats. Isopods are most abundant in calcareous soils, especially *A. vulgare*, which is largely confined to calcareous grassland. However, other widely distributed species such as *Trichoniscus pusillus*, *Oniscus asellus* and *Porcellio scaber* are less restricted by soil reaction and were more abundant in Dutch deciduous woodland on acid than on calcareous soils (van der Drift, 1962).

2.6.2 Abundance and biomass

Isopods reach their greatest abundance in unmanaged temperate grasslands where populations are frequently in the range 500–1000 m^{-2} (Table 2.4). The maximum population recorded for any species was 7900 m^{-2} for *T. pusillus* in scrub grasslands in Britain (Sutton, 1972). Populations are favoured by the ample litter and sheltered environment available in rough grassland and under favourable conditions biomass can exceed 5 g m^{-2} (Table 2.4). Mean annual biomass exceeded 7 g m^{-2} in some years in dune grassland in Yorkshire (Davis and Sutton, 1978), while those authors calculated a value exceeding 8 g m^{-2} for Californian coastal grassland based on data given by Paris and Pitelka (1962). Marked seasonal fluctuations in abundance and biomass have been reported, reflecting the life histories of the species present. Reproduction typically occurs from April to September and peak density occurs during the breeding season after the recruitment of juveniles into the population (Dunger, 1958; Sutton, 1972). Post-breeding female mortality and high juvenile mortality during the first 6 months of life bring a population decline, with minimum populations being reached in spring before the new breeding season. Marked seasonal changes in biomass also occur and these may reflect the different life-history patterns of the various species present, as illustrated by the example in Figure 2.2. Assuming no net migration, changes in biomass reflect the differences between weight gains due to tissue and embryo growth and losses due to mortality. In the

Table 2.4 Population density (numbers m^{-2}) and, in parentheses, biomass (g m^{-2}) of isopods in four grassland sites

	Rough grassland (Wytham, Oxon, England; Sutton, 1972)	Dune grassland (Spurn Head Yorkshire, England; Davis and Sutton, 1978)	Heath grassland (East Anglia, England; Al Dabbagh and Block, 1981)	Coastal grassland (California, USA; Paris and Pitelka, 1962)
Armadillidium vulgare	6 (0.18)	184 (3.0)	400–500 (4.7–6.6)*	538 (8.1)*
Trichoniscus pusillus	1270 (0.78)			
Philoscia muscorum	103 (0.63)	322 (2.0)		
Porcellio scaber		59 (0.5)		
Total isopods	1580 (1.6)	565 (5.5)		

*Biomass values estimated by Davis and Sutton (1978).

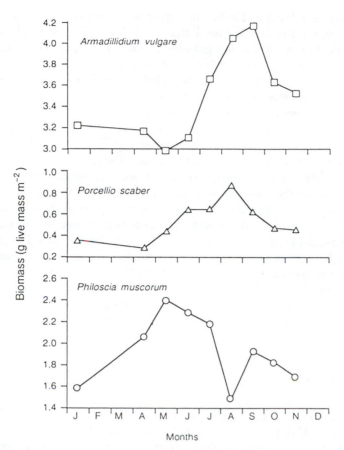

Figure 2.2 Seasonal changes in biomass of three isopod species in dune grassland, Spurn Head, Yorkshire, England. (From Davis and Sutton, 1978.)

example considered, the biomass increase in slowly developing *A. vulgare* (22 months to maturity) throughout the summer was attributed to growth of juveniles, whereas in the more rapidly developing *P. muscorum* (10 months to maturity) maximum growth occurred in the spring prior to the breeding season. Heavy mortality of young and spent females caused a sharp decline in biomass in late summer, with some recovery due to growth of surviving young in the autumn. *P. scaber* showed intermediate trends.

2.6.3 Feeding habits

Isopods feed primarily on dead plant material, although dung, animal remains and some living plant material may also be consumed (Sutton,

1972; Paris and Pitelka, 1962). They may show a preference for certain plant species: Paris (1963) found that *A. vulgare* prefers dead tarweed (*Picris echioides*) in Californian grassland, but concluded that other kinds of vegetation are equally suitable. Rushton and Hassall (1983b) reported that *A. vulgare* prefers litter from dicotyledonous plants from a grassland heath in East Anglia to that from monocotyledons and bryophytes, but preference for monocotyledonous litter increases as this decays. Higher growth and fecundity rates were also found on dicotyledonous plant material, while feeding on monocotyledenous litter decreased growth, fecundity and survival (Rushton and Hassall, 1983a). According to Sutton (1972), leaching and bacterial conditioning of some litters may be required before they are taken, especially if the tannin content is high. Isopods appear to be unable to acquire their copper requirements from litter when it is organically bound, and there is evidence that it must first be made available by microbial activity (Wieser, 1968, 1978). The practice of coprophagy was believed to be a behavioural adaptation that facilitated this; however, Hassall and Rushton (1982, 1985) concluded from feeding studies with *P. scaber* that copper is not normally a critically limiting nutrient for isopods, but that coprophagy is necessary in order to increase the overall nutrient status of the faeces through enhanced microbial activity.

2.6.4 Ecosystem role

Where abundant, isopods can be among the main invertebrate decomposers. In the Yorkshire dune grassland ecosystem studied by Hassall and Sutton (1978), isopods consumed about 1300 kJ m^{-2} yr^{-1}, equivalent to about 10% of the total annual input of litter. Like most invertebrate decomposers their assimilation efficiency is low and their direct contribution to litter oxidation is limited: only 162 kJ m^{-2} yr^{-1} energy was dissipated through their metabolic activity in the study quoted. However, their indirect contribution to litter metabolism through conditioning litter for microbial activity may be significant (Chapter 6). Also, their role in mineral cycling may be greater than that in energy release since Reichle (1967) reports that they may absorb as much as 87% of mineral nutrients from litter.

2.7 MYRIAPODA

2.7.1 Pauropoda

These small myriapods (up to 1 mm in size) are true soil dwellers with only 10% occurring in the surface 15 cm of a pasture studied by Salt *et al.*, (1948). They are also found living superficially under stones, in leaf litter and moss, etc. Their biology and feeding habits are poorly understood. Starling (1944) reported that *Pauropus silvaticus* breeds mainly in early summer in US woodland and development from newly emerged larva to adult takes 3–4 months. They are reported to feed on decaying plant material, carrion and fungi, and some species may be predatory (Verhoeff, 1934).

Populations in excess of 2000 m^{-2} have been reported from deciduous woodland in Tennessee, USA (McBrayer, Reichle and Witkamp, 1974). Densities in grassland soils generally seem to range from a few hundred to in excess of a thousand per m^2. Salt *et al.* (1948) recorded 600 m^{-2} from permanent pasture in Cambridgeshire, England, Curry (1969a) 1023 m^{-2} in old grassland in Ireland, and Belfield (1956) 1022 m^{-2} in West African pasture. Andrén and Lagerlöf (1983) reported around 3500 m^{-2} from permanent pasture while Lagerlöf and Scheller (1989) reported mean population densities of 1100 and 1200m^{-2} from grass and lucerne ley, respectively, in Sweden. Because of their small size and relative scarcity pauropods are generally assumed to be of little significance in the functioning of the grassland soil ecosystem.

2.7.2 Symphyla

Symphylids are widely distributed in temperate and tropical regions, preferring moist organic loam soils of open texture (Edwards, 1958). They are often more abundant deeper in the soil than in the surface layers. Pronounced seasonal vertical migrations in the profile appear to be a feature of symphylid behaviour. Edwards (1958, 1959) reported that high surface temperature and desiccation induces downward migration in summer, with upward migration in spring and autumn when conditions in the surface soil are favourable. Seasonal vertical migrations also appear to reflect changes in feeding, moulting and ovipositional cycles (Raw, 1967).

Oviposition in *Scutigerella* commences in spring and continues throughout summer and early autumn (Anglade, 1967), and developmental time from hatching to adult is 12–36 weeks (Edwards, 1955). Highest populations have been recorded from glasshouses (up to 22 000 m^{-2} and cultivated soils (up to 75 000 m^{-2}) (Edwards, 1958). Edwards (1958) recorded populations of 100–1750 m^{-2} in British grasslands. Curry (1969a) recorded 800 m^{-2} from the top 7.5 cm of an Irish grassland, Athias (1976) reported mean populations of over 2000 m^{-2} in burned savannah at Lamto, Ivory Coast, and Belfield (1956) 10 800 m^{-2} from West African pasture.

Symphylids as a group are polyphagous, with the larger Scutigerellidae preferring living plant tissue and the smaller Scolopendrellidae feeding mainly on decaying plant and animal material and dead microorganisms (Edwards, 1974a). On the basis of gut contents Walter, Moore and Loring (1989) concluded that grassland symphylids (*Symphylella* sp.) in the central USA should be regarded as omnivorous predators. Most gut contents were comprised of soft-bodied microarthropods (Acari, Collembola and Symphyla) and nematodes.

2.7.3 Chilopoda

Chilopods (centipedes) are widespread in most habitats in tropical and temperate regions. The Scolopendromorpha and Scutigeromorpha are restricted to the tropics, while the Lithobiomorpha and Geophilomorpha are widely distributed in temperate regions. Being susceptible to desiccation, they are predominantly associated with moist woodland habitats, but are also common in grassland. The lithobiomorphs are unable to burrow and are mainly surface-soil and litter dwellers, while the geophilomorphs are strong burrowers and are essentially subterranean in occurrence.

Breeding mainly occurs in the spring and autumn, and development from egg to adult extends over 3–4 years. Morris (1968) recorded a population density of 387 centipedes m^{-2} in ungrazed chalk grassland in England, but estimates from other grassland sites are considerably lower: 50 m^{-2} in West African pastures (Belfield, 1956), 63 m^{-2} in German meadows (Albert, 1982) and 120 m^{-2} in Irish grassland (Curry, 1969a).

Centipedes are primarily carnivorous, feeding on many groups of soil and litter invertebrates, although significant amounts of plant material have been reported from the gut in some instances. They are active predators and when abundant probably are of some significance in the natural regulation of soil invertebrates.

2.7.4 Diplopoda

Like centipedes, diplopods (millipedes) are primarily woodland animals, but are also common in rough grassland. The Polydesmoidea (flat millipedes) are fairly inactive and are restricted to the superficial layers, while the Iuliformia (snake millipedes) such as *Blaniulus guttulatus* and Pentazonia (pill millipedes) such as *Glomeris marginata*, both of which are common in grassland, are active soil burrowers. Millipedes have a worldwide distribution, being particularly common in the tropics although little information is available on the tropical fauna. In temperate regions, egg laying may occur in spring, summer or autumn, depending on the species, and eggs are often protected within specially constructed nests. The length of the life cycle varies from 1 to 3 years.

Highest populations have been reported from deciduous woodland

and unmanaged grassland. Dunger (1958) recorded 210–700 m^{-2} in woodland on calcareous soils while Morris (1968) recorded 114 millipedes m^{-2} (almost entirely Polydesmoidea) from ungrazed chalk grassland in England compared with only 22 from grazed grassland. Curry and Momen (1988) recorded a mean population density of 350 Polydesmoidea m^{-2} (mainly *Brachydesmus* and *Polydesmus* spp.) from unmanaged ley grassland on reclaimed peat in Ireland compared with fewer than 10 m^{-2} in adjacent managed grassland. Millipedes are scarce in tropical grasslands – Lamotte (1947) reported 5 m^{-2} from savannah in Guinea – and are virtually absent from moorland. The Iuliformia are more resistant to desiccation than are the Polydesmoidea and can be found in drier habitats. Thus, Baker (1978a) described how the southwestern European species *Ommatoiulus moreletii*, which has become established in the Eyre Peninsula region of South Australia, has reached pest outbreak levels, constituting a severe nuisance in many areas. Behavioural and physiological adaptations that enable this species to survive dry summer weather have been mentioned earlier (Chapter 1). Davis and Sutton (1978) cite biomass values for millipedes of 0.5 g m^{-2} in coastal sand dunes in Yorkshire and 12.5 g m^{-2} in rough grassland at Wytham Wood, Oxfordshire, England.

Millipedes are phytophagous to some extent and some species (notably *Brachydesmus* and *Polydesmus* spp.) may cause considerable root damage to the seedlings of crop plants; however, as a group they are predominantly saprophagous and when abundant they can consume significant amounts of plant litter. Striganova (1971) estimated that the millipede population in woodland in the Caucasus Mountains area could consume practically all the leaf fall during their spring and autumn activity, so that by early summer the soil was covered with a 0.5–1-cm layer of their excrement. Kubiena (1955) and Blower (1956) likewise show that millipede faeces can be an important constituent of moder-type humus in woodland soils.

There is some doubt about the direct role of millipedes in the decomposition of organic matter. They are known to be among the macroarthropod groups that have some cellulase activity (Nielsen, 1962; Hartenstein, 1982), and Striganova (1971) reported that they assimilate 30–40% of the litter they consume. On the other hand, van der Drift (1975) estimated that *Glomeris marginata* assimilated only 9% of oak litter ingested. Other studies indicate that while some chemical breakdown of litter and some humification may occur on passage through the gut, the main role of millipedes is in the fragmentation and disintegration of litter, thereby promoting and enhancing subsequent microbial decay (Blower, 1955, 1956; Marcuzzi, 1970). Because of the large quantities of litter that they process, millipedes are often among the most significant soil invertebrates in litter decomposition in moist, undisturbed habitats.

2.8 ARACHNIDA

The arachnids of most significance in grasslands are the Araneae (spiders) and Acari (mites). Related groups, the Scorpiones (scorpions) and Solifugae (sun spiders) are widely distributed in warm, dry tropical and subtropical habitats. They are voracious and highly mobile and may be important predators of surface invertebrates and small vertebrates in desert ecosystems. The Chelonethi (pseudoscorpions) are common in damp, decaying litter in forest soils where they feed on small litter-dwelling invertebrates. Opiliones (harvestmen) are also common in damp situations where there is an accumulation of leaf litter; they are also predatory and feed readily on a range of insect larvae and other invertebrates (Phillipson, 1960). While harvestmen are commonly caught when sampling the grassland fauna, they are mainly scarce and probably have little influence on the functioning of the community.

2.8.1 Araneae

(a) Distribution and abundance

Spiders are ubiquitous in distribution and are a significant component of the fauna of temperate grasslands that can support communities rich in species and individuals (Table 2.5). Spiders' webs are a common feature of grassland, particularly the orb webs of the Araneidae and Tetrangnathidae in the vegetation and the surface sheet webs of the Linyphiidae. The latter group are particularly prevalent in cool temperate grasslands. They contributed 58 of the 141 species recorded from limestone grassland (Duffey, 1962) and over 70% of all spiders present in mountain moorland in England (Cherrett, 1964); they constitute over 95% of the spider fauna in intensively mown and grazed grasslands in Ireland (Purvis, 1978). A feature of the biology of many species is the ability of juveniles, and small adults in some cases, to disperse by ballooning. This adaptation facilitates rapid recolonization of disturbed habitats by linyphiids. Unmanaged grassland typically supports the greatest number of species (Table 2.5), but a rich spider fauna may also develop under monoculture: for example, Wheeler (1973) recorded 78 species from alfalfa plots in New York state, USA.

Spiders typically show pronounced horizontal and vertical stratification related to the structure of the vegetation. Duffey (1962) reported that dense *Festuca* tufts with large numbers of fine leaves and stems provide a habitat supporting high populations of spiders (up to 842 m^{-2}) in limestone grassland compared with the less-dense *Brachypodium* sward. Aggregations of linyphiids also occurred in moorland vegetation tussocks (Cherrett, 1964). Kajak (1965) reported a linear relationship between the

Table 2.5 Population density and biomass of spiders in various grasslands

Author	Site	Numbers (m^{-2})	Biomass (mg m^{-2} dry mass)	No. of species
Kajak (1971)	Natural meadow, Poland	52	78.3	
Kajak (1971)	Managed meadow, Poland	4.4	5.2	
Delchev and Kajak (1974)	Mountain sheep pasture, Poland	<7		
Morris (1968)	Ungrazed chalk grassland, England	253		
Morris (1968)	Grazed chalk grassland, England	22		
Duffey (1962)	Limestone grassland, England	842 max.		141
Cherrett (1964)	Moorland, England	482 max.		71
Curry and Momen (1988)	Grass leys on reclaimed peat, Ireland	110–204		
Curry and O'Neill (1979)	Grass and clover leys, Ireland	73		} 13
Curry and O'Neill (1979)	Old pasture, Ireland	144		
Purvis (1978)	Conserved leys, Ireland	481 max.		20
Purvis (1978)	Grazed leys, Ireland	104 max.		
Curry (1986)	Grass and lucerne leys, Sweden	70–131		
Persson and Lohm (1977)	*Agropyron* grassland, Sweden	197	39	19
Palmgren (1972)	Moist meadow, Finland	150–250		
Palmgren (1972)	Mesic meadow, Finland	100–120		
Van Hook (1971)	Old field, Kentucky, USA	9.5 max.	219 max.	
Dondale (1971)	Meadow, Ontario, Canada	53		150
Turnbull (1966)	Overgrazed pasture, Canada		146	
Hutchinson and King (1980)	Sheep pasture, Australia	5–11		

diversity of the vegetation and the populations of web-spinning spiders in Polish grassland. Diversity depends on the admixture of grasses, sedges and dicotyledons of differing growth habits and structure, creating a mosaic with uneven, multilayered surfaces. Grazing and cutting drastically simplify the habitat, and managed grasslands typically support far fewer species than unmanaged grassland, with the web-spinning spiders characteristic of the vegetation layer being largely eliminated. Thus, Kajak (1971) reported that the six dominant species in natural meadow in Poland (*Theridion bimaculatum, Tibellus maritimus, Singa hemata, Singa heri, Araneus quadratus* and *Cheirocanthium erraticum*) are largely confined to the unmown margins of mown and fertilized meadow. The spider fauna of managed grassland in Ireland consists almost entirely of small ground-dwelling Linyphiidae, notably *Erigone atra, E. dentipalpus, Oedothorax fuscus* and *Bathyphantes gracilis* (Purvis, 1978; Curry and O'Neill, 1979).

Spiders in unmanaged grasslands often number several hundred per m^2 (Table 2.5). There are usually fewer than 100 m^{-2} in short grazed swards, but numbers increase dramatically when swards are allowed to grow long. Peak populations are frequently reported in summer and autumn (Duffey, 1962; Cherrett, 1964; Purvis and Curry, 1981) when large numbers of juveniles are present.

(b) Feeding habits

Spiders are carnivorous; the web-building species use their webs to capture prey while the hunting species either ambush or outrun their prey depending on the group. Examination of insect catches in webs permits an assessment of prey preferences and consumption by web-building species. Thus, Kajak (1967) found that the food of *Araneus quadratus* in a Polish meadow consisted mainly of Acridoidea (grasshoppers, 58% by weight), with Diptera constituting a further 13%, whereas in the case of *Araneus cornutus* Diptera were the major component of the diet (49%), followed by Auchenorrhyncha (planthoppers etc.; 23%) and Acridoidea (21%). Taking the spider community as a whole, Diptera constituted over 80% of the prey captured in the meadow, and spiders captured 30–50% of all Diptera emerging in July–September (Kajak *et al.*, 1972). Spiders are also known to be cannibalistic, and Breymeyer (1980) suggested that lycosid spiders in moist peat bog meadows feed mainly on the large numbers of young of their own species. The role of spiders in the regulation of prey populations and their potential as biocontrol agents for pest species has been reviewed by Riechert and Lockley (1984), while Sunderland, Fraser and Dixon (1986) demonstrated that they can have a significant impact on aphid populations in cereal crops early in the season.

2.8.2 Acari

(a) Abundance and biomass

The Acari (mites) often constitute the most abundant arthropod group in temperate grasslands and have been the subject of many ecological studies. Mites found in the soil and vegetation mainly belong to four orders – Crytostigmata, Mesostigmata, Prostigmata and Astigmata. Mean densities and biomass in a range of grassland soils are given in Table 2.6. Information on populations in the vegetation layer is less extensive but Table 2.7 shows that significant numbers may inhabit this zone in ungrazed swards.

Acarine populations tend to be higher in forest than in non-wooded sites, with temperate coniferous forests on mor soils having highest mean densities (Petersen and Luxton, 1982). Populations in old grassland soils can exceed 100 000 m^{-2}, with highest numbers being recorded where there is an accumulation of organic matter on the surface. Lowland agricultural grasslands on mineral soils typically support fewer mites, notable exceptions being the permanent pasture studied by Salt *et al.* (1948) in Cambridgeshire, England, and the grassland site studied by McMillan (1969) in New Zealand, where a maximum population density approaching 600 000 mites m^{-2} was recorded on one sampling occasion. Generally lower populations are found under more arid tropical and steppe conditions than in mesic temperate soils.

Biomass estimates approaching 7 g dry mass m^{-2} have been recorded from deciduous woodland in Tennessee (McBrayer, cited by Petersen and Luxton, 1982), but most sites reviewed by these authors fall within the range 50–1000 mg. Most of the biomass estimates given in the studies cited in Table 2.6 fall within the range 100–200 mg dry mass, with rather higher values for mountain meadow in Norway (Østbye *et al.*, 1975), old pasture on peat in Ireland (Whelan, 1978) and moorland grassland in England (Block, 1965, 1966).

Cryptostigmata (oribatid mites) generally constitute the main proportion of the Acari, particularly when there is an organic surface layer; they are especially characteristic of woodland habitats and are often the dominant group in grassland as well. Prostigmata are also generally abundant in grasslands, and are sometimes the dominant group under conditions as disparate as temperate moorland, peat and South African steppe (Table 2.6). Mesostigmata are generally most abundant in mineral soils while the Astigmata are normally the scarcest component of the acarine fauna. Two notable exceptions are the New Zealand grassland on peat studied by Luxton (1982, 1983) and the English mineral grassland studied by Salt *et al.* (1948). In the former case *Tyrophagus similis*, a species often found in great density associated with hay (Hughes, 1976), was particularly abundant.

Table 2.6 Population density and biomass of mites in various grassland soils

Author	Site	Numbers × 10³ (m⁻²)					Biomass (mg m⁻²)*
		Cryptostigmata	Mesostigmata	Prostigmata	Astigmata	Total	
Davis (1963)	Permanent pasture, England	6.7	2.9	0.5	4.0	14.1	
Dhillon and Gibson (1962)	Old meadow, England	0.2	7.6	14.7	0.5	23	
Salt et al. (1948)	Permanent pasture, England	104	56.4	7.7	58	234	
Curry (1969a)	Old grassland, Ireland	42.1	28.2	32.8	3.0	106.1	
Bolger and Curry (1980)	Permanent pasture, Ireland					21.6	
Curry and Momen (1988)	Grass leys on reclaimed peat, Ireland					14.5–66	
Sheals (1957)	Old grassland, Scotland	19.4	8.8	0.4	5.4	34	
Whelan (1978)	Permanent pasture on peat, Ireland	26.2	16.2	6.4	3.9	52.7	1891(F)
Macfadyen (1952)	*Molinia* fen, England					16–132	
Wood (1967a)	Moorland grassland, England	40–68	13–24	61–75	5–10	121–176	
Block (1965, 1966)	Moorland grassland, England	18–66	9.4–9.9	0.9–6.9	0.6–2.4	28.7–77.8	890–1850(F)
Haarlov (1960)	Pasture on sandy soil, Denmark	143.7	20.6	14.7		179	
Ricou (1976)	Permanent pasture, France	18.8	11.3	2.7	1.6	36	
Andrén and Lagerlöf (1983)	Permanent pasture, Sweden	30.6	11.3	36.4	0.9	79.2	
Persson and Lohm (1977)	Fen grassland, Sweden	39.2	7.6	50.5	14.6	111.9	131(D)

Reference	Location						
Törmälä (1979)	Abandoned field, Finland					60–100	
Østbye et al. (1975)	Mountain meadows, Norway					32–119	
McMillan (1969)	6-year-old pasture, New Zealand	70.9	4.4	41.1	2.3	118.8	153–867(D)
Adams (1971)	Hill pasture, New Zealand	63				86	
King and Hutchinson (1976)	Sheep pasture, Australia					3.4–20.6	
Luxton (1982)	6-year-old pasture on peat, New Zealand	68.9	12.9	58.3	59	199.1	
Luxton (1983)	10-year-old pasture on peat, New Zealand	49.3	16.9	51.4	28.1	145.7	
Chernova et al. (1975)	Tundra, Russia	5	8				
Bunnell, MacLean and Brown (1975)	Tundra, Alaska					26.2	110(D)
Ryke and Loots (1967)	Steppe, South Africa	0.5–21	0.05–2.8	3.2–38.5	0.05–3.2	5.4–50	
Willard (1974)	Prairie, Canada					6.2	18(D)
Crossley, Proctor and Gist (1975)	Prairie, USA						186(D)
Salt (1952)	Pasture, East Africa					27.8	
Belfield (1956)	Pasture, West Africa					6.8	
Block (1970)	Pasture, Uganda	6.1	3.5	0.6	2.8	13.0	
Athias (1976)	Unburned savannah, Ivory Coast	8.2	3.1	8.0	0.8	20.2	
Athias (1976)	Burned savannah, Ivory Coast	4.5	2.3	7.2	0.5	14.5	

*F = fresh mass; D = dry mass.

Table 2.7 Population density and biomass of mites in the herbage layer of various grasslands

Author	Site	Numbers (m^{-2})					Biomass (mg m^{-2})
		Cryptostigmata	Mesostigmata	Prostigmata	Astigmata	Total	
Whelan (1978)	Permanent pasture on peat, Ireland	9423	1507	3407	427	14 764	286.6(F)
Curry and Tuohy (1978)	Old pasture, Ireland					16 394	
Purvis and Curry (1980)	Grass ley, Ireland	2465	548	22 186	2191	27 390	
Solhöy (1972)	Mountain meadows, Norway					5766–6664	
Kauri, Holdung and Solhöy (1969)	High mountain grassland, Norway					10 154	
Törmälä and Raatikainen (1976)	Abandoned field, Finland					9164	

The mite fauna of the vegetation in managed grassland is frequently dominated by small Prostigmata, notably Tarsonemidae and Pyemotidae, and Astigmata, notably *Tyrophagus* spp., while in rank, unmanaged vegetation Cryptostigmata may predominate (Whelan, 1978; Curry and Tuohy, 1978; Purvis and Curry, 1980).

(b) Factors affecting distribution

Cryptostigmata are strongly associated with surface organic accumulations that provide suitable living conditions and an abundant supply of food. Non-food aspects of greatest significance probably include the microhabitat complexity afforded by organic layers in various stages of decay and the high degree of moisture and temperature stability afforded by well-developed surface organic horizons. Adequate moisture is of paramount importance, for the Cryptostigmata as a group require conditions approaching saturation (Vannier, 1970). Nevertheless, a wide variation is found within the group in their ability to withstand desiccation. The presence of a waterproof, cuticular layer in species inhabiting the upper litter layer and epigeal habitats is an adaptation facilitating the colonization of more exposed habitats (Madge, 1964a,b,c). Behavioural mechanisms are also involved and diurnal and seasonal migrations undertaken by some species between the litter layer and the vegetation are probably at least partially in response to changing temperature and humidity. The other groups are probably less dependent on saturated conditions. The Prostigmata can be abundant in arid soils (Loots and Ryke, 1966) and are often well represented in the vegetation layer by large herbivorous and predatory forms as well as by small microbivorous species. Many predatory Mesostigmata are active on the soil surface. Some Astigmata, which can become very abundant in decaying plant residues and in rank vegetation, can survive adverse conditions in hypopial form.

Soil microarthropods in general tend to be concentrated in the surface organic layers or hemiedaphic zone of undisturbed soil. This surface concentration is particularly marked in the case of podzolic soils with marked surface organic zones; for instance, Wood (1967b) reported that 87–91% of the mites and collembolans occurred in the upper 4 cm of peaty podzolic moorland soils in England. In mull soils, where the organic matter is more evenly distributed in the profile, the fauna is less vertically compressed. Thus, Wood (1967b) found a lower percentage (76–79%) of microarthropods in the top 4 cm in moorland brown earth and mull-like rendzina soils than in peat and peaty podzols. Vertical distribution reflects pore size distribution in the profile to a degree, with larger species being restricted to the surface layers where larger-sized cavities occur (Weis-Fogh, 1948). Haarløv (1960) found agreement between the vertical distribution of soil microarthropods in Danish pastures and the development of soil cavities, but he concluded that other factors such as humidity and food supply must also have an influence.

True soil-dwelling (euedaphic) grassland mites comprise mainly Rhodacaridae (Mesostigmata), small Prostigmata of the families Nanorchestidae, Alicorhagiidae and Tydeidae, and Cryptostigmata of the family Brachychthoniidae (Haarløv, 1960; Wood, 1967b; Curry, 1971; Whelan, 1985). Most grassland species are predominately hemiedaphic, but many hemiedaphic forms such as *Platynothrus peltifer, Hemileius initialis, Minunthozetes semirufus, Punctoribates punctum, Liebstadia similis* and *Trichoribates incisellus* (Cryptostigmata) are also frequently abundant in the vegetation layer (Purvis and Curry, 1980; Whelan, 1985). Predominantly epigeal (vegetation layer) mites include *Tyrophagus* spp. (Astigmata); *Eupodes viridis, Siteroptes* spp., Bdellidae, Eriophyidae, Tarsonemidae and Tetranychidae (Prostigmata); and Phytoseiidae and Epicriidae (Mesostigmata). Vertical migrations between the vegetation layer and superficial soil horizons appear to be common. In some cases, these movements seem to be associated with the reproductive cycle: for example, Whelan (1985) found that *M. semirufus* migrates into the herbage in summer and reproduces there.

Some Cryptostigmata appear to undertake diurnal movements between soil and vegetation (Krull, 1939; Ibarra, Wallwork and Rodriguez, 1965) and these movements are believed to be of significance in the transmission of anoplocephalid tapeworms in cattle and sheep, which have as their intermediate hosts various species of grassland Cryptostigmata (Wallwork and Rodriguez, 1961). Krull (1939) considered that adults might move into herbage in search of food or to escape deleterious conditions such as waterlogging after heavy rain on the litter layer. Many authors have reported changes in the vertical distribution patterns of mites within the soil profile throughout the year (e.g. Strickland, 1947; Belfield, 1956; Stockli, 1957; Curry, 1971). These are especially pronounced in soils with a strongly seasonal climate: for example, Strickland (1947) and Belfield (1956) reported marked downward shifts in soil microarthropod populations in tropical soils coinciding with the dry season and a great increase in the density of the fauna in the surface layers with the advent of the wet season. Such pronounced seasonal migrations have generally been interpreted as responses to changing temperature and humidity conditions. Under more mesic conditions, where the overriding effect of climate is less pronounced, many factors including food supply and biological rhythms may be involved.

(c) Life cycles and phenology

Reproduction is predominantly sexual, but parthenogenesis (mainly arrhenotoky) is relatively common (e.g. Luxton, 1981). Basically, the developmental stages consist of egg, hexapod larva, one to three nymphal stages and adult, but there is considerable variation between groups. In the Mesostigmata, there are usually two nymphal stages (protonymph and deutonymph). Developmental time varies with the species and is influenced by environmental factors such as food supply, temperature and humidity. The Prostigmata is a particularly heterogenous group with regard to life cycles. Among the Pyemotidae, Scutacaridae and Tarsonemidae there is usually only one nymphal stage, whereas among other common families such as the Bdellidae and Tydeidae three nymphal stages are usual. Multiple generations in the year are common, particularly among the smaller epigeal and hemiedaphic species.

Many of the common astigmatic species in the families Acaridae and Anoetidae have one hypopal stage in the life cycle. This facilitates survival during adverse conditions and dispersal by attachment to the bodies of insects. It is likely that those species can complete several generations a year under favourable conditions in the field, although the norm for most species is probably one or two generations a year (Luxton, 1981).

Three nymphal stages occur in the Cryptostigmata – the protonymphal, deutonymphal and tritonymphal stages. One generation a year appears to be usual in temperate grasslands (Haarløv, 1960; Block, 1965; Whelan, 1976), but some species such as *Tectocepheus velatus* can have up to five generations a year in woodland habitats (Lebrun, 1965).

Seasonal changes in microarthropod populations have frequently been recorded, with peak density usually occurring during autumn or winter and a marked summer decline in numbers (Weis-Fogh, 1948; Macfadyen, 1952; Haarløv, 1960; Block, 1966; McMillan, 1969). Generally, only Cryptostigmata among the mites show marked seasonal periodicity. Seasonal variations have been variously attributed to climatic factors, availability of food and reproductive cycles. It is possible that marked seasonality in population trends is largely imposed by weather synchronizing to some extent the phenology of the populations; where seasonality in weather patterns is not marked seasonal population trends are not so apparent (Dhillon and Gibson, 1962; Curry, 1971; Whelan, 1985).

(d) Feeding habits

A wide range of feeding habits is represented among the Acari. Most of the Gamasina (Mesostigmata) are predatory, feeding on small invertebrates in soil and litter. Predatory Mesostigmata, which commonly occur in grassland, include members of the families Parasitidae, Veigaiidae, Eviphididae, Rhodacaridae, Ascidae and Phytoseiidae, and, associated with dung, Macrochelidae. As a group they have a wide range of prey, including nematodes, collembolans and other apterygotes, eggs and young larvae of insects and immature mites although individual groups and species may have a more restricted prey range. Many Gamasina feed avidly on nematodes. *Alliphis halleri* is strictly nematophagous while several species (e.g. *Rhodacarus roseus* and *Rhodacarellus silesiacus*) will accept small arthropods such as collembolans (*Tullbergia*) and immature mites in captivity, but have a preference for nematodes (Sardar, 1980). Some prairie species such as *Gamasellodes vermivorax* and *Macrocheles* sp. laid eggs in culture only when fed on nematodes, while *Cosmolaelaps vacua* developed more quickly on a nematode than on an arthropod diet (Walter, Hunt and Elliott, 1987). *Hypoaspis aculeifer* readily eats a range of prey but various aspects of its biology are affected by the choice of prey (Sardar, 1980). In the field, it is likely that the choice of prey is largely determined by availability.

A few Gamasina are non-predatory; these include some Phytoseiidae, which have been reported as feeding on fungi, pollen and leaf sap of higher plants (Chant, 1959; Porres, McMurtry and March, 1975), *Proctolaelaps hypudaei* (Ascidae), which has been reared on a fungal diet (Mathys and Tencalla, 1959), and some Ameroseiidae, which live in plant inflorescences and feed on nectar and pollen (Krantz and Lindquist, 1979).

The Uropodina (Mesostigmata) are mainly mycophagous and saprophagous, often coprophagus (Wallwork, 1970). There are some records of predation by members of this group with nematophagy being fairly prevalent (Krantz and Lindquist, 1979; Mueller, Beare and Crossley, 1990).

The Prostigmata are a highly diverse group as regards feeding habits. Several families are exclusively phytophagous – the Tetranychidae (spider mites), Tenuipalpidae (false spider mites) and Eriophyidae. Phytophagous members also occur among the Nanorchestidae, Eupodidae, Penthalodidae, Stigmaeidae, Tydeidae, Pygmephoridae, Pyemotidae, Tarsonemidae and other families (Krantz and Lindquist, 1979). Phytophagus prostigmatic species commonly found in grassland include *Bryobia* spp., *Steneotarsonemus* spp., *Halotydeus destructor* (red-legged earth mite), which can sometimes cause severe wilting of grasses, and *Siteroptes* spp., although there is some question as to whether *Siteroptes* can feed on healthy plants (Jepson, Keifer and Baker, 1975). These are vectors of fungal pathogens in cereals and grasses and it may be that symptoms such as 'white top' attributed to the mites may be primarily due to the pathogens (Suski, 1973). The eriophyid species *Abacarus hystrix* transmits ryegrass mosaic virus in Europe (Plumb, 1978).

The feeding habits of many common grassland Prostigmata are poorly understood, particularly the smaller species. *Alicorhagia fragilis* fed on nematodes, algae and fungi in culture but did not thrive when nematodes were excluded from the diet (Walter, 1987a). *Nanorchestes* spp. were reported feeding on algae (Schuster and Schuster, 1977), and *Daidalo-*

tarsonemus on algae and lichens (Suski, 1972); while microfloral feeding has been widely reported among Eupodidae (Baker and Wharton, 1952; Walter, 1987a; Mueller, Beare and Crossley, 1990), Tydeidae (Momen, 1986; Walter, 1987a), Tarsonemidae (Suski, 1972; Walter, 1987a), Scutacaridae and Pyemotidae (Wallwork, 1967) and other families. Predatory Prostigmata of the families Bdellidae, Cheyletidae, Cunaxidae, Rhagidiidae, Trombidiidae and Erythraeidae are well represented in soil and vegetation.

Most free-living Astigmata are primarily saprophagous or microbivorous, although some (e.g. *Tyrophagus* spp.) which are often abundant on grasses, may feed on higher plants (Hughes, 1976). Some *Tyrophagous* spp. feed and reproduce on a range of food types, including nematodes, fungi, lichen, algae, dead grass and dead and injured mites in laboratory culture (Walter, Hudgens and Freckman, 1986). Walter and Kaplan (1990) described *Schwiebia bakeri* as an omnivorous grazer, feeding on decaying root material, fungi and small invertebrates such as nematodes and protozoans. Many Anoetidae appear to derive their nutrition from the liquified products of plant putrefacation rich in microorganisms (Wallwork, 1967), although *Histiostoma bakeri* feeds primarily on decaying root material (Walter and Kaplan, 1990).

The Cryptostigmata are predominantly mycophagous or saprophagous (Krantz and Lindquist, 1979). Feeding on higher plants is usually confined to decaying tissue, although many species consume plant pollen (Behan-Pelletier and Hill, 1983) and the larvae and nymphs of *Minunthozetes semirufus* burrow into the stems of grasses (Michael, 1884–8). Most species are unspecialized in their feeding habits, but some such as common-grassland *Oppia* spp. and *Brachychthonius* spp. feed primarily or even exclusively on microflora (Schuster, 1956; Luxton, 1972). Most species ingest a wide range of fungi, bacteria, yeasts and algae but often show distinct preferences for different groups and species or even for specific parts such as fungal spores. Several grassland oribatid species will ingest the eggs of anoplocephalid cestodes, which occasionally are swallowed whole and complete their development within the mite. Bulanova-Zachvatkina (1967) listed 54 species associated with one or more of 13 species of parasites. In common with many other predominantly mycophagous invertebrates, at least some species of oribatid mites avidly attack and ingest nematodes with fungal hyphae (Walter, 1987a,b).

(e) Ecosystem role

Luxton (1972) demonstrated the presence of a wide range of carbohydrases in 12 oribatid species assayed. Cellulases were generally accompanied by xylanases or pectinases among species that ingest remains of

higher plants, indicating some ability to decompose structural polysaccharides. Litter consumption by oribatid mites can be considerable in
woodland (Berthet, 1963), but oribatids tend to be less abundant in
grassland where the more active predatory Mesostigmata are usually the
dominant acarine group in terms of energy metabolism. Thus, Persson
and Lohm (1977) reported that cryptostigmatic respiration accounted for
only 23% of the total acarine respiration of 12.3 kJ m^{-2} yr^{-1} in a Swedish
grassland while Gamasina contributed 50%. Macfadyen (1963) reported
an even greater dominance of predators in an English grassland: here, the
metabolism of oribatids was only 42.7 kJ m^{-2} yr^{-1} compared with 266
for parasitids. Predatory mites probably have a significant effect on their
prey populations: for example, bdellid mites appear to play an important
part in the natural regulation of the 'lucerne flea' *Sminthurus viridis*
(Collembola) in Australian pasture (Wallace, 1967, 1974). The contribution of mites to total invertebrate respiratory metabolism was 1–1.6% in
moorland grassland in England (Heal and Perkins, 1976), 1.2% in
Swedish grassland (Persson and Lohm, 1977), 1.9% in Australian sheep
pasture (Hutchinson and King, 1980) and 6.5% in a grazed English
meadow (Macfadyen, 1963). Estimated consumption by saprophagous
mites in Swedish grassland was 41.8 kJ m^{-2} yr^{-1}, about 0.3% of the
total litter input (Persson and Lohm, 1977).

2.9 COLLEMBOLA

2.9.1 Distribution and abundance

Collembolans are world-wide in distribution; they are sometimes equal in
abundance with mites in grassland soil and are often the most abundant
arthropods in the herbage layer. Their tolerance for low temperatures
makes them particularly significant under arctic/subarctic conditions and
many species remain active during winter in temperate latitudes: Lienhard (1980) observed very high densities of collembolans in alpine
meadows during winter when the soil was totally frozen. Usher and
Edwards (1984) reported that collembolans constituted over 80% of the
soil arthropod fauna under *Deschampsia antarctica* on Lynch Island,
Antarctica (Table 2.8).

Mean population densities in the soil typically range from a few
thousand per square metre in tropical grasslands to over 100 000 m^{-2} in
some temperate grasslands (Table 2.8). Biomass values exceeding 2 g
fresh mass m^{-2} have been recorded for tundra grassland in Russia, but
the average for grassland is considerably less and probably rarely exceeds
1 g fresh mass m^{-2} (Petersen and Luxton, 1982). Herbage population
densities in the range 1000–1600 m^{-2} appear to be the norm for agri-

cultural leys with considerably higher populations under less-disturbed conditions. Highest numbers have been recorded from natural grasslands at high altitude in Norway (Table 2.9).

Like the Acari, the Collembola have a marked vertical distribution pattern in the soil profile. Surface concentration is particularly marked in soils with a discrete organic layer; in mull grasslands, surface concentration is less well marked but numbers do show a progressive decrease with depth (Dhillon and Gibson, 1962; Davis, 1963; Curry, 1971). Weis-Fogh (1948), Haarløv (1960) and others have investigated the relationship between depth distribution and soil structure and have pointed to the significance of decreasing soil porosity with depth. Collembolans are unable to burrow and the larger species are therefore restricted to the superficial layers where soil cavities are sufficiently large.

Typical herbage-dwelling collembolans in grassland include many members of the Symphypleona, such as *Sminthurus* spp., *Deuterosminthurus* spp., *Sminthurinus* spp., *Sminthurides* spp., *Bourletiella* spp. and *Dicyrtoma* spp. (Curry and O'Neill, 1979; Purvis and Curry, 1980). An important adaptation in this group is the possession of a tracheate respiratory system. This allows greater waterproofing of the cuticle and hence the colonization of more exposed habitats than is possible with members of the Arthropleona, which respire through the cuticle and require a nearly saturated atmosphere. Many of the larger species of Arthropleona – notably, Isotomidae such as *Isotoma viridis* and Entomobryidae such as *Entomobrya* spp. – are also commonly sampled from vegetation (Edwards, Butler and Lofty, 1976; Curry and O'Neill, 1979; Curry *et al.*, 1980; Purvis and Curry, 1980) although these groups are best represented in the superficial soil and litter layers. Species with a predominantly hemiedaphic distribution in grassland include *Isotoma viridis*, *I. notabilis*, *Folsomia quadrioculata*, *Lepidocyrtus* spp., *Entomobrya* spp., *Hypogastura* spp., *Sminthurinus* spp. and *Sminthurides* spp., while euedaphic species characteristic of the mineral soil layers include *Folsomia candida*, *F. montigena*, *Tullbergia* spp. and *Neelus* spp. (Haarløv, 1960; Dhillon and Gibson, 1962; Hale, 1966; Curry, 1971). As is the case with Acari, vertical distribution patterns are not static and vertical movements between vegetation and topsoil and between topsoil and deeper soil layers are common. Diurnal migrations may occur in response to changes in humidity (Joosse and Groen, 1970), while seasonal migrations have also been reported (Glasgow, 1939; Strickland, 1947; Persson and Lohm, 1977). Downward population shifts in summer and in winter have been interpreted in terms of escape from unfavourable conditions in the surface layers; however, under more equable climatic conditions it is difficult to see any clear relationship between such migrations and climatic and edaphic variables (Dhillon and Gibson, 1962; Curry, 1971).

Table 2.8 Population density and biomass of collembolans in various grasslands

Author	Site	Numbers $\times 10^3 (m^{-2})$	Biomass $(mg\ m^{-2})$*
Dhillon and Gibson (1962)	Permanent pasture, England	27	
Davis (1963)	Permanent pasture, England	11.1	
Curry (1969a)	Old *Agrostis/Festuca* grassland, Ireland	105.4	
Bolger and Curry (1980)	Permanent pasture, Ireland	28.3	
Bolger and Curry (1984)	Sandy estuarine soil, Ireland	46.0	
Curry and Momen (1988)	Grass leys on reclaimed peat, Ireland	39.8–73.8	
Salt et al. (1948)	Permanent pasture, England	68.6	
Sheals (1957)	Old grassland, Scotland	22.8	
Hale (1966)	Moorland limestone grassland, England	52.9	
Hale (1966)	Moorland alluvial grassland, England	43.7	
Wood (1967a)	Moorland grassland, England	46–73	
Macfadyen (1952)	*Molinia* fen, England	24–25	
Haarløv (1960)	Danish pasture	109	
Persson and Lohm (1977)	Old grassland on peat, Sweden	108.7	142(D)
Ricou (1976)	Permanent pasture, France	19.3	
King and Hutchinson (1976)	Sheep pastures, Australia	9.8–24.5	
Adams (1971)	New Zealand hill pasture	24.7	
Luxton (1982)	6-year-old pasture on peat, New Zealand	18.4	

Reference	Location		
Luxton (1983)	10-year-old pasture on peat, New Zealand	81.4	
McMillan (1969)	New Zealand pasture	40	
Törmälä (1979)	Abandoned field, Finland	30	
Solhöy (1972)	Wet mountain meadow, Norway	33.3	
Solhöy (1972)	Dry mountain meadow, Norway	62	
Østbye et al. (1975)	Mountain meadows, Norway	39–80	168–287(D)
Crossley, Proctor and Gist (1975)	Prairie, USA		14(D)
Belfield (1956)	West African pasture	4.1	
Block (1970)	Pasture, Uganda	0.5	
Block (1970)	Grazed bush, Uganda	2.2	
	Elephant grass, Uganda	1.7	
Lamotte (1947)	Savannah, Guinea	40	
Athias (1976)	Savannah, Ivory Coast	1.4–2.6	
Salt (1952)	East African pasture	6.7	
Bunnell et al. (1975)	Tundra meadow, Alaska	96.8	484(D)
Chernova et al. (1975)	Herb grassland, Russian tundra	31.8	
Usher and Edwards (1984)	Deschampsia grassland, Lynch Island, Antartica	610	2220(F)

*F = fresh mass; D = dry mass.

Table 2.9 Numbers of collembolans in the herbage layer of various grasslands

Author	Site	Numbers (m^{-2})
Curry and O'Neill (1979)	Permanent pasture, Ireland	1 211
Curry and O'Neill (1979)	Grass/clover leys, Ireland	1 589
Purvis and Curry (1980)	Ley pasture, Ireland	1 279
Edwards, Butler and Lofty (1976)	Pasture, England	1 040
Solhöy (1972)	Wet mountain meadow, Norway	6 168
Solhöy (1972)	Dry mountain meadow, Norway	6 129
Törmälä and Raatikainen (1976)	Reserved meadow, Finland	1 298

2.9.2 Life cycles and phenology

Reproduction in collembolans is considered to be predominantly sexual (Mayer, 1957), but parthenogenesis is common, especially among euedaphic species (Petersen, 1978). Breeding appears to occur throughout the year when climate allows (e.g. Hutson, 1981). Under the rigorous climatic conditions of Pennine moorland at Moor House, England, Hale (1965) found that most species produced eggs in spring and early summer but not during winter. Developmental rates and life-cycle duration vary with species and climate. Many species have several generations a year in temperate climates: *Folsomia quadrioculata* had four generations a year in Danish pasture (Haarløv, 1960); *Sminthurus viridis* had four generations a year during the active season in Australia (Davidson, 1934) and *Isotoma notabilis* had several generations a year in reclaimed land in Northumberland (Hutson, 1981). Under the subarctic conditions at Moor House, one generation a year seems to be the norm for most species of Isotomidae, but some species, such as *Isotomiella minor, Friesea mirabilis, Onychiurus tricampatus, Tullbergia krausbaueri, Folsomia* spp. and Sminthuridae, may have two (Hale, 1966). *Hypogastrura tullbergi* requires 3–5 years to complete its life cycle under high arctic conditions, compared with 2–12 months in more temperate regions (Addison, 1977).

Marked fluctuations in abundance throughout the year are commonly observed. Population maxima have often been reported in spring and autumn in temperate latitudes, with minima in winter and summer (e.g. Glasgow, 1939; Haarløv, 1960; Dhillon and Gibson, 1962; Hale, 1966; McMillan, 1969; Curry, 1971). Maxima often coincide with the appearance of large numbers of juveniles of the dominant species in the population when weather conditions are favourable. However, most studies reveal considerable variation in times of occurrence of maxima and minima for individual species, reflecting the differing phenologies of the species concerned.

2.9.3 Feeding habits

Information on the feeding habits of non-phytophagous collembolans and other small arthropods is largely based on laboratory culture studies, or on examination of the gut contents of specimens taken from the field. In either case there is considerable risk of misinterpretation since food readily eaten in the laboratory may not be available in the field, and gut contents will tend to overestimate material that is resistant to digestion and has a recognizable structure. Most accounts agree that the Collembola as a whole have a very broad range of feeding habits (e.g. Christiansen, 1964; Harding and Stuttard, 1974), with bacteria, fungal hyphae and spores, algae, lichens, faeces, carrion, decaying plant material, pollen and, in some cases, living plant material commonly listed among their foods. Larger, surface-dwelling Isotomidae and Entomobryidae ingest significant quantities of litter while pollen grains can comprise a significant portion of the diet of some species (Dunger, 1956; Christiansen, 1964; McBrayer and Reichle, 1971). However, most collembolans appear to be mainly microherbivorous and often have a high proportion of fungal material in their guts. Many of the smaller species commonly feed on the excreta of litter-ingesting invertebrates, and bacteria can be an important food source for many species (Christiansen, 1964).

Collembolans are able to utilize dead plant material directly only to a limited extent. Dunger (1956) states that collembolans will feed on fresh leaves but that leaves attacked by microorganisms are more readily eaten, while Hale (1967) suggests that plant material is eaten primarily because of the fungal mycelia and other microbes present. Thus collembolans appear to be mainly secondary decomposers, deriving their nutrition largely from microbial biomass or from plant material conditioned by microbial activity.

Some species are known to be at least partially carnivorous. Wallwork (1970) states that *Friesea* spp. consume rotifers, proturans and tardigrades, and also includes the ubiquitous grassland species *Isotoma viridis* and *Pseudosinella alba* among species that may feed on living prey and eggs. *I. viridis*, although normally vegetarian, was observed to kill and devour *I. notabilis* in the laboratory (Poole, 1959). *Hypogastrura scotti* in prairie grassland readily consumes nematodes, in addition to algae and fungi (Walter, 1987b).

Several species of Symphypleona can feed on living plant tissue. The best-known example is the lucerne flea *Sminthurus viridis*, which was considered by Davidson (1934) to be one of the most important pests of forage crops in Australia. Walters (1964) listed over 90 host plants belonging to 19 families for this species in South African pastures but it prefers lucerne, clover and young grass. *Sminthurus viridis* is omnivorous, with fungi, algae and mosses also being important components of its diet in South Africa. It also feeds readily on pollen, and consumed cast skins, carrion and eggs of its own and other species.

2.9.4 Ecosystem role

While *Sminthurus viridis* can be a grassland pest of economic importance in drier climates the main role of collembolans in ecosystems relates to the decomposition of organic matter. Their direct contribution to decomposition as measured by their respiratory metabolism is fairly modest. This was estimated to be about 8 kJ m^{-2} in moorland mineral soils in England (Coulson and Whittaker, 1978), 25.5 kJ m^{-2} in Swedish grassland (Persson and Lohm, 1977), and 154 kJ m^{-2} for the dominant species in Japanese grassland (Tanaka, 1970). Higher figures have been reported for lightly grazed sown pasture in Australia (466 kJ m^{-2}, Hutchinson and King, 1980) and for lightly grazed old grassland in England (640 kJ m^{-2}; Macfadyen, 1963), although it has been suggested that the latter figure is probably an overestimate (cf. Persson and Lohm, 1977). Collembolan respiration represented 0.4% and 3.1%, respectively, of total invertebrate respiration in the English moorland and Swedish grasslands, and 26% and 13.4%, respectively, of that in Australian and English lowland pastures. In addition, collembolans are believed to promote the decomposition and mineralization of organic matter through comminution of organic residues, through dissemination of fungal spores and by generally facilitating microbial activity (Macfadyen, 1964; Harding and Stuttard, 1974). Larger species selectively feeding on nutritious fungal hyphae may stimulate organic matter mineralization, while smaller species living in fine-textured soils may contribute to humification by non-selective scavenging and mixing of organic material and mineral soil particles (van Amelsvoort, van Dongen and van der Werff, 1988). Some collembolan species such as *Isotoma notabilis* are among the earliest colonizers of decaying grass residues (Curry, 1969c, 1973) and under these conditions their role in the release of nutrients immobilized in microbial biomass may be of considerable significance in enhancing further decomposer activity (Chapter 6).

2.10 OTHER APTERYGOTA

2.10.1 Protura

Proturans appear to be scarce or absent from many grassland soils; they are said to favour moist, organic soils that are not too acid (Wallwork, 1970). Their piercing mouthparts suggest fluid-feeding, and *Acerentomon* spp. and *Eosentomon* spp. have been reported to feed by sucking on the outer coating of fungal hyphae (Sturm, 1959). Recorded mean population densities include 41 m^{-2} in Swedish fen grassland (Persson and Lohm, 1977), 200 m^{-2} in tropical savannah with a maximum of 1625 m^{-2} (Athias, 1976), 662 m^{-2} in an old field soil in Michigan (Engelmann, 1961), 6700 m^{-2} in an Irish grassland (Curry, 1969a) and over 8000 m^{-2} in an English pasture (Raw, 1956). Petersen and Luxton (1982) cite a study in a Norwegian spruce forest where populations ranging from 28 000 to 51 000 m^{-2} were recorded.

2.10.2 Diplura

Diplura have been recorded even less frequently than Protura from grassland. Some members of the two commonest families, the Campodeidae and Japygidae, are considered to be predatory (Wallwork, 1970) while Carpenter (1988) observed *Campodea staphylinus*, the commonest British species, feeding on grass roots, scavenging on dead organic matter and animals such as slugs, and preying on small fauna. Persson and Lohm (1977) recorded 25 *Campodea* m^{-2} from Swedish grassland and Engelmann (1961) reported 389 Japygidae m^{-2} from an old field in Michigan. Belfield (1956) reported 244 Diplura m^{-2} from West African pasture while Athias (1976) found 657 and 859 m^{-2} in burned and unburned savannah respectively, with a seasonal max of 2625 m^{-2} in one unburned plot. The highest population density recorded from grassland was 6605 m^{-2} in an English pasture (Salt *et al.*, 1948).

2.11 ISOPTERA

2.11.1 Distribution

Isoptera (termites) are often the dominant invertebrate group in tropical and subtropical habitats. There are about 7000 known species of these social insects (Sands, 1977) and their distribution is largely confined to tropical and subtropical regions where their mound-like terminaria are often a conspicuous feature of the landscape. They can be grouped into arboreal species, which have their nests within or attached to the exterior of trees and shrubs; epigeal species, which form surface mounds; and subterranean species, which have their nests underground. Termites are polymorphic insects, with caste systems of varying complexity. There are reproductive and sterile worker and soldier castes, which differ in morphology, physiology and behaviour. Caste differentiation appears to depend on pheromones given off by the reproductive and soldier castes, and the proportions are controlled to some extent by selective cannibalism (Lee and Wood, 1971b). Development of a termite colony takes several years to complete, with a progressive increase in size and numbers.

The success of termites in the tropics can largely be attributed to their social organization and nest systems, which allow a degree of microclimatic control, storage and (in the case of fungal-growing species) cultivation of food, and a measure of protection against predators. Various materials are used in the construction of nests and associated structures, including soil, plant remains, excreta and saliva. The nests of some species are entirely subterranean. Surface mounds may range from small structures a few centimetres high to the massive structures several metres high that are characteristic of some African Macrotermitinae. Subterranean galleries extending out from the mounds are constructed for foraging purposes, while termites nesting on the outside of trees have external covered runways for access to the ground. Some colonies are very long-lived: Grassé (1949) cites examples of *Bellicositermes* and *Nasutitermes* mounds 80–100 years old.

2.11.2 Abundance and biomass

Population densities of termites are difficult to estimate because of their colonial organization. Many of the estimates given in Table 2.10 are probably underestimates because the soil cores on which they were based were often insufficiently deep to sample subterranean species, typically the most abundant forms in savannah grasslands. Wood and Sands (1978) cite termite densities of up to 4000 m^{-2} and biomass up to 11 g m^{-2} from several studies reviewed and suggest that the upper limits are about 15 000 individuals and 50 g biomass m^{-2} for any ecosystem. Data on mound-building species are fairly scarce. Large mounds of Macrotermitinae usually number less than 10 ha^{-1} but may contain several million termites per colony (Lee and Wood, 1971b). The large lignivorous and fungus-growing *Macrotermes* spp. prefer dense forest habitats and are scarce in open African savannah (Goffinet, 1976) where small mound-building humivores such as *Cubitermes* spp. and litter feeders and graminivores such as *Trinervitermes* spp predominate. About 5000 *Cubitermes* mounds ha^{-1} were reported from degraded savannah in Zaire by Goffinet (1976), while Ohiagu and Wood (1976) state that several hundred mounds of *Trinervitermes geminatus* ha^{-1} may often be found in West African savannah.

Table 2.10 Population density and biomass of termites in various tropical and subtropical grasslands

Author	Habitat	Species or category	Numbers (m^{-2})	Biomass $(g m^{-2}$ fresh mass)
Salt (1952)	East African pasture	Soil dwellers	1563	
Harris (1963)	Various, Congo	Soil dwellers	1800	
Lee and Wood (1971b)	N. Australian native steppe	Soil dwellers	2000	
Sands (1965a,b)	Northern Guinea savannah, Nigeria	*Trinervitermes geminatus*	110–2860	
Wood and Sands (1978)	Southern Guinea savannah, Nigeria	Soil and mound dwellers	4402	11.1
Ohiagu (1979)	Southern Guinea savannah, Nigeria	*Trinervitermes geminatus*	737	3.08
Josens (1972)	Derived savannah, Lamto, Ivory Coast	Soil and mound dwellers	1100	2.4
Lepage (1974)	Sahel savannah, Senegal	Soil and mound dwellers	200–300	0.15–0.3
Bouillon (1970)	Savannah, Zaire	*Cubitermes exiguus*	612–701	1.3–1.9
Goffinet (1975, 1976)	Degraded savannah, Zaire	All mound dwellers	Several thousand	
		Cubitermes spp.		9
Bodine and Ueckert (1975)	Semiarid grassland, N. America	Soil dwellers	2139	5.2

2.11.3 Feeding habits

Termites as a group derive their food from a wide range of organic materials, including living woody and non-woody plants, standing dead plants, plant debris and litter, dung, stored foodstuffs, soil organic matter, and fungi. Most lower termites and some Termitidae are wood feeders, although few species attack living wood. Some wood feeders are polyphagous and will consume a wide variety of plants and plant products (Lee and Wood, 1971b). It is usual to divide termites into major feeding categories such as foragers, humivores and fungal growers, although these groupings are not entirely mutually exclusive. Foraging termites gather their food from the soil surface and plants and carry it to their nests to be eaten or stored. These can be divided into harvesters, which prefer to cut down fragments of standing grasses and herbs, and scavengers, which collect plant material from the soil surface. Of particular importance in tropical grassland ecosystems are the grass-harvesting termites, including members of the genera *Hodotermes*, *Bellicositermes* and *Trinervitermes* in Africa. Generally, dead or dry grass material is preferred to green tissue, and foraging is most active during the dry season or winter (Sands, 1961; Nel and Hewitt, 1969). However, species such as *Trinervitermes trinervicus*, which do not store grass, forage all year round, the intensity of foraging being related to seasonal weather patterns (Lee and Wood, 1971b).

Ohiagu and Wood (1976) describe grass harvesting by *Trinervitermes geminatus*, the most widely distributed and abundant member of this genus in West Africa. Each colony has several mounds, a large one used for breeding and several smaller mounds from which foraging expeditions can be made to areas remote from the breeding colony. Grass collected from the surrounding areas is stored in all mounds. Foraging parties of workers and soldiers travel from the mounds via subterranean galleries to the soil surface and forage within a 2–3 m radius of the exit holes. Workers climb standing grass and cut it down; as many as 500–800 workers may invade a single grass tussock. Grass is also picked up from the soil surface. Pieces 2–20 cm long are carried back to the nests. A distinct preference is shown for dry grass, and *Andropogon gayanus*, the most abundant grass in the area, is the species most often attacked in the field. Sands (1961) carried out preference tests with *Trinervitermes geminatus* and found that few grasses were consistently avoided but smaller, finer grasses were preferred. He also found that some species accepted by one colony were avoided by others and concluded that selection was affected by familiarity with grasses normally available near the mound.

Humivorous species consume large quantities of mineral soil, deriving their food from soil organic matter. These are predominantly subterranean and were about three times more abundant than foragers at

Lamto, Ivory Coast (Josens, 1972). *Cubitermes* spp. are particularly prominent among humivores in Africa, although not at the Lamto site.
Cubitermes exiguus was the subject of an intensive study by Hébrant
(1970), who reported 510 nests ha^{-1} of this species in a Zaire savannah
site and calculated that it needed to ingest 17 t soil ha^{-1} yr^{-1} to satisfy its
energy requirements.

Fungal cultivation occurs exclusively in the Macrotermitinae. Fungal
growers are scarce in arid areas and are most abundant under humid
tropical conditions. They were the dominant group in humid savannah at
Lamto (Josens, 1972), where four subterranean nesting species occurred.
They ingest large quantities of dead wood, dried grass and sometimes
living grass and construct special fungal gardens primarily from faecal
material, which they tend. Fungal activity probably enhances the quality
of a litter diet by degrading lignin, increasing the N content of the diet
and providing vitamins (Lee and Wood, 1971b).

A feature of termite feeding is the efficiency with which they can derive
energy from the breakdown of polysaccharides, especially cellulose,
hemicellulose and some lignin. This is attributable to symbiotic protozoans and bacteria in the digestive tract.

2.11.4 Ecosystem role

Termites are sometimes referred to as being the tropical analogues of
earthworms in their effects on soil turnover and their ingestion and
decomposition of orgnic matter. There are however, some significant
differences between their activities and those of earthworms, as will be
discussed in Chapter 6.

There have been few studies on the energetics of entire termite populations in particular sites. Hébrant (1970) estimated that a population of
600–700 *Cubitermes exiguus*, a humivorous species, in Zaire savannah
assimilated 56 kJ m^{-2} yr^{-1}. Assuming that 68% of ingested food was
assimilated (Wood and Sands, 1978), this represents an ingestion rate of
86 kJ m^{-2} yr^{-1}.

Josens (1973) constructed partial energy budgets for two species at
Lamto – *Trinervitermes geminatus*, a grass cutter (density 160 m^{-2}, and
Ancistrotermes cavithorax, a fungus grower (density 190 m^{-2}). Ingestion
rates were 35 and 749 kJ m^{-2} and the ratios of tissue production to
ingestion were 0.09 and 0.018, respectively. The high energy ingestion
rate for the fungal grower is typical of this group, while Peakin and
Josens (1978) suggest that the low energy flow characteristic of grass-
cutting species is well adapted to the conditions of low primary production typical of drier savannah.

A few studies have attempted to quantify grass consumption by termites. Some fungus-growing Macrotermitinae may consume significant

amounts of grass; for example, Lepage (1972) estimated that *Bellicositermes bellicosus* consumed at least 5.4% of grass production in Senegal and, when particularly abundant, up to 49.2%. However, most information is available on the specialized grass-harvesting groups. Josens (1972) estimated a relatively modest annual consumption of 3–5 g m^{-2} by an average population of 160 foraging termites m^{-2} at Lamto, Ivory Coast. Consumption by *Trinervitermes geminatus* (mean population 200 m^{-2}) was estimated by Ohiagu and Wood (1976) in southern Guinea savannah, Nigeria, at 11–83 mg dry grass g^{-1} termite m^{-2} d^{-1} during the dry season. Total consumption at this site was less than 4% of grass production, but further north, where populations were as high as 1250 m^{-2}, up to 30% of annual grass production could be consumed by this species (Sands, 1977).

Hodotermes mossambicus is abundant in the dry steppe regions of South Africa. Nel and Hewitt (1969) found that during the winter of 1966, when numbers were high, 27–61% of the canopy of marked vegetation was removed, equivalent to an average consumption of about 36 g m^{-2} yr^{-1} or 33% of total grass production. Following a crash in population, the consumption rate dropped to about 1% of grass production in 1967. Average rate of grass collection by experimental colonies was 1.6 g per colony per day (Nel, Hewitt and Joubert, 1970), a foraging rate of about 5.0 mg g^{-1} termite biomass d^{-1}.

Several examples of significant grass consumption and damage to pasture by other termite species in various regions are given by Sands (1977). Several species of *Anacanthotermes* often remove up to 20% of the grass cover throughout the Middle East and Asia while *Gnathamitermes tubiformans* consumes around 20% of the standing crop of grass, an amount similar to that used by cattle, in dry Texan grassland. *Cornitermes cumulans* can damage pasture in Brazil, while its mounds reduce the area available for grazing (Amante, 1962). According to Sands (1977) serious damage to pasture is most likely when pasture is overgrazed. Termites may then eat the remainder, including root stocks, with resulting denudation of soil and soil erosion, particularly on steep hillsides in areas of high rainfall.

2.12 ORTHOPTERA

This order of insects includes predatory mantids and omnivorous crickets, and grasshoppers and locusts (Acridoidea), which can cause major damage to grasslands and croplands especially in areas with warm dry summers. There is no clear distinction between grasshoppers and locusts; at one extreme are the solitary species that never swarm and at the other the gregarious locusts that do. In between, there are many species that often aggregate in large numbers and can cause considerable damage to crops and pastures, but do not exhibit the marked morphological and physiological changes associated with the swarming phase of the true locusts. Of the 10 000 species of Acridoidea, only about 12 can be considered locusts and these occur in the warmer regions of the world (Table 2.11).

2.12.1 Biology

All species require a habitat that provides a mosaic of bare ground for oviposition and vegetation for food and shelter. Weather has an important effect on distribution and abundance. Moisture is required for egg development, but nymphs and adults need warm, dry, sunny conditions. The important influence of temperature on acridid activity and behaviour is well established (Uvarov, 1928). Many species have the ability to withstand night temperatures below freezing: after a period of basking in the sun they very quickly thaw and recover. Fecundity is reduced by cool damp weather. High-density populations are associated with mosaic patterns of plant cover, with bare ground and sparse and dense vegetation providing a range of microclimates so that all stages can find their requirements (Dempster, 1963).

The swarming phase in locusts develops from the solitary phase through multiplication, concentration and gregarization (Chapman, 1976). Gregariousness can be induced in the laboratory by crowding; in the field environmental factors that tend to concentrate solitaries provide the first step in the process of phase change. Once forced into a group, locusts tend to aggregate spontaneously. Adult swarms may extend over tens of square kilometres, and a single swarm may contain more than 10^9 individuals (1.5×10^6 kg), which can eat their own weight of vegetation daily. Swarms may migrate over hundreds of kilometres, depending on the wind; it is assumed that this adaptation enables the locust to exist in regions where the food supply is very seasonal and irregular (Chapman, 1976).

2.12.2 Abundance and biomass

While adult densities in settled swarms of the desert locust *Schistocerca gregaria* are typically in the range 30–150 m^{-2} (Chapman, 1976), most population estimates for non-swarming acridids in temperate grasslands are less than 10 m^{-2}. Typical examples are 2–3 m^{-2} in English grasslands (Richards and Waloff, 1954; Morris, 1968), 3.3–7.7 m^{-2} in Polish meadows (Breymeyer, 1971) and 2.8 m^{-2} in old field in Michigan (Wiegert, 1965). Rather higher population densities have been reported from a Finnish meadow (up to 30 *Chorthippus parallelus* m^{-2}) by Gyllenberg (1969); 18 m^{-2} were recorded from alfalfa in the USA (Wiegert, 1965) and

Table 2.11 Distribution of locust species (from Chapman, 1976)

Species	Common name	Distribution
Schistocerca gregaria	Desert locust	Northern Africa, Arabia, Indian subcontinent, from about 10 to 35° N
Schistocerca americana americana		Central America
Schistocerca americana paranensis		South America
Anacridium melanorhodon	Sahelian tree locust	Across Africa south of Sahara, from about 10° to 20° N but extending to equator in the east
Anacridium wernerellum	Sudanese tree locust	Slightly further south than *A. melanorhodon*
Nomadacris septemfasciata	Red locust	Africa, mainly south of equator to about 30° S
Patanga succinata	Bombay locust	South-west Asia
Melanoplus spretus	Rocky mountain locust	North America, now extinct?
Chortoicetes terminifera	Australian plague locust	Australia
Locusta migratoria	Migratory locust	Different subspecies in southern Europe, Africa south of the Sahara, Malagasy Republic, Southern Russia, China, Japan, Philippines, Australia
Locustana pardalina	Brown locust	South Africa
Dociostaurus macroccanus	Moroccan locust	Middle East and Mediterranean countries

10–30 m^{-2} from Arizona rangeland (Nerney, 1960). Lamotte (1947) reported 1.25 acridids m^{-2} in Guinea savannah and Gillon and Gillon (1973) found fewer than 1 m^{-2} in dry thornbush savannah in northern Senegal. Biomass data have been provided by Sinclair (1975), who recorded a maximum of 120 mg dry mass m^{-2} during the wet season in the Serengeti long-grass savannah, Tanzania, while Van Hook (1971) recorded a maximum biomass of 705 mg fresh mass m^{-2} for *Melanolpus sanguinipes* in an old field in Kentucky, USA.

2.12.3 Food preferences

Grasshoppers and locusts are mainly polyphagous, but definite preferences are shown for certain categories of plants and most species studied show some degree of selectivity. Mulkern (1972) classified grasshoppers of the north-central Great Plains area of the USA into monophagous, oligophagous, pleophagous and polyphagous species and reported that survival, growth, development and fecundity were influenced by the plants fed on. *Schistocerca* spp. are polyphagous, and Bernays and Chapman (1978) reported that *S. gregaria* readily ate 160 species in 53 families of plants. *Melanoplus* spp. are also polyphagous, feeding mainly on broad-leaved weeds but also commonly on grass. *Locusta, Locustana, Chorthippus, Chortoicetes* and *Nomadacris* are essentially grass feeders, rarely feeding on broad-leaved plants (Chapman, 1976; Bernays and Chapman, 1978). Several studies have indicated preference for particular grass species among the grass feeders; for example, Bernays and Chapman (1970) found that *Chorthippus parallelus* preferred *Agrostis* while Gyllenberg (1969) reported preference for *Agrostis tenuis* and *Poa pratensis* by this species. Richards and Waloff (1954) recorded preference for *Holcus* by five grasshopper species in *Festuca–Agrostis* grassland in England.

2.12.4 Ecosystem role

Devastation of crops and grassland associated with locust swarms is well known. Chapman (1976) estimated that settled swarms of the desert locust *Schistocerca*, at adult densities of 30–150 m^{-2}, would consume 45–225 g m^{-2} d^{-1}. Direct losses in primary production attributable to consumption by non-outbreak grasshopper populations have been estimated in several studies; these range from less than 1% of primary production to more than 10%. Knutson and Campbell (1976) recorded little damage to tall-grass prairie over a period of 16 years, while at the other extreme grasshopper populations of 18 late-instar nymphs or adults m^{-2} caused 20–70% yield reductions in range grasslands in the USA, with highest levels of damage being associated with heavily grazed, sparse vegetation (Nerney, 1958). Wiegert (1965) estimated consumption by a maximum grasshopper population of 2.8 m^{-2} in an old field in Michigan, USA, to be less than 0.5% of shoot production, and that of a maximum population of 18 m^{-2} in alfalfa to be about 1.3% of alfalfa shoot production.

A population of up to 30 *Chorthippus parallelus* m^{-2} in a Finnish meadow consumed 1.4–2.7% of above-ground primary production (Gyllenberg, 1969). Andrzejewska and Wojcik (1970) arrived at rather higher estimates of damage due to grasshoppers in Polish meadows, pointing out that direct consumption may account only for a small proportion of the actual damage, which is mainly due to withering of the leaf above the point fed on. They estimated that destruction of relatively short-leaved grasses in a *Stellario–Deschampsietum* meadow was on average six times greater than consumption, whereas in a tall sedge (*Caricetum*) association this factor was 15. Loss of primary production in the meadow due to 10 hoppers m^{-2} feeding for 4 months was approximately 24 g dry matter m^{-2}, or 14% of net primary production. They estimated that 44.1 g m^{-2} was consumed or destroyed in tall sedge meadow with an average population of 7 m^{-2} feeding for 3 months, reducing primary production by 8%. Mitchell and Pfadt (1974) give lower destruction/consumption ratios of 0.5–0.99 for three short-grass prairie species but conclude that the main function of grasshoppers in this ecosystem is making litter rather than assimilating food.

Grasshoppers are the main invertebrate consumers in the Serengeti savannah grassland in Tanzania. Sinclair (1975) estimated that the grasshopper consumption was 7.6% of net primary production in long grassland and 4.1% in short grassland on a yearly basis: the corresponding figures for ungulates were 18.8% and 34%. During the wet season when conditions were favourable for rapid population increase the impact of grasshoppers was considered to be at least equal to that of the resident ungulate population. White (1974, 1979a) studied the impact of grasshoppers on alpine tussock grasslands in New Zealand and concluded that their overall consumption was no more than 1–2% of annual primary production on average but under extreme grazing pressure this could be up to 18%. In addition, much of their feeding was selective on important ground-cover species of low biomass, and consumption over several years could account for up to 11% loss of total ground cover with serious consequences for the stability of this fragile ecosystem.

While most studies have stressed the negative effects of grasshopper feeding on plant growth, a few have pointed to some positive effects. Andrzejewska and Wojcik (1970) concluded that grasshopper feeding could stimulate more intense grass growth during the phase of maximum growth in early summer and Dyer and Bokhari (1976) reported a similar effect, possibly due to the injection of growth-promoting substances into the plant during chewing.

Less attention has been given to the role of crickets in grassland eco-systems. These are generally thought of as being mainly detritivorous but some species can be sporadically serious crop pests (Hill, 1987). Field crickets of the family Gryllidae including the cosmopolitan *Gryllus* spp. and several species of *Acheta*, which are distributed throughout southern Europe, Africa, southern Asia, the USA and South America, can defoliate grasses and destroy roots and seedlings. The migratory mormon cricket, a grassland pest in Canada and the USA, is actually a grasshopper (family Tettigoniidae). Cowan and Shipman (1947) measured food consumption by this species and concluded that a migrating band of 12 m^{-2} would consume 120 t dry mass of rangeland vegetation in 4 months, equivalent to the forage needed to sustain 44 head of cattle.

2.13 THYSANOPTERA

Thysanoptera (thrips), a group of small insects, is well represented in grassland and can sometimes cause economic losses to grass seed crops.

2.13.1 Biology

Female thrips are always diploid and males haploid (i.e. derived from unfertilized eggs; Lewis, 1973). Arrhenotoky is the prevalent form of reproduction in many species, while in others males are scarce and the-lytoky is the norm. There are indications that temperature affects the sex ratio in nature, and it may be that cyclical changes in the type of repro-duction occur in some species, with thelytoky predominating in summer and arrhenotoky or sexual reproduction in the autumn (Lewis, 1973). Eggs are placed in incisions made with the ovipositor in plant tissue in the case of most Terebrantia, and on plant surfaces in the cases of Tubulifera. There are usually four or five immature stages in the life cycle, including two feeding 'larval stages' and two resting 'pupal stages'. Developmental time depends on temperature: Sharga (1933) reported that the total life cycle (egg laying to adult emergence) took 29–35 days for *Limothrips cerealium* and 40 days for *Aptinothrips rufus* in Scottish summers. In warm climates, there are several generations a year, but in cool temperate regions one or two is the norm. Sharga (1933) concluded that *Limothrips cerealum* is univoltine in the Edinburgh area. Here, only adult females survive the winter, hibernating in grasses and other shel-tered situations and breeding during the summer. *Aptinothrips rufus*, by contrast, can breed all year round in mild seasons and in Sharga's study, gravid females and larvae occurred throughout the year.

Some species may go into diapause or become quiescent during unfavourable periods of the year either as fully grown second-stage larvae or as adults. Overwintering is most common in adult females in soil, dense turf or hedgerow litter or under bark of trees. Mass flight migrations of winged grass and cereal thrips from ripening crops towards the end of the growing season are comnon in Europe and Central Asia (Lewis, 1973): these are usually confined to a few days when meteorological conditions are favourable.

2.13.2 Species richness and abundance

The most species-rich thrips communities of temperate regions occur in herb-rich seminatural grasslands where there may be 30–40 species per site (von Oettingen, 1942). Grass and cereal crops tend to have a thrips fauna dominated by a few species, such as *Limothrips denticornis* in barley and *Chirothrips* spp. in grass seed crops (Lewis, 1973). Cereal crops tend to support several thousand thrips m^{-2} (Lewis, 1973); populations in grasslands are usually considerably lower. Solhöy (1972) recorded 400 m^{-2} from mountain meadow in Norway; Tansky (1961) recorded 351 m^{-2} and 254 m^{-2}, respectively, from reverted grassland and virgin steppe in the former USSR; and Persson and Lohm (1977) recorded 720 m^{-2} in Swedish fen grassland. Population densities in mown leys and permanent grassland in Ireland were somewhat lower (42–112 m^{-2}), probably due to repeated removal of breeding and feeding sites in the upper leaf sheaths and inflorescences (Curry and O'Neill, 1979). Mean population densities in Swedish grass and lucerne ley plots were 229 and 126 m^{-2}, respectively, with *Frankliniella tenuicornis* being the dominant species (Curry, 1986).

2.13.3 Feeding habits

Most species of thrips are phytophagous, sucking juices from leaves, flowers and shoots of plants; some are moss and fungal feeders. Some are predatory on small arthropods, including other thrips. Many flower-dwelling species swallow pollen grains or suck their contents. Lewis (1973) distinguished between penetrating feeding, where the walls of epidermal and mesophyll cells are ruptured by the stylets and plant juices oozing out are sucked up with the mouth cone, and shallow feeding, where only epidermal cells and a few mesophyll cells are ruptured. Sharga (1933) described how *Limothrips cerealium* feeds on protected areas of leaf tissue in wheat seedlings first, and later on the developing inflorescence. Infested grass and cereal inflorescence withers and turns silvery-white, giving the characteristic 'silvertip' or 'whitehead' appearance. Grasses provide many suitable crevices in inflorescence and leaf sheaths for thrips and are hosts to more species than any other family of plants (Lewis, 1973).

2.13.4 Ecosystem role

Few studies have provided data on the role of thrips in community energetics: an exception is that of Persson and Lohm (1977), who computed an annual respiration of 1.2 kJ m^{-2} for this group in Swedish grassland, comprising 4.4% of total above-ground invertebrate respiration.

Several hundred species have been recorded as crop pests, and economic losses to cereals and grass seed crops have frequently been reported, notably due to *Chirothrips* spp., but also to *Anaphothrips obscurus* in the case of grass seed crops in Oregon (Kamm, 1972). Kamm noted that relatively small numbers of overwintering thrips can produce silvertop by feeding on the inflorescence within the leaf sheath, thereby reducing yield in bent grass. After the heads emerge from the leaf sheaths these thrips move to the protected environment of the lower leaves and can destroy up to two-thirds of the leaves under favourable conditions, with consequent reduction of quality and yield of seed. Losses ranging from 10% to 40% in seed yield of grasses such as meadow foxtail, cocksfoot, timothy and ryegrass have been largely attributed to *Chirothrips* spp. in New Zealand, Sweden, Finland and Germany (Hukkinen, 1936; Johansson, 1946; Doull, 1956; Wetzel, 1964).

Thrips probably play only a minor role in disease transmission in grassland, and also have a role in pollination of seed crops but are probably only of significance when large insect pollinators are absent. Predatory species usually feed on small soft-bodied insects such as other thrips, aphids, scale insects, and on eggs of mites and of various insects. These are usually polyphagous and can survive temporarily on plant juices and pollen (Lewis, 1973).

2.14 HEMIPTERA

The Hemiptera (plant bugs) are a large and diverse group of insects comprising some 50 000 species. They are common in grassland where some representatives are associated with roots and basal parts, others with aerial plant structures. Although a wide range of feeding habits occurs in the group, bugs are mainly phytophagous and include many important pests of agricultural crops. In addition to direct damage caused by their feeding they are the main vectors of viral and mycoplasmal plant diseases. Mean populations in the herb layer usually range from fewer than 100 to over 500 m^{-2} during the growing season (Table 2.12), but may reach several thousand m^{-2} during periods of maximal abundance. The order is divisible into two distinct suborders, the Heteroptera and the Homoptera. The Homoptera are best represented in grassland by the Auchenorrhyncha, comprising leafhoppers (Cicadellidae), planthoppers (Delphacidae), froghoppers (Cercopidae), and aphids (Aphidoidea).

2.14.1 Heteroptera

(a) Distribution, abundance and biology

The Heteroptera are best represented in unmanaged, seminatural grassland and are usually scarce in managed grass leys. Thus, Morris (1969, 1973, 1979) recorded 28 and 35 species, respectively, from two areas of chalk grassland and 42 from limestone grassland in the UK. Population densities were low in grazed areas (Table 2.12) and were generally reduced by cutting (Morris, 1979). Heteropterans were scarce in grass and clover leys sampled in Ireland (Curry and O'Neill, 1979), but *Lygus rugulipennis* was abundant in lucerne leys in Sweden (Curry, 1986).

Many of the Holarctic heteropterans are univoltine, but several species found in grassland in Britain, have two generations a year including *Notostira elongata*, *Anthocoris nemorum* and *Lygus rugulipennis* (Southwood and Leston, 1959).

(b) Feeding habits and damage

Feeding habits in heteropterans range from entirely predacious to entirely herbivorous, including monophagous and polyphagous forms. The Miridae, the commonest heteropteran family, is represented in grassland by phytophagous species including grass-feeding Stenodemini such as *Stenodema*, *Notostira*, *Trigonotylus* and *Leptoterna* spp. Most mirids belonging to the subfamilies Phylinae, Orthotylinae and Dicyphinae, and several of the Mirinae, are at least partial predators, while Deraeocorini are entirely predacious (Southwood and Leston, 1959). Predacious damsel bugs of the Family Nabidae such as *Nabis ferus* and *N. rugosus* are common in grassland and feed on a variety of prey, including mirids, aphids, leafhoppers and caterpillars.

The host plants and feeding habits of herbivorous grassland heteropterans are varied. Thus, the dominant species in chalk grassland in Bedfordshire, England, include a moss-feeding species *Acalypta parvula* and a sedge-feeder (*Agramma laeta*); *Mecomma dispar*, a possible sedge-feeder; *Berytinus signoreti*, associated with *Lotus* and *Vicia* and possibly other host plants in grassland; *Stygnocoris pedestris*, a common species of unknown feeding habits; and several grass-feeding Stenodemini, of which *Leptoterna ferrugata* and *Notostyra elongata* are the most abundant (Morris, 1968).

Lygus rugulipennis feeds on a wide variety of herbaceous plants and is sometimes a crop pest (Southwood and Leston, 1959). The chinchbug *Blissus leucopterus* is an important pest of cereals and many grasses in North America. It has recently become established in Brazil, where serious damage to pasture has been recorded involving populations of up to

Table 2.12 Mean population densities of hemipterans in various grasslands

Author	Habitat	Group	Population density (numbers m^{-2})	
			Mean	Maximum
Morris (1969, 1971b)	Grazed chalk grassland, England	Ground-dwelling Heteroptera	24	
		Herbage-dwelling Heteroptera	7	
		Auchenorrhyncha	17	
	Ungrazed chalk grassland, England	Ground-dwelling Heteroptera	167	
		Herbage-dwelling Heteroptera	85	
		Auchenorrhyncha	245	
Curry and O'Neill (1979)	Leys and old grassland, Ireland	Total Hemiptera	63–91	
Purvis (1978)	Grazed and mown grass plots, Ireland	Auchenorrhyncha	32–401	c.1000
		Heteroptera	1–6	
		Aphidoidea	48–349	c.900
Breymeyer (1971)	Three grasslands, Poland	Homoptera	16–41	
Persson and Lohm (1977)	Fen grassland, Sweden	Homoptera	289	931
Törmälä (1982)	Abandoned field, Finland	Heteroptera	29	161
		Auchenorrhyncha	814	1324
		Heteroptera	203	693
		Aphidoidea	207	861
Curry (1986)	Fescue ley, Sweden	Total Hemiptera	267	1233
	Lucerne ley, Sweden	Total Hemiptera	168	502

20 000 m^{-2} (Samways, 1979). In Brazil, it is confined to one species of grass, *Brachiaria radicans*.

2.14.2 Homoptera – Auchenorrhyncha

(a) Distribution and abundance

The Auchenorrhyncha (leafhoppers, planthoppers and froghoppers) as a group are very characteristic of grassland, where numbers sometimes may exceed 1000 m^{-2} at times of peak nymphal abundance (Waloff, 1980). However, with the exception of an abandoned field in Finland, where populations sampled in June, July and August over a 6-year period averaged over 800 m^{-2}, mean population densities rarely exceed 200 m^{-2} (Table 2.12).

Communities of planthoppers reach their greatest diversity in areas of grassland rich in plant species and where the architecture of the sward is varied. Thus, 63 species were recorded, 42 of them from regularly breeding colonies, in an area of acid grassland dominated by *Agrostis tenuis* in England (Waloff and Solomon, 1973), while in a more uniform area dominated by *Holcus* 49 species were recorded but only 10 bred there regularly (Waloff, 1980). Andrzejewska (1965) found that different species of Auchenorrhyncha in Polish meadow tend to occupy definite layers in the habitat, with vertical movements of different populations occurring during the growing cycle. Tall grasslands typically support a richer community than do grasslands where the sward is kept uniformly short. Morris (1971b, 1973, 1981a), Morris and Lakhani (1979) and Morris and Plant (1983) showed that grassland that is managed by cutting or grazing has fewer species and individuals and a lower species diversity than grassland that is ungrazed or unmown. However, managed grasslands have their own characteristic fauna such as *Macrosteles laevis* and *Psammotettix cephalotes*, which are adapted to short grass (Andrzejewska, 1962; Morris, 1973).

(b) Biology

Waloff (1980) outlined the life histories of 24 leafhopper species common in acidic grassland in southern England. Ten have one generation a year, 13 have a second generation and 1 has a partial second generation a year. Most Delphacidae hibernate as nymphs and hence predominate in number at the beginning of the growing season, while most Cicadellidae hibernate in the egg stage and tend to build up in numbers later. Alternating waves of abundance of delphacids and cicadellids throughout the growing season were observed. The number of generations a year depends on the length of the growing season: Kontkanen (1954) compared a site in eastern Finland with a district in western Germany and found that of the species that the two sites have in common three have only one generation a year both in Finland and Germany and seven have only one generation annually in Finland but two in Germany.

(c) Feeding habits

Many grassland Auchenorrhyncha are oligophagous, with most species being confined to a single grass genus or tribe (Whitcomb *et al.*, 1974 cited by Waloff, 1980). The group includes mesophyll and vascular feeders, with the Typhlocybinae (Cicadellidae) being mostly mesophyll feeders, most Cercopidae, Cicadidae and Cicadellidae feeding predominantly on xylem sap and Delphacidae being mostly phloem feeders (Waloff, 1980; Raven, 1983). Because of the low concentration of solutes in xylem sap, xylem feeders must ingest large volumes of sap to obtain their nutritional requirements: daily sap intake can be several hundred times the body weight (Andrzejewska and Gyllenberg, 1980) and heavy infestation of species such as the leafhopper *Cicadella viridis* can cause drying up and withering of affected plants (Andrzejewska, 1967). Phloem sap frequently contains suboptimal levels of N and some elements required for growth, and the level of soluble N influences parameters such as instar duration, food ingestion rates, efficiency of N utilization and oviposition rates (Prestidge, 1982a). Maturation and reproduction of alates and egg production are often geared to periods when feeding sites with high levels of soluble N are available (Waloff, 1980). Prestidge and McNeill (1983a) showed that the occurrence of some oligophagous leafhoppers is strongly correlated with particular N levels in a range of grasses and switching of host species occurs as N levels change, while other species such as *Macrosteles laevis* show little host plant specificity and little association with plant N levels.

(d) Economic importance

Auchenorrhyncha are rarely implicated in serious crop damage in Europe. Several species are capable of transmitting virus diseases (Ossiannilsson, 1966), but only *Javesella pelucida* is of any consequence as a virus vector in Europe. It transmits European wheat striate mosaic and oat sterile dwarf virus (Plumb, 1978). This species is very abundant in Finland where the climate is favourable and its preferred host plants (oats and timothy) are widely grown (Raatikainen, 1967). In parts of North America and some tropical countries, planthoppers can be serious pests of grassland and crops. Weaver and Hibbs (1952) reported that the meadow spittle bug (*Philaenus spumarius*) can be a serious pest of alfalfa and red clover throughout the northeastern USA where yield increases of up to 1.5 t hay ha^{-1} occurred following insecticidal treatment. The two-lined spittle bug *Prosapia bicincta* is an important pest of improved pasture grasses throughout the southeastern USA (Fagan and Kuitert, 1970), while Hawkins *et al.* (1979) reported average yield increases of 18% in coastal Bermuda grass in Louisiana after suppression of leafhopper and planthopper populations with insecticides.

2.14.3 Homoptera–Aphidoidea

(a) Distribution and abundance.

Aphids are widely distributed in temperate regions and include many important pests of agricultural crops. Over 40 species are associated with Gramineae in Europe, seven of these being commonly found on cereals and grasses (Vickerman and Wratten, 1979). Considerable attention has been given to aphids in cereal crops (Vickerman and Wratten, 1979; Carter *et al.*, 1980) but their effects on grass growth and yield have not been extensively studied. *Rhopalosiphum padi* and *Metopolophium festucae* are usually the most abundant aphids in northern European grassland, while the root-feeding species *Aploneura lentisci* can sometimes be abundant in grass leys (Purvis and Curry, 1981). Aphid populations in grassland do not usually reach the kind of levels sometimes experienced in cereal crops: typically, seasonal mean populations are less than 200 m^{-2} but may reach several thousands per m^{-2} in occasional years (Vickerman, 1978).

(b) Biology

Monoecious species that spend the whole year on grasses and cereals include *Sitobion avenae*, *Schizaphis graminum*, *Metopolophium festucae* and *Rhopalosiphum maidis*, while *Rhopalosiphum padi*, *R. insertum* and *Metopolophium dirhodum* are heteroecious, alternating between their primary woody hosts (family Rosaceae) and their secondary herbaceous hosts (grasses and cereals). Although the egg is the main overwintering stage most species can survive the winter as nymphs or parthenogenetic adults on grasses and cereals in areas with milder climates. Aspects of aphid biology that contribute to their status as perhaps the most successful of all phytophagous insects include (1) their ability to produce a rapid succession of generations of apterous viviparous females under favourable environmental and food conditions, (2) the production of alate morphs in response to overcrowding followed by dispersal to new host plants and (3) the production of a sexual phase that gives rise to diapausing fertilized winter eggs (van Emden *et al.*, 1969; Vickerman and Wratten, 1979; Carter *et al.*, 1980).

(c) Effects on host plants

Some aphids derive their nutrition from parenchymous tissue, but most are believed to feed predominantly on phloem sap (van Emden *et al.*, 1969). The direct effects of aphid feeding on plants include depletion of plant carbohydrates, distortion of plant tissues, local cellular damage around the feeding site, and physiological disorders of various kinds due to toxins in the injected aphid saliva. The effects of heavy aphid infestations on yield and quality of cereal grains can be serious (Vickerman and Wratten, 1979). Most attention has been given to species that feed on the aerial part of the plant; the effects of root-feeding aphids such as *Rhopalosiphum insertum* and Pemphagidae are unknown. Aphid damage to grass has rarely been quantified, although there are several general reports of injury (Carter *et al.*, 1980).

Aphids are the commonest vectors of plant viruses but transmit only two to grasses – the barley yellow dwarf virus (BYDV) and cocksfoot streak virus (CfSV) (Plumb, 1978). CfSV is widespread and decreases yield of seed crops, but infected plants rapidly disappear from mixed swards because of competition from healthy plants (Catherall, 1971). BYDV has a wide host range and is widespread in ryegrass fields although the characteristic discoloration seen in cereal crops is not manifested. Significant losses in grass yields have been attributed to the virus in Sweden (Lindsten and Gerhardson, 1969) and in the UK, where Catherall (1971) recorded losses in dry matter of up to 70% in individual infested plants and simulated losses of up to 20% in ryegrass–clover swards. Forage legumes, particularly lucerne, are susceptible to attack by several aphid species. These include the spotted alfalfa aphid *Therioaphis maculata* and the blue-green lucerne aphid (*Acyrthosiphon kondoi*), which are major pests of lucerne in North America (Nielson *et al.*, 1976) and parts of Australia (Turner and Franzmann, 1979).

2.14.4 Other grass-infesting Hemiptera

Other Hemiptera reported as damaging grass include the grass mealybug *Heterococcus graminicola*, which occurs on a wide range of grass species in the USA (Dietz and Harwood, 1960).

2.14.5 Energy utilization by Hemiptera

Van Hook (1971) estimated that Homoptera consumed 42 kJ m^{-2} yr^{-1} in a *Festuca/Andropogon* old field in Kentucky. This represented about 9% of total herbivore consumption and less than 1% of net primary production in that site. Plants bugs were considerably more important in the energetics of a coastal salt marsh in Georgia, USA, where the energy flow through the leafhopper population was estimated at 1150 kJ m^{-2} yr^{-1} (Wiegert and Evans, 1967). The annual energy flow through a meadow spittlebug population (*Philaneus spumarius*) was 162 kJ m^{-2} in an alfalfa field in Michigan and 2.5–5 kJ m^{-2} in an old field (Wiegert, 1964); 354 kJ m^{-2} were ingested in the alfalfa field, and Wiegert calculated that this was equivalent to a loss of 1757 kJ m^{-2} in plant energy. Hinton (1971) estimated that a low-density population of *Neophilaenus lineatus* ingested 9.5 kJ m^{-2} yr^{-1} in the form of xylem sap and assimilated 3.8 kJ m^{-2} of this in grassy heathland in Britain. Annual respiration by the Hemiptera in a Swedish grassland was 7 kJ m^{-2} and their estimated energy consumption was 29 kJ m^{-2} yr^{-1} (Persson and Lohm, 1977). Hemiptera were estimated to respire 6–16 kJ m^{-2} in grass ley plots and 8–18 kJ m^{-2} in lucerne plots in central Sweden (Curry, 1986).

2.15 COLEOPTERA

2.15.1 Distribution and biology

The Coleoptera (beetles) comprises the largest of all insect orders, with more than 300 000 known species. Although many species have become adapted to aquatic habitats, this is essentially a terrestrial order being best represented on the soil surface and in low vegetation. Some species are true soil dwellers for all or part of their life cycle. The size range is enormous, ranging from the minute Ptiliidae, which are less than 0.5 mm long, to giant tropical Scarabaeidae such as *Goliathus regius* and *Dynastes hercules*, and Cerambycidae such as *Macrodontia cervicornis*, which are up to 155 mm long. Characteristic features are the hard cuticle and tough elytra that allow them to exploit a wide range of habitats and environmental conditions. Adults are quite resistant to desiccation and can, therefore, live in drier conditions than most insects.

Detailed knowledge of the life cycle and biology is available only for a relatively small number of grassland species of economic importance. In many families (e.g. Carabidae, Staphylinidae, Silphidae and Scarabaeidae) the univoltine life cycle is most common. The larval stage is often prolonged, extending over 4–5 years in *Agriotes* spp. (Elateridae) and from 1 to 5 years in many members of the Scarabaeidae. The duration of the life cycle may vary depending on environmental conditions. Thus, Stewart (1979) reported that while a 1-year life cycle is typical for *Costelytra zealandica*, the grass grub, throughout New Zealand, in some areas, particularly in the South Island, a 2-year cycle is common. He found that the incidence of the 2-year cycle rises with increasing latitude and altitude, suggesting that the main environmental determinant is temperature, but other factors such as drought, which affect rate of development, can modify the dominance of each cycle.

2.15.2 Abundance and biomass

Because of the great variation in size, habits and activity patterns few reliable data are available for total beetle populations in grassland. Published estimates range from fewer than 10 to over 4000 m^{-2} (Table 2.13), but much of this intersite variability may be attributable to differences in methods of sampling. For example, large surface-dwelling species are inadequately sampled by soil coring and heat extraction whereas the smaller, soil-dwelling species are often underestimated by hand sorting. Most estimates fall in the range of one to several hundred individuals per m^2. The highest estimate (4397 m^{-2}), for permanent pasture in Britain, was based on the Salt and Hollick (1944) flotation extraction technique developed especially for wireworm (Elateridae) sampling, and this group accounted for 66% of all beetles extracted.

2.15.3 Feeding habits and roles in grassland ecosystems

Coleoptera have a wide range of feeding habits. For example, there are phytophagous forms, which feed on woody and herbaceous plants, organic-matter feeders such as dung beetles, and predators of soil and surface-dwelling invertebrates such as members of the families Carabidae and Staphylinidae. Considerable differences in feeding habits may be found within the various beetle families, even between closely related species. For example, although most of the Carabidae (ground beetles) are predators, many *Amara* and *Harpalus* species are phytophagous (Tischler, 1965). Most Staphylinidae are predators, but some *Aleochara* larvae are ectoparasites of dipteran puparia (Palm, 1972). Some Tachyporinae and many small Aleocharinae such as *Oxypoda* spp. and the widely distributed *Amischa analis* and *Atheta fungi* are probably mainly saprophagous (Tischler, 1965), but many mycophilous species are probably also predatory (Newton, 1984). The family Scarabaeidae includes saprophagous and necrophagous forms as well as coprophages and phytophages (Raw, 1967).

(a) Phytophagy

Phytophagous forms of economic importance include root-feeding wireworms (*Agriotes* spp.), Scarabaeidae such as *Melolontha melolontha* (cockchafer) and *Phyllopertha horticola* (garden chafer) in Europe (Richter, 1958), and *Costelytra zealandica* (grass grub) and *Heteronychus arator* (black beetle) in New Zealand (Kain, 1979; East, King and Watson, 1981). Many scarabaeids feed mainly on dead organic matter, while some species such as *Sericesthis nigrolineata* subsist on dead organic matter in soil but prefer and grow better on living roots (Ridsdill-Smith, 1975). Many weevils

Table 2.13 Population density and biomass of beetles in various grasslands

Author	Habitat	Group	Numbers (m^{-2})	Biomass ($g\ m^{-2}$)*
Luff (1966)	Tussock grassland, England	Surface dwellers	256	
Salt et al. (1948)	Permanent pasture, England		4397	
Edwards and Lofty (1975a)	Permanent pasture, England	Larvae	Up to 900	
Curry (1969a)	Old grassland, Ireland		608	
Curry and Momen (1988)	Grass leys on reclaimed peat, Ireland		751–1422	
Curry and O'Neill (1979)	Leys and old grassland, Ireland	Surface dwellers	38–69	
Morris (1968)	Chalk grassland, England			
	Intensively grazed		82	
	Ungrazed		244	
Morris and Rispin (1987)	Limestone grassland, England		435	
Persson and Lohm (1977)	Fen grassland, Sweden		1404	2.9 (D)
Törmälä (1979)	Abandoned field, Finland		663	0.5 (D)
Nowak (1976)	Polish meadows		21–35	
Luxton (1982)	6-year-old grassland on peat, New Zealand		39	
Luxton (1983)	10-year-old grassland on peat, New Zealand		218	
Hutchinson and King (1980)	Sheep pastures, Australia	Adults	35–50	0.6–1.0(F)
		Scarabaeidae larvae	47–61	4.2–20.3(F)
		Other larvae	14–49	0.1–2.0(F)
Nakamura (1965)	Japanese grassland	Scarabaeidae	20–500	
Lloyd et al. (1973)	Short-grass prairie, USA		178	0.3(D)
Gillon and Gillon (1973)	Thornbush savannah, Senegal		<2	0.004(D?)
Athias, Josens and Lavelle (1974)	Savannah, Lamto, Ivory Coast	Small beetles	230	0.3(F)
Chernova et al. (1975)	Savannah, Lamto, Ivory Coast	Larger beetles	2–11	0.1–1.0(F)
	Herb grassland, Russian tundra			107(F)

*F = fresh mass; D = dry mass.

(Curculionidae) feed on grassland plants. Adults of the Argentine stem weevil (*Listronotus bonariensis*) are leaf feeders while the larvae tunnel in grass tillers (Kelsey, 1958) and legume-feeding species, such as the white-fringed weevil *Graphognathus leucoloma*, *Sitona* spp., *Apion* spp. and the alfalfa weevil (*Hypera* spp.), can damage clover and lucerne swards in various parts of the world. The larva of *Dascillus cervinus* (Dascillidae) is considered to be a root feeder (Kühnelt, 1961); this species occurred in densities of up to 1560 m^{-2} in grassland on fen peat in Ireland (Baker, 1981). Herbivory is common among the Chrysomelidae, which is represented in grassland by species such as *Oulema melanopa*, the cereal leaf beetle, and flea beetles. Some weed-feeding *Longitarsus* spp. appear to have potential for the biological control of pasture weeds, such as heliotrope (*Heliotropium europaeum*) and ragwort (*Senecio jacobaea*) in Australia (Waterhouse, 1979).

(b) Predation

Many Coleoptera are predatory – notably ground beetles (Carabidae), which are common on the soil surface and feed on a range of surface-dwelling prey. Soil-burrowing species such as *Scarites, Dyschirius* and *Clivinia* spp. feed on soil insects. Predatory rove beetles (Staphylinidae) are also common in grassland, feeding on a wide range of small arthropods. Their larvae live in the upper soil layers where they are partly predacious and partly feeders on decaying organic matter. The list of prey consumed by Carabidae and Staphylinidae is comprehensive. Thus, Mitchell (1963) reported fragments of collembolans, small insects, mites, pseudoscorpions and earthworms from the gut of *Bembidion lampros*. Sunderland (1975) examined the gut contents of predatory species caught in pitfall traps in cereal crops and found remains of a wide range of prey such as collembolans, flies, beetles and aphids, although individual species varied in their degree of prey specialization. *Pterostichus melanarius*, *Harpalus rufipes* and *Nebria brevicollis* contained a wide range of food, whereas *Notiophilus biguttatus, Loricera pilicornis* and *Lamycetes fulvicornis* fed primarily on collembolans. Other predatory beetles common in grassland include members of the families Cantharidae and Histeridae.

(c) Carrion and dung feeding

Most grassland beetles are saprophagous, associated with decaying organic materials of various kinds. The carrion feeders are a rather specialized group, being prominent during all stages of carrion decomposition. A succession of species associated with various stages in decomposition can be distinguished (Raw, 1967), with silphids such as the burying beetles *Necrophorus* spp. being prevalent in the early stages, and dermestids, clerids and tenebrionids later on. Many species of beetles are coprophagous, notably some members of the families Geotrupidae and Scarabaeidae. *Geotrups* spp. (dor beetles) excavate burrows under dung pats and provision them with dung for their larvae. Two subfamilies of the Scarabaeidae, the Aphodiinae and Coprinae, are predominantly dung feeders although not exclusively so: some *Aphodius* spp. are phytophagous and can be important root-feeding pests in Australian grassland (Waterhouse, 1979). Most Coprinae (and some other Scarabaeidae) are dung buriers. They dig tunnels beneath pats, line them with dung and store balls of dung in them. The adults live on this dung for short periods before flying off in search of new pats. The females lay eggs inside the balls of dung, which serve as food for the larvae. *Aphodius* spp. do not bury dung as a food store for their progeny but live and breed in the dung pats while they are still on the surface.

The significant role of dung beetles in the decomposition and mineralization of dung is well recognized. Bornemissza (1970) found that *Onthophagus gazella*, an Afro-Asian species, buried dung so rapidly in an insectary study that four insects per 100 cm^3 dung completely buried cow pats within 30–40 h. Breymeyer (1974) estimated that *Aphodius* and *Onthophagus* spp. buried about 10% of sheep dung in Carpathian mountain pasture. Holter (1979) concluded that, although the proportion of dung metabolized by dung feeding beetles is negligible, *Aphodius* spp. substantially accelerate rates of dung disappearance by mixing dung with soil, stimulating microbial activity, aerating dung, etc. Gillard (1967) outlined how, in the absence of a coprophagous fauna adapted to the faeces of large ungulates, dung from domestic animals remains undisturbed on Australian pastures for up to 1 year, resulting in fouling of pasture and immobilization of plant nutrients. Under these conditions, dung-breeding nuisance flies such as *Musca vetustissima* proliferate.

(d) Energy utilization

Few studies have attempted to assess the total contribution of beetles to energy flow in grassland ecosystems. Persson and Lohm (1977) calculated the annual respiration of the soil Coleoptera in a Swedish fen grassland to be 167 kJ m^{-2}, a figure that is considerably higher than Macfadyen's (1963) estimate of 39 kJ m^{-2} for a grazed meadow in England. Hutchinson and King (1980) give the following values for annual respiration by Coleoptera in lightly grazed Australian temperate pasture: adult beetles 79.5 kJ m^{-2}, Scarabaeidae larvae 416 kJ m^{-2} and other beetle larvae 183 kJ m^{-2}. Nakamura (1965) calculated an energy budget for the larval population of *Anomala orientalis* (Scarabaeidae) in Japanese grassland and estimated that this species consumed over 300 g dry mass of dead plant material m^{-2} yr^{-1}.

2.16 DIPTERA

2.16.1 Distribution, abundance and feeding habits

Larvae of Diptera (true flies) are widely distributed in grassland soils, being best represented in moist organic habitats and least in arid sites. Population densities in mesic grassland soils range from a few hundred per m^2 to almost 10 000 per m^2 (Table 2.14). In addition, several hundred shoot-mining larvae may inhabit the vegetation, while 100–300 adults m^{-2} are commonly encountered in the herbage layer (this number may exceed 1000 in long grass). Olechowicz (1971) estimated that Diptera represented 48–56% of all insects emerging from Polish meadows.

(a) Nematocera

Most terrestrial Diptera generally belong to the sub-order Nematocera; for example, this group accounted for 80% of the dipterous larvae in Swedish fen grassland (Persson and Lohm, 1977). The families Tipulidae, Bibionidae, Mycetophilidae, Cecidomyidae and Chironomidae are generally most abundant. Many Nematocera are saprophagous (Raw, 1967), but some Bibionidae (*Bibio* and *Dilophus* spp.) may also feed on grass roots (D'Arcy Burt and Blackshaw, 1991) and some tipulids (crane flies; leather jackets) such as *Tipula paludosa* can be important crop pests in heavy soils. Tipulid larvae frequently form an important component of the soil faunal biomass in wet soils, and especially in the relatively simple ecosystems of alpine and arctic regions (Hofsvang, 1972, 1973). A study of the tipulid fauna of Moor House Nature Reserve, Northumberland, England, yielded 66 species, but none of these inhabit both mineral and peat soils (Coulson and Whittaker, 1978). One large species (*Tipula subnodicornis*) and two small ones (*Melophilus ater* and *Pedicia immaculata*) occurr abundantly in peat, highest densities being recorded in *Juncus squarrosus* moorland and lowest in waterlogged *Sphagnum* peat, while three species are abundant in mineral soils (*Tipula varipennis*, *T. paludosa* and *T. pagana*). Coulson (1962) reported high mortality of eggs and first-instar larvae associated with desiccation. During dry periods in 1955, *Tipula subnodicornis* died out in many areas or was severely reduced; only in areas such as *Sphagnum* flushes that retained water was density similar to that of previous years. Such areas do not normally support high densities of larvae but act as reservoirs from which the surrounding areas can be repopulated.

Mycetophylidae (fungus gnats) are mainly fungivorous. Larvae of Cecidomyidae (gall midges) have a wide range of feeding habits: some are herbivorous and are associated with gall formation and loss of seed production (Hukkinen, 1936; Barnes, 1946), while soil-dwelling forms may be saprophagous, coprophagous or predacious. Larvae of many families, including Chironomidae (non-biting midges), Scatopsidae, Psychodidae (owl midges), Sciaridae (also fungus gnats), Bibionidae (fever flies; St Mark's flies) and Cecidomyidae, commonly occur in animal dung (Mohr, 1943; Laurence, 1954; Curry, 1979) but are probably not primarily coprophagous. Larvae of Ceratopogonidae (biting midges), Culicidae (mosquitoes) and Simuliidae (black flies) are aquatic, but 'biting' adults can cause severe irritation to humans and animals while mosquitoes and, to a lesser extent, black flies can be important vectors of disease.

(b) Brachycera

Soil-inhabiting representatives of the suborder Brachycera are mostly predacious. Members of the families Stratiomyidae, Empididae, Dolichopodidae and, in wet habitats, Tabanidae, are common in grassland. The Tabanidae (horse flies) contain several blood-sucking species that can be a nuisance for humans and livestock. Brachycerans tend to be most abundant in organic habitats where their prey includes saprophagous nematoceran larvae and other small invertebrates. One species of Stratiomyidae, *Inopus rubriceps* (Australian soldier fly), feeds on plant roots and has become a major pest of sown pastures in New Zealand (Robertson *et al.*, 1979).

(c) Cyclorrhapha

Grassland representatives of the suborder Cyclorrhapha are varied in their biology and feeding habits. Families such as Lonchopteridae, Borboridae, Sepsidae, Calliphoridae, Sapromyzidae and Muscidae are well represented among the saprophagic fauna, where organic matter accumulates in the form of dung or rotten vegetation. Olechowicz (1974) reported that 33% of invertebrates colonizing sheep dung were Diptera. She distinguished a primary coprophagic dipteran fauna consisting of members of the cyclorrhaphan families Anthomyidae, Scathophagidae, Sepsidae and Borboridae that colonized freshly deposited dung and that had an important influence on the rate of decomposition. Some dung-breeding muscids may constitute a severe nuisance for humans and livestock: these include *Musca autumnalis* (face fly), *Musca vetustissima* (Australian bush fly), *Haematobia irritans* (horn fly), *Musca domestica* (the common house fly) and *Stomoxys calcitrans* (stable fly) (Hughes and Walker, 1970; Schmidtmann, 1985; Wright, 1985). Larvae of many Calliphoridae (blow flies) and Sarcophagidae (flesh flies) occur in carrion, notably *Calliphora* spp. which are early colonizers of decaying flesh. Some Tachinidae, Phoridae and Calliphoridae are parasitic on earthworms and

Table 2.14 Mean population density of dipterans in various grasslands

Author	Habitat	Numbers (m^{-2}) Larvae in soil	Shoot-mining larvae	Adults in herbage
Salt et al. (1948)	Permanent pasture, England	672		
Curry (1969a)	Old grassland, Ireland	1280		
Curry and O'Neill (1979)	Grass and clover leys, Ireland			152
Curry and O'Neill (1979)	Old pasture, Ireland			240
Curry and Momen (1988)	Grass leys on reclaimed peat, Ireland	2635–6091		
Purvis (1978)	Grazed and cut grass leys, Ireland			84–297
Clements, Chapman and Henderson (1983)	Ryegrass leys, England		300–400	
Heard and Hopper (1963)	Grass clover leys, England		< 800	
	Range of 29 leys, England		96	
Mowat (1974)	Mown grassland, Northern Ireland		200–400	
Nabialczyk-Karg (1980)	Polish meadow	434		
Persson and Lohm (1977)	Fen grassland, Sweden	5500		
Curry (1986); Delettre and Lagerlöf (1992)	Grass ley plots, Sweden	< 3500		371
Curry (1986); Delettre and Lagerlöf (1992)	Lucerne plots, Sweden	< 9700		515
Tömälä (1979)	Abandoned field, Finland	1500		262 max
Breymeyer (1971)	Meadow, Poland			157–296
MacLean (1973)	Wet Arctic tundra	< 250 Tipulidae		
Coulson and Whittaker (1978)	Moorland, England	> 3000 Tipulidae		
Luxton (1982)	6-year-old pasture on peat, New Zealand	2019		
Luxton (1983)	10-year-old pasture on peat, New Zealand	4492		

other soil invertebrates (Raw, 1967), while some Muscidae such as *Fannia* spp. are predacious on other insect larvae in dung. Several families (Agromyzidae, Anthomyzidae, Opomyzidae, Chloropidae and Anthomyidae) contain phytophagous members and some species, notably shoot-mining Chloropidae, can damage grass leys.

2.16.2 Life cycles

The length of the dipteran life cycle varies enormously. Cecidomyidae have one to three generations a year in Britain (Jones and Jones, 1984), while in suitable habitats such as commercial mushroom beds *Sciara* spp. complete their larval stage in about 3 weeks (Raw, 1967). *Smittia aterrima*, the dominant chironomid species in Swedish grass and lucerne leys, has two generations a year (Delettre and Lagerlöf, 1992). At the other extreme, larval development in arctic crane fly (*Pedicia hannai antenatta*) extends over at least 4 years. (MacLean, 1973). This species, which is able to survive the short growing season and rigorous climate of the tundra, appears to be able to maintain a high larval biomass through a prolonged life cycle and overlapping generations. By contrast, the marsh crane fly *Tipula paludosa*, often the most abundant species in lowland mineral grassland, is univoltine, with the larval stage typically extending from September to June, pupation occurring in July–August and the adult craneflies emerging in August–September. The frit fly *Oscinella frit* is multivoltine in Britain, the number of generations varying with the temperature. In southern England, there are normally three generations a year whereas in northeastern England there are usually only two (Vickerman, 1980). Emergence of species of stem-boring Diptera appears to occur in sequence and oviposition is extended so that larvae are found in grass throughout the year (Vickerman, 1980; Clements, Chapman and Henderson, 1983).

2.16.3 Ecosystem role

The main role of grassland Diptera is in the decomposition of organic matter, and moist organic residues of all kinds support high densities of a range of species. Coprophagic species accelerate dung decomposition, their role in tunnelling through and ingesting fresh dung being particularly important in facilitating microbial activity. Olechowicz (1974) estimated that dipterous larvae consumed 16% of the sheep dung on mountain pasture in Poland.

The marsh crane fly *Tipula paludosa* can cause economic damage to lowland grasslands on heavy soils in northern Europe (French, 1969; Blackshaw, 1984), while damage to grass swards caused by *Dilophus febrilis* has been reported (Port and French, 1984). Several gall midges can affect seed crops (Jones and Jones, 1984). Infestations of *Contarinia merceri* and *Dasineura alopecuri* can reduce seed production in meadow foxtail *(Alopecurus pratensis)* in Britain, while *Sitodiplosis dactylidis, Contarinia dactylis* and *Dasineura dactylis* infest cocksfoot *(Dactylis glomerata)* and *Contarinia geniculati* infests meadow foxtail and cocksfoot. Clover is attacked by a range of species, including *Dasineura leguminicola and D. trifolii* (clover leaf midges); *Campylomyza ormerodi*, the red clover gall grub, infests the tap root and plant apex, resulting in die off of heavily infested young plants. *Contarinia medicaginis*, the lucerne flower midge, can reduce seed production in lucerne. Several shoot-mining species, notably *Oscinella frit* but also *Opomyza* spp. and *Geomyza* spp., infest a range of grasses and can cause economic loss (Mowat, 1974; Henderson and Clements, 1977a; Clements, Chapman and Henderson, 1983).

Few studies have attempted to calculate the total respiratory activity of dipterans in grassland. Persson and Lohm (1977) give an estimate of 128.5 kJ m^{-2} yr^{-1} for Swedish fen pasture where flies were second only to beetles in importance in terms of soil-arthropod respiration. Tipulid respiration amounted to 335–498 kJ m^{-2} yr^{-1} in moorland soils at Moor House Nature Reserve, Northumberland, England (Coulson and Whittaker, 1978), accounting for 17–22% of total invertebrate respiration.

2.17 HYMENOPTERA

This enormous order of insects contains well over 100 000 known species, including bees, wasps, ants, gall wasps and a wide range of parasitic forms.

The suborder Symphyta (sawflies) are herbivorous and include several important pests, mainly of coniferous trees. Larvae of the family Cephidae are stem borers and several species are injurious to grasses. These include the wheat-stem sawfly *Cephus cinctus*, a native-American grass feeder and serious pest of wheat in the USA (Metcalf, Flint and Metcalf, 1962), and the related European species *Cephus pygmeus*.

The second major subdivision of the order, the suborder Apocrita, is divided into the Parasitica, which contains the gall wasps and a vast number of parasitic forms, and the Aculeata, which comprises the bees, wasps and ants.

2.17.1 Parasitic and gall-forming Hymenoptera

The more important superfamilies of Parasitica include the Ichneumonoidea, Chalcidoidea, Proctotrupoidea and Cynipoidea. Ichneumonids (ichneumon wasps) mainly parasitize insect larvae, notably butterfly and moth caterpillars; also sawflies and beetle larvae. Many species are hyperparasites. The Braconidae (braconid wasps) have essentially similar habits, and the subfamily Aphidiinae contains some species of aphid parasites, notably *Aphidius* spp., *Praon* spp., *Lysiphlebus fabarum, Ephedrus plagiator, Trioxys auctus, Monoctonus cerasi, Diaeretiella rapae* and *Aphelinus varipes* (Vickerman and Wratten, 1979). The role of these parasites in the natural regulation of aphid numbers in grass is not well understood but there are indications that it could be substantial; for example, Jones and Dean (1975) recorded up to 24% parasitism in *Sitobion avenae* collected from oats and barley in 1972 and up to 45% in *Metopolophium dirhodum* apterae.

Members of the Cynipidae (gall wasps), the main family of Cynipoidea, are phytophagous and are associated with gall formation in a range of plants. Most Chalcidoidea (chalcid wasps) are parasites or hyperparasites with a few exceptions which are phytophagous, feeding on seeds or causing gall formation on other parts of plants. These include the wheat jointworm, *Harmolita tritici*, an important pest of wheat in the USA, and related species associated with various grasses.

Curry and O'Neill (1979) recorded a mean population of 46–62 parasitic Hymenoptera m^{-2} over the growing season from grass/clover leys and old pasture in Ireland while Vickerman (1978) recorded 37–173 m^{-2} from grass leys in West Sussex. Mean population densities in Swedish grass and lucerne plots were 83 and 256 m^{-2}, respectively (Curry, 1986).

2.17.2 Ants

(a) Distribution and abundance

There are about 10 000 species of ants (Formicidae) in the world. All are social, but there is considerable variation in colony size and complexity. Temperature is an important factor controlling their occurrence and high populations are most characteristic of drier, warmer climates, but ants are widely distributed and occur up to and above the treeline in the Arctic and most of the world's highest mountains (Brown, 1973). Their social organization and the diversity of their nest construction allows ants to regulate their own environment to some extent and to occupy many different habitats.

Ant nests vary in size and complexity. Typically, there is a central part occupied by the queen for brood rearing and a peripheral set of galleries and chambers occupied by workers; in some species, nests are also used for food storage, fungus growing and aphid rearing. Colony size varies from a few dozen in more primitive species of Ponerinae and Myrmicinae to several million individuals in some tropical Dorylinae and European Formicinae (Pisarski, 1978). In arid regions, soil-dwelling species predominate. Species of Ponerinae common in Australia construct almost completely subterranean nests, while some common European *Lasius* and *Formica* species construct distinctive mounds that may be more than 1 m high.

Colony densities, numbers of individuals per colony and mean populations per m^2 are given in Table 2.15 for a range or grassland and related habitats. Colony density in general is inversely proportional to colony size (Brian, 1965). Nest densities exceeding $1\ m^{-2}$ have been recorded but this is unusual. *Lasius flavus* is the main species in neutral grassland in Britain, and Brian (1965) calculated densities of 6–15 000 and biomass approaching $15\ g\ m^{-2}$ for this species based on Waloff and Blackith's (1962) maximum density of 0.6 mounds m^{-2}. High densities of ants have been reported from dry, sparsely vegetated sandy soils. Nielsen and Jensen (1975) reported over 5000 *Lasius alienus*, a small soil-inhabiting species, per m^2 in Danish heathland. Typically, highest numbers of species are recorded from dry steppe and prairie grasslands of Europe and North America and from tropical savannah (Pisarski, 1978), and under these conditions ants form a greater proportion of invertebrate numbers and biomass than they do in mesic European grasslands. Ants constitute up to 15% of the soil-invertebrate biomass in North American desert grassland (Lewis, 1971); they constitute up to 3% of the invertebrate biomass in Polish meadow soil (Petal, 1980) but this figure is usually less than 1% in European grassland. Species diversity and population densities are generally lower in cultivated intensively managed grassland than in uncultivated old pasture.

(b) Feeding habits

Ants as a group have a wide range of food and feeding habits. They are basically omnivorous but there is considerable variation between species and within species at different times of the year, depending on food availability and the requirements of the colony. There is a high demand for protein during the larval stage while workers have high energy requirements (Stradling, 1978). The most primitive ants are predominantly hunter carnivores, with members of the subfamilies Ponerinae, Dorylinae and many Myrmicinae feeding mainly on other ants, insect larvae, spiders and other invertebrates. An extreme form of hunter carnivory is practised by the driver ants (Dorylinae) of the tropics, which raid their surroundings in vast armies, attacking vertebrates and invertebrates along their path.

Herbivores and granivores predominate in the natural grasslands of the Americas and Africa (Pisarski, 1978). Granivorous species of the genera *Messor*, *Veromessor* and *Pogonomyrmex* are common in desert and arid grasslands, where seeds constitute a major portion of the harvested material. Their capacity to store seed in their nests permits their survival in harsh environments where food is only seasonally available. Rogers

Table 2.15 Population density and biomass of ants in various grasslands

Author	Habitat	Group	Colony density	Number per colony (× 10³)	Number (m⁻²)	Biomass (gm⁻² fresh mass)
Waloff and Blackith (1962)	Neutral grassland, England	Lasius flavus	0.6 m⁻² (max.)			
Brian (1965)	Acid grassland, Scotland	4 species	0.7 m⁻²		700	3
Petal et al. (1971)	Natural meadows, Poland				225–300	
Petal (1976)	Fertilized meadow, Poland				31–188	
Gaspar (1971)	Grassland, Belgium		1.13 m⁻²			
Baroni-Urbani (1968)	Alkaline meadow, Italy		0.5 m⁻²			
Baroni-Urbani (1969)	Festuca grassland, Appenines, Italy		0.65 m⁻²			
Nielsen (1972)	Heathland, Denmark	Lasius alienus		9.7–18		
Nielsen and Jensen (1975)	Heathland, Denmark	Lasius alienus			5068	1.6
Hutchinson and King (1980)	Pasture, New South Wales, Australia				108–332	0.4–0.5
Golley and Gentry (1964)	Old field, South Carolina, USA	Pogonomyrmex badius	27 ha⁻¹	4.6		
Schumacker and Whitford (1976)	Chihuahu desert, USA		0.2–0.4 m⁻²			

Rogers, Lavigne and Millers (1972)	Shortgrass prairie, USA	*Pogonomyrmex occidentalis*			
	Ungrazed to moderately grazed		23–31 ha^{-1}	2676	
	Heavily grazed		3 ha^{-1}		
Talbot (1953)	Old field, Georgia		1 m^{-2}		
Lamotte (1947)	Savannah, Guinea				
Lamotte (1979); Lamotte *et al.* (1979)	Savannah, Lamto, Ivory Coast		0.3–0.7 ha^{-1}	500	250
Cherrett, Pollard and Turner (1974)	Natural savannah, Guyana	*Atta laevigata*	0.59 ha^{-1}		
	Natural savannah, Guyana	*Acromyrmex landolti*	2.98 ha^{-1}		
	Sown savannah, Guyana	*Atta laevigata*	1.9 ha^{-1}		
	Sown savannah, Guyana	*Acromyrmex landolti*	95 ha^{-1}		
Labroder, Martinez and Mora (1972)	*Panicum* pasture, Venezuela	*Acromyrmex landolti*	5930 ha^{-1} (max.)		1

(1974) found that seeds constituted 39% of forage collected by western harvester ant (*Pogonomyrmex occidentalis*) in short-grass prairie, with plant litter accounting for a further 24% and the rest consisting of dead insects, insect prey, faeces and mineral material for mound construction. Tevis (1958) reported that seeds constitute 68–92% of the material harvested by *Veromessor pergandei*, depending on the season. Estimates for seed removal range from 2 to 10% in desert and prairie habitats (Whitford, 1978). Whitford concluded that harvester ants do not appear to affect total seed reserves, but could affect the success and distribution of preferred species.

Several species of leaf-cutting ants (Myrmicinae) of the genera *Atta* and *Acromyrmex* harvest grass in South American savannah; they use vegetable material as a substrate for growing fungi on which they feed. Some species, such as *Acromyrmex landolti* in Guyana, can be important pests of tropical pasture (Cherrett, Pollard and Turner, 1974). *Acromyrmex* spp. also harvest considerable amounts of grass in the southern parts of the USA: Werner (1973) reported that Gramineae account for over 50% of material foraged by *Acromyrmex versicolor* during July–August, while at other times grasshopper faeces and leaves of the leguminous tree *Prosopis juliflora* are major components of the foraged material, depending on availability.

The ants of European grasslands are mainly eurytopic species of forest origin, feeding mainly on other small invertebrates and aphid honeydew (Petal, 1978; Pisarski, 1978). The exact composition of the food depends on food supply and on the requirements of the colony: during the period of intense development of larvae, mainly animal food is collected, while more honeydew is foraged when young workers are maturing (Petal, 1978). The dominant ants in Polish meadows are *Myrmica* spp. and *Lasius niger* (Petal, 1980), and 50% of their food is of animal origin. *Myrmica* spp. are mainly predacious, but *Lasius niger* is a scavenger, feeding on insect bodies, secretions of root and stem aphids and on seeds, nectar and floral parts of flowers (Talbot, 1953; Petal, 1978, 1980). *Myrmica* and *Lasius* species, particularly *Lasius niger*, also carry to their nests considerable amounts of plant material (Petal, 1974).

Plant exudates and sap are consumed extensively by most ants, with aphid honeydew being widely exploited as a rich source of sugar, amino acids, amides and minerals (Stradling, 1978). Aphids are cultivated by many species for their honeydew.

(c) Ecosystem role

Ants can have significant roles in the grassland ecosystem as primary consumers of vegetation and as predators, and they can also modify their habitat through their activities in nest building and food storage. Because of their high rates of metabolism and low production efficiencies ($\leqslant 2\%$) ants have food requirements that are disproportionately greater than their biomass. Petal (1974) estimated that *Myrmica* spp. and *Lasius niger* consumed up to 0.11% of primary production in sheep-grazed Polish pasture, while in eight meadow sites studied over 11 years these ants consumed 0.19–7.26 g dry mass m^{-2} yr^{-1} – equivalent to 0.04%–2.76% of shoot production (Petal, 1980).

Nielsen (1972) calculated the energy flow through a population of *Lasius alienus* in Danish heathland to be 262 kJ m^{-2} yr^{-1}, somewhat higher than Golley and Gentry's (1964) estimate of 59–201 kJ m^{-2} yr^{-1} for the harvester ant *Pogonomyrmex badius* in old field, South Carolina, USA. This latter species stored 5–10% of the seeds produced in its habitat. Ant respiration amounted to 38–53 kJ m^{-2} yr^{-1} in Australian sheep pastures (Hutchinson and King, 1980).

Grass-cutting ants can destroy newly planted grass swards and can cause severe defoliation of established pastures in South America. Cherrett, Pollard and Turner (1974) calculated that *Acromyrmex landolti* cuts 566 g vegetation ha^{-1} day^{-1}, while *Atta capigura* may remove as much as 52.5 kg grass ha^{-1} day^{-1} (Amante, 1967). Also, the large nests built by *Atta* spp. represent a considerable nuisance in managed grassland.

Ants are often the most voracious surface predators in meadows: Petal *et al.* (1971) estimated that a *Myrmica* population of mean density 225–300 m^{-2} consumed 40–80 g fresh mass of prey per m^2 per year in a Polish meadow. Spiders (mainly migrating juveniles), planthoppers (Auchenorrhyncha) and Diptera accounted for 78–88% of the food consumed by *Myrmica laevinodis*. The proportion of available prey consumed by ants was estimated to vary between 0.42% and 39% (Petal, 1980), with total predation being greatest in mid-May to mid-July when the production of prey was greatest (Kajak *et al.*, 1972). Kajak *et al.* calculated that 43% of Auchenorrhyncha, 32% of adult Diptera and 49% of wandering lycosid spiders were consumed.

Ants also have the capacity to modify their environment through their nest-building activities. Nests vary in size and complexity, depending on the species; their influence on soil properties will be discussed in Chapter 6.

2.17.3 Wasps and bees

Other Hymenoptera of some significance in grassland include the true wasps (Vespidae), social and solitary, which feed their young on insect larvae and other prey. Sand wasps (Sphecidae), the largest group of solitary wasps, are common in sandy soil, especially desert, and construct their brood chambers in the soil. Other 'diggers' of sandy habitats include members of the family Scoliidae, which have strongly developed legs and dig in the soil in search of prey. The bees (Apoidea), social and solitary, rear their young on pollen and nectar and are the main agents in plant pollination; their role in this context is of greatest significance in legume seed crops. Solitary bees of the family Adrenidae build subterranean nests, which can be a nuisance in lawns and sports turf.

2.18 LEPIDOPTERA

2.18.1 Abundance, feeding habits

The order Lepidoptera (butterflies and moths) contains more than 100 000 known species. The caterpillars are almost exclusively plant feeders, many being important crop pests and some having a role in the biological control of weeds. Populations in grassland are generally low, usually less than 10 m^{-2} (Morris, 1968; Persson and Lohm, 1977; Nabialczyk-Karg, 1980; Curry and Momen, 1988), but several species can cause considerable damage when abundant.

(a) Pasture damage

The most potentially damaging lepidopteran species belong to the families Hepialidae, Pyralidae and Noctuidae. Soil-dwelling larvae of the Hepialidae (ghost moths) can damage pasture and a range of other crops: in Britain, the swift moths (*Hepialus* spp.) are sometimes pests, while in New Zealand porina caterpillars (*Wiseana* spp.) cause widespread pasture loss, especially when the clover content of the sward is high (Harris, 1969; Harris and Brock, 1972). Harris (1969) calculated that a population of 10 larvae m^{-2} would reduce yield by 3.2 kg ha^{-1} day^{-1}.

Among the Pyralidae, the grass moth, *Agriphila straminella*, occasionally damages grass by feeding at the base of the plants in Britain, while related webworms (*Hednota* spp.) can damage pasture in Australia (Waterhouse, 1979). Seven to ten species of sod webworms (*Crambus* spp.) can seriously damage grasslands in the USA and Canada (Hill, 1987). Pass (1966) reported serious damage to bluegrass (*Poa pratensis*) turf in central Kentucky over several years; the species involved were the bluegrass webworm *Crambus teterrelus*, the larger sod webworm *C. trisectus* and the striped sod webworm *C. mutabilis*.

Several species of Noctuidae can damage grassland and a range of crops. The armyworm (*Mythimna unipunctata*) can destroy much of the vegetation over large areas during epidemics in the USA and Canada while the fall armyworm, *Spodoptera frugiperda*, occasionally causes similar damage in the southern USA and tropical America (Metcalf, Flint and Metcalf, 1962; Byers, 1967; Hill, 1987). *Apameae sordens*, the rustic shoulder knot moth, is widely distributed in the British Isles; the caterpillars feed chiefly on grasses (*Agropyron* sp.) and cereals and may hollow the grain in early autumn and destroy the stems in spring (Jones and Jones, 1984). Caterpillars of *Mesapamea secalis*, the common rustic moth, are essentially grass feeders, their usual hosts including *Dactylis glomerata*, *Poa annua* and *Festuca elatior*. Caterpillars of the antler moth *Cerapteryx graminis* are common in hill pastures in Britain, feeding on all kinds of grasses but especially on *Nardus strita*. Cutworms, noctuid caterpillars that attack plants at soil level, are general soil pests which can damage newly established leys.

(b) Weed feeders

Many lepidopteran caterpillars are weed feeders and some have a useful role in the natural control of pasture weeds. The cinnabar moth, *Tyria jacobaeae*, is widely associated with ragwort (*Senecio jacobaea*) and often causes severe defoliation, although attempts to use this species for biological control have had only limited success (Dempster, 1971). An outstandingly successful example of the use of Lepidoptera for biological control of pasture weeds concerns the introduction and establishment of a cactus-feeding caterpillar (*Cactoblastis cactorum*) for the control of *Opuntia* spp. (prickly pear), which had infested some 25 million ha of Australian grassland by 1925 (Harris, 1973).

2.19 OTHER INSECTS

Other insect groups commonly found in pasture but usually at low abundance include the orders Dermaptera, Psocoptera and Neuroptera.

2.19.1 Dermaptera

The Dermaptera (earwigs) is a small order of about 1000 species; they are mainly subtropical in distribution and are at the northerly edge of their range in northern Europe. They are omnivorous in feeding habits, feeding on flower petals, carrion and living insects. Sunderland and Vickerman (1980) found that 28% of *Forficula auricularia* examined from cereal fields in West Sussex contained aphid remains in their guts, suggesting that it can be an important aphid predator. However, it is only abundant in rough grassland and its activity is restricted to the vicinity of field boundaries in cultivated fields and managed grassland.

2.19.2 Psocoptera

Psocoptera (plant lice) are frequently encountered in litter and vegetation samples. These insects feed on fragments of animal or vegetable matter, particularly on fungi, unicellular algae and lichens (Imms, 1977).

2.19.3 Neuroptera

Neuroptera (lacewings) are predatory, with the aphid-feeding Chrysopidae being of some significance in the natural control of crop pest species (Rautapää, 1977; Vickerman and Wratten, 1979).

2.20 SUMMARY

The composition of the grassland fauna is extremely variable, being influenced by factors such as climate, soil characteristics, the nature of the sward and its management, and time of year. Table 2.16 summarizes population and biomass data for a range of sites representing different climatic zones and soil types. Invertebrate biomass ranges from less than 1 g m^{-2} in arid sites to well over 100 g m^{-2} in base-rich mesic temperate sites. At the drier end of the range, microfauna (Protozoa and Nematoda) inhabiting the water film on soil particles comprise the dominant component of the soil invertebrate biomass, while under favourable soil and climatic conditions earthworms predominate.

Protozoans are rarely included in faunal studies: biomass estimates range from less than 0.5 g m^{-2} in arid soils to over 5 g m^{-2} in mesic temperate soils. Nematode population densities vary from about 0.2 million to 10 million or more (*c.* 4 g biomass) m^{-2} over a similar range of conditions. Mean earthworm density frequently reaches several hundred per m^2 with biomass often exceeding 100 g m^{-2} in fertile temperate grasslands. Earthworms can also be abundant in humid tropical soils; they tend to be scarce or absent from arid soils and from wet, acidic moorland and tundra although substantial populations have been recorded from some steppe and tundra sites in the area of the former USSR.

Enchytraeid worms reach greatest abundance in acid moorland soils while the other main groups of mesofauna, the Acari and Collembola, are most abundant in grasslands with a well-developed surface organic mat.

Termites and, to a lesser extent, ants, are often the dominant invertebrates in terms of biomass in tropical soils. Coleoptera and Diptera are generally the most significant of the remaining soil faunal groups, with tipulid larvae (Diptera) often being particularly abundant in wet soils.

Generally speaking, the faunal groups that are dominant in terms of biomass in different grassland soils also tend to be dominant in terms of

Table 2.16 Mean population density and biomass of invertebrates in various grasslands

Habitat	Soil Herbage	Nematodes (× 10⁶)	Enchytraedae (× 10³)	Earthworms	Acari (× 10³)	Collembolans (× 10³)	Other arthropods (× 10³)	Biomass (g m⁻²)*
		Numbers (m⁻²)						
Old pasture, Co. Kildare, Ireland (Curry, 1969a)	S	–	–	364	106.1	105.4	10.6	c.88(F)
Permanent pasture, Co. Meath, Ireland (Curry, 1976b; Curry and Tuohy, 1978; Curry and O'Neill, 1979; Bolger and Curry, 1980)	S H	– –	– –	311	21.6 8.0	28.3 1.2	0.4 0.6	>100(F) 0.5(F)
Moorland alluvial grassland, England (Coulson and Whittaker, 1978)	S	3.3	120	390	36	40	0.5	35.5(D)
Juncus moorland on peat, England (Coulson and Whittaker, 1978)	S	3.9	200	4	45	23	3.1	7.2(D)
Old grassland on fen peat, Sweden (Persson and Lohm, 1977)	S H	–	23.8	133	106.9 5.0	108.7	7.4 1.0	10.6(D) 0.2(D)
Mown meadow, Poland (Breymeyer, 1978)	S	–	–	154	46.9	73.1	11.8	10(D)
Natural meadow, Poland (Breymeyer, 1978)	S	–	–	97	105.9	52.5	36.1	6.5(D)

Table 2.16 (continued)

Habitat	Soil Herbage	Nematodes (×10^6)	Enchytraeidae (×10^3)	Earthworms	Acari (×10^3)	Collembolans (×10^3)	Other arthropods (×10^3)	Biomass (g m^-2)*
Lightly grazed sheep pasture, NSW, Australia (King and Hutchinson, 1976; Hutchinson and King, 1980)	S	0.3	6	74	24.5	20.6	0.5	37.5(F)
Dry mountain meadow, Norway (Solhöy, 1972)	S	–	–	–	25.4	33.3	–	
	H				5.8	6.1	1.0	
Herb grassland, Russian tundra (Chernova et al., 1975)	S	1.0	20.5	150	13	31.8	0.3	100.4(F)
Tundra meadow, Alaska (Bunnell, MacLean and Brown, 1975)	S	0.7	56.7	0	26.2	96.8	0.1	c.3.0(D)
Savannah pasture, Panama, wet season (Breymeyer, 1978)	S	–	–	35	36.8	1.3	1.5	6.3(F)
	H						<0.1	0.1(F)
Humid Savannah, Ivory Coast (Athias, Josens and Lavelle, 1974)	S	–	0.6	230	17.5	17.8	6.8	54.7(F)
Prairie, USA (Lloyd et al., 1973)	S	–	–	20	–	–	0.6	
Arid steppe, Russia (Zlotin, 1970)	S	–	–	–	–	–		1.2(F)

*F = fresh mass, D = dry mass.

energy utilization. Normally invertebrates living on the soil surface and in the vegetation comprise only 1–2% of the total invertebrate biomass, but their activities can be very important in terms of the overall grassland ecosystem. This category includes a wide range of taxonomic and feeding groups such as sap-feeding aphids and other hemipterans, leaf-feeding caterpillars and beetles, shoot-mining Diptera, thrips, grasshoppers and other herbivores in the herbage layer; surface-foraging termites and ants gathering living and dead plant material, litter-feeding isopods and millipedes, microbivorous collembolans and amites, predatory carabid and staphylinid beetles and mites; parasitic wasps and omnivorous earwigs.

3

The grassland invertebrate community

3.1 GENERAL FEATURES OF THE COMMUNITY

The term community is used to describe the assemblage of species populations that occupy a designated area or habitat. The term may be applied very broadly to encompass the biota of major biomes such as grassland and deciduous forest, or it may be applied in a more restricted sense to the species populations associated with individual plants or microhabitats such as cowpats, or to assemblages of species that exploit a common resource in a similar way (guilds). Regardless of the scale of resolution, communities are generally described and compared in terms of features such as species composition, species richness and diversity, dominance structure and trophic interactions.

While the grassland invertebrate community broadly defined is variable in composition, it does have certain characteristic features that derive from the general nature of the grassland habitat. Grassland is relatively simple in structure compared with the multilayered structure of mixed deciduous woodland, for example, and the range and variety of invertebrate species present in the vegetation are correspondingly lower. Not surprisingly given the preponderance of Gramineae, a high proportion of grassland herbivores are adapted for feeding on grasses. Grassland leaves being narrow and predominantly vertical in orientation afford less-favourable microhabitat conditions than do those of deciduous trees. Sucking insects in particular, which can occur in great variety on the sheltered undersurfaces of deciduous leaves, tend to be less common in grassland. Many grassland invertebrates tend to be elongate and narrow in form or to have other adaptations appropriate to the predominantly vertical structure of the habitat. The litter layer of the forest floor generally supports a much more varied and abundant invertebrate fauna than that found in grassland. On the other hand, grassland soils tend to have higher and more evenly distributed organic matter levels than do forest soils and this is reflected in a correspondingly richer euedaphic soil fauna.

In this chapter, some of the main characteristics of grassland invertebrate communities are considered. Aspects of particular interest include the ways in which communities are structured and the factors that influence this structure; the degree to which communities are stable and the factors that determine this stability; the ways in which communities change over different timescales; and the degree to which different assemblages of species are associated with different types of grassland.

3.2 TROPHIC STRUCTURE AND ENERGY FLOW

3.2.1 Biomass distribution

The way in which biomass is allocated among the major functional units provides one basis for describing community structure and for making comparisons between different habitats. In Table 3.1, for example, four contrasting grassland types are compared on the basis of their plant biomass to faunal biomass ratios. The ratios are general indicators of habitat favourableness, being narrowest under the most favourable conditions and widening progressively as conditions become harsher.

A difficulty in describing the structure of grassland invertebrate communities in terms of feeding or trophic relationships is that many groups may occupy a number of trophic levels: for example, many molluscs and insect larvae may feed on living roots and dead organic matter, while centipedes and some staphylinid beetles may be partially plant feeders and partly predacious. This difficulty may be overcome by assigning fractional trophic-level values to the various taxa (e.g. Hutchinson and King, 1980), although these values are often rather arbitrary because of lack of reliable quantitative information on feeding habits.

Table 3.2 presents an analysis of the distribution of invertebrate bio-

Table 3.1 Ratios of plant to faunal biomass in a range of IBP sites (after Breymeyer, 1980)

Grassland type	Plant biomass
	Faunal biomass
Meadows and pasture, Poland	89–113
Wet meadows and steppe meadows, Russia	221–313
North American prairie	737–2071
Semi-desert and desert, Russia	2115–3785

Table 3.2 Invertebrate biomass and its trophic composition in various grasslands

Author	Site	Total biomass (g m⁻²)*	Distribution		Trophic composition (%)		
			Above ground (%)	Below ground (%)	Decomposers	Herbivores	Carnivores
Persson and Lohm (1977)	Fen grassland, Sweden	10.8(D)	2	98	69.6	25.1	5.3
Macfadyen (1963)	Grassland, England	189.5(F)			85.7	9.2	5.1
Macfadyen (1963)	Upland limestone grassland, England	191.1(F)			80.2	19.1	0.7
Breymeyer (1978)	Meadows, Poland	6.5–12.3(D)	3–6	94–96	91–96	2–4	1–6
Breymeyer (1978)	Savannah, Panama	0.3–1.7(D)	6–31	69–94	60–92	8–29	0–19
Breymeyer (1978)	Prairie, Colorado, USA	0.8(D)	13	87			
Breymeyer (1978)	Wet meadows, Asia†	54(F)	0.5	99.5	97.6	1.9	0.5
Breymeyer (1978)	Arid steppe, Asia†	1.0(F)	24.3	75.7	43.4	51.7	4.9
Breymeyer (1978)	Desert, Asia†	0.2(F)	23.4	76.6	53.8	34.3	11.8

*F = fresh mass; D = dry mass.
†Calculated from Zlotin (1970).

mass and its trophic composition in various grassland habitats. In all cases, the vast bulk of the invertebrate biomass (70–98%) occurs below ground and decomposers are the dominant trophic group in all except arid habitats.

3.2.2 Energy flow

Just as biomass distribution gives a measure of community structure, the rate of energy flow and the efficiency of energy transfer between trophic levels are indicators of community function. The question of how energy budgets for populations and communities are compiled will be deferred to Chapter 6: for present purposes the discussion will be confined to two examples that illustrate contrasting patterns of energy utilization (Figure 3.1). The example considered in Figure 3.1a is that of an English grassland extensively grazed by cattle (Macfadyen, 1963). Of the total energy fixed annually in net primary production, about 16% is consumed by the grazing cattle, 18% is consumed by other herbivores and 56% is returned to the soil as dead plant material, where it supplies the energy source for the decomposer community. Thus, less than 40% of available sward production is exploited by herbivores, and, of this, less than half is utilized by cattle. The second example (Figure 3.1b) concerns an old field site unexploited by humans in Tennessee, USA (Van Hook, 1971). Herbivores (mainly the grasshoppers *Melanoplus sanguinipes* and *Conocephalus*

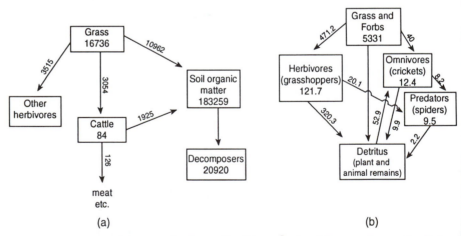

Figure 3.1 Annual energy budgets (in kJ m^{-2}) for (a) grazed grassland in England and (b) unutilized old field in Tennessee. The values inside the boxes represent mean standing crop in the case of the grazed grassland and net tissue production per year in the case of the old field; those on the arrows represent annual energy fluxes. Energy losses through respiration are not considered. ((a) After Macfadyen, 1963; (b) after Van Hook, 1971.)

fasciatus) and omnivorous crickets (*Pteronemobius fasciatus*) consumed 9.6% of net primary production, while predatory spiders (*Lycosa* spp.) consumed 21% of herbivore and omnivore production. Such energy budgets are inevitably reductionist and highly simplified, but they can provide a useful basis for making general comparisons between ecosystems in terms of their structure and functioning. They can also provide a preliminary basis for assessing the role of different groups of organisms within ecosystems, and for evaluating ecosystem response to human management (Chapter 6).

3.2.3 Trophic relationships in soil

Under the more stable conditions that prevail in soil, food chains tend to be longer and food webs more complex than those above ground (Coleman, 1985). Figure 3.2 illustrates, in much simplified form, the many interconnections between different groups of organisms in a short-grass prairie soil. While significant numbers of root herbivores can occur in some soils, nearly all soil organisms belong to the detritus food web. Three major decomposer trophic systems may be recognized in soil, broadly separated on size (Heal and Dighton, 1985). The microtrophic system is largely confined within the water film around pieces of organic matter, on root surfaces, and around soil particles; it is based on the utilization by bacteria or yeasts of more readily available carbon sources.

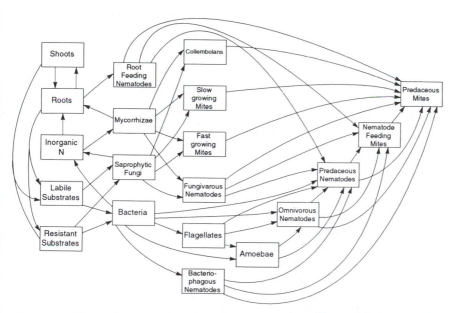

Figure 3.2 The detrital food web in a short-grass prairie. (After Hunt *et al.*, 1987.)

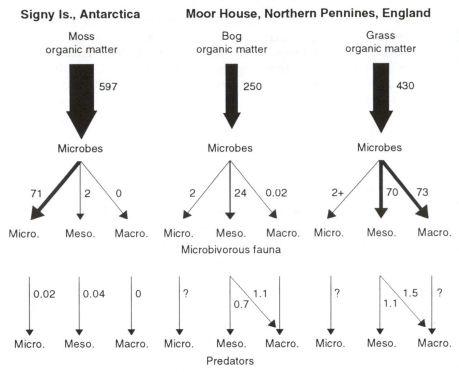

Figure 3.3 Trophic structure of three soil ecosystems differing in climatic severity and in resource quality. The data represent assimilation (g m^{-2}) by successive trophic groups. (After Heal and Dighton, 1985.)

Microflora are consumed by protozoans and nematodes, which may also be preyed on by predatory members of these groups. The mesotrophic system is based on the utilization by fungi of a wide range of substrates, including intractable dead organic matter. The fungal biomass is exploited by the mesofauna, mainly collembolans and mites, which inhabit the aerobic soil–litter pore system and are limited by the size of the spaces available in the soil–litter system. Their predators are also mesofauna, with some degree of predation by surface-dwelling pseudoscorpions, linyphiid spiders, etc. The macrotrophic system comprises invertebrates that are large enough to disrupt the physical structure of the soil or litter, such as earthworms and millipedes; these ingest both organic matter and associated microflora and fauna.

The relative importance of these three trophic systems will be influenced by site factors such as climatic severity, physicochemical conditions including moisture content and pH, and the quality of the food resources available (Figure 3.3).

3.3 SPECIES RICHNESS, DOMINANCE AND DIVERSITY

3.3.1 Diversity

Grassland habitats can support large numbers of invertebrate species. For example, Evans and Murdoch (1968) recorded 1584 insect species from the herbage (field layer) and 112 from the surface (ground zone) of an old field grassland in Michigan over several years, although many were sporadic in occurrence. The most recorded in any one year was 840 field-layer species, of which 245 were found regularly in each year.

Species richness (number of species recorded) is the simplest measure of community diversity, but its usefulness as an index for making comparisons between communities is limited because the number of species recorded depends on the size of the area sampled and because equal weight is given to scarce and abundant species. Some alternative measures of diversity have been proposed that reflect both species richness and the evenness of distribution of individuals among species. Indices derived from information theory such as the Shannon–Wiener function (H') are widely used (Shannon and Weaver, 1949). The related Brillouin index was found useful by Morris and his coworkers in their studies on the effects of management on Hemiptera and Coleoptera in calcareous grassland (Morris and Lakhani, 1979; Morris and Plant, 1983; Morris and Rispin, 1987; cf. Chapter 5). Figure 5.7 illustrates relationships between diversity, species richness and numbers of individuals in one such study. Reviews and critiques of a range of diversity indices are given by Hurlbert (1971), Whittaker (1972), Peet (1974), Pielou (1975) and others. Difficulties of interpretation arise because parameters such as species richness, evenness, the numbers of individuals and the area sampled are compounded when single indices are used.

3.3.2 Dominance--diversity relationships

An alternative approach that avoids the difficulties associated with indices of diversity is the use of graphical displays of both species richness and evenness. The curves are usually drawn such that the ordinate represents the log of the percentage abundance (dominance), which is plotted against ranked species sequence, and the shape of these curves may be interpreted in terms of various species-abundance models (Figure 3.4). Of the three models illustrated, the geometric series only fits communities poor in species, while the log series and the lognormal models appear to be widely applicable (Preston, 1962; Williams, 1964). The geometric series often applies to early successional communities with few species, the log series to immature communities with relatively few abundant and many scarce species, while the lognormal distribution may

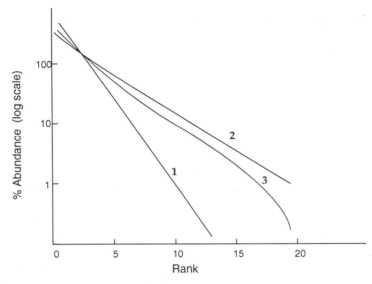

Figure 3.4 Dominance–diversity curves for different underlying species abundance models (1) geometric series, (2) log series and (3) log normal. (After Whittaker, 1972.)

often fit complex communities with large numbers of species, most being of intermediate abundance (May, 1975; Whittaker, 1975). Figure 3.5 shows species abundance curves for soil mites occurring in samples taken from a cereal field, from grass-ley pastures and from old permanent pasture in Ireland. The steepness of the cereal field curve reflects low species richness and a high concentration of dominance among a few species; the curve for the permanent pasture reflects a species-rich, mature community; while the grass-ley curves occupy intermediate positions between the cereal field and the permanent pasture. However, there are many situations in which species abundance models do not fit ecological data very well and, even when there is a good fit, the ecological interpretation may not always be apparent (Gray, 1987).

3.3.3 Species–area relationships

An analogous approach is to consider the relationship between species richness and area. Purvis and Curry (1980) found highly significant relationships between the numbers of arthropod species in soil and vegetation and the $\log_{10}$ area sampled in a maturing grass-ley pasture in Ireland (Fig. 3.6). Furthermore, the slope of the regression lines increased dramatically between November 1975 and November 1976, indicating

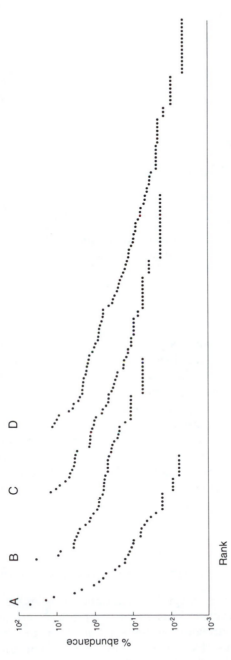

Figure 3.5 Rank-relative abundance curves for soil mites in Lyons Estate, Co. Kildare, Ireland. A, Cereal field; B, second-year grass ley; C, third-year grass ley; D, old permanent pasture. (From Curry, 1968; Emmanuel, Curry and Evans, 1985.)

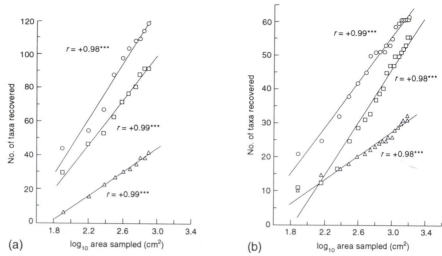

Figure 3.6 Cumulative species/area curves with correlation coefficients (r) for arthropods in (a) the soil and (b) the herbage layer of maturing grass leys at Celbridge, Co. Kildare, Ireland. △, 1975; □, 1976; ○, 1977; ***probability that r differs from $+1 \leqslant 0.001$. (After Purvis and Curry, 1980.)

rapid colonization during this period. The rate of accumulation of new species continued to increase between 1976 and 1977, but at a slower rate in the case of soil, while there was a significant decline in the rate of accumulation of new species over this period in the case of vegetation, suggesting different patterns of colonization of soil and vegetation.

More generally, a species–area relationship of the form $S = CA^Z$ has been applied, where S is the number of species, A is the area sampled, C is a constant measuring the number of species in the first unit of area and Z is a constant measuring the slope of the line relating S to A. It has been shown to hold for a wide range of plant and animal examples both on islands and in large land masses (MacArthur and Wilson, 1967; Williams, 1964) and appears to be valid in the case of soil fauna as well (Usher, 1985).

3.4 FACTORS INFLUENCING COMMUNITY STRUCTURE

While patterns of diversity, species abundance, species–area and related attributes offer empirical evidence that communities are organized in a structured way, they do not offer any explanation as to what the reasons for this structuring are. However, a number of recent lines of research offer pointers as to what the major features shaping community structure may be. These features, as they relate to grassland herbivores in particular, can be considered under the following headings:

1. Host-plant characteristics.
2. Habitat characteristics.
3. Interspecific interactions within the community.

3.4.1 Host-plant characteristics

The ways in which plant physical and chemical characteristics influence their attractiveness and food quality for herbivores will be considered in Chapter 4; the present discussion will be restricted to some general aspects of how features such as size and structural complexity (plant architecture), geographical distribution, physical and chemical variability, and seasonal phenology influence the species composition and abundance of the herbivore community.

(a) Plant architecture

The concept of plant architecture includes a variety of attributes including size and growth form, seasonal development and persistence and· the variety of structural parts (Lawton, 1983). Large and structurally complex plants such as trees and woody shrubs support a richer fauna than do smaller and simpler plants such as herbs and grasses. There is a strong correlation between plant and herbivore biomass in meadows, with multilayered plant structure promoting the development of a more diversified phytophagous community (Andrzejewska, 1979a). Plant-hoppers and leafhoppers (Auchenorrhyncha) show pronounced stratification in the meadow habitat, with different species tending to occupy definite layers (Andrzejewska, 1965). Some species undergo vertical movements related to their life-cycle dynamics, thus utilizing different parts of the sward canopy during the growing season. Closely grazed or cut swards typically have an impoverished fauna compared with long swards (Morris, 1971b, 1979, 1981a; Purvis and Curry, 1981). Using the number of different types of grass structures as a measure, Stinson and Brown (1983) found that plant architecture accounted for 79–86% of the variance in leafhopper species richness and total abundance in a fallow field undergoing secondary succession in Britain. A strong positive correlation also existed between homopteran diversity and plant structural diversity (estimated on the basis of height) in three abandoned forb-grass fields in Michigan, USA, although a complication in this case was the fact that the fields also differed in plant species diversity (Murdoch, Evans and Peterson, 1972).

Structural diversity has a major influence on the assemblages of sap-feeding insects (Homoptera) inhabiting two different grasses that dominate the vegetation of salt marshes along the northeastern coast of the USA (Denno, 1977). *Spartina patens* forms a dense, persistent thatch composed of dead plants from previous growing seasons, while dead culms of *S. alterniflora* form a loose lattice that decomposes rapidly during the following year. *Spartina patens* supports a much more diverse assemblage of sap-feeding insects (mainly Delphacidae, Cicadellidae, Issidae and Miridae) than does *S. alterniflora*. Removal of the thatch resulted in reduced species diversity and evenness and Denno concluded that the complex microstructure and thatch of *Spartina patens* provide a more heterogenous and protective environment, which supports a more diverse and specialized fauna.

(b) Seasonal phenology

Plant architecture changes considerably throughout the year depending on the phenology of the individual plant species and management; these changes are paralleled by changes in the insect community. A continuous turnover of insect species throughout the summer was recorded in a Michigan old field (Evans and Murdoch, 1968). Insect turnover was influenced particularly by the phenology of flowering plants as 73% of the adult species recorded were flower feeders. Pronounced changes in the composition of the leafhopper fauna in acidic grassland in Britain also occurred with successive waves of abundance of Delphacidae and Cicadellidae as the season progressed (Waloff and Solomon, 1973).

(c) Geographical range

The importance of geographical range in determining the diversity of insects associated with individual plants was highlighted by Strong (1974), who demonstrated a highly significant species–area relationship between insect species richness and the distribution of their host trees in Britain.

(d) Chemical variability

In addition to the marked structural variability within plants, there is also a considerable degree of chemical variability and this has important implications for the herbivore community. Krischik and Denno (1983) note the great variations in allelochemistry and morphology between leaves within plants, between seasons and between plants, and review evidence that supports the hypothesis that the evolution of this heterogeneity is at least partially due to herbivory. Defoliation experiments indicate that some leaves (e.g. terminal and midseason leaves) are more valuable to plants than others (e.g. basal and early- and late-season leaves), and there is some evidence that these valuable leaves are better defended than others. Within-plant variation may be an important defensive adaptation against rapidly evolving pests and parasites that are unlikely to be adapted to all parts of the plant at once, so that no one species is likely to reach epidemic proportions (Whitham, 1983). The major effect of such chemical variation on non-adapted insects may be to reduce the frequency of successful colonization by decreasing the chances of a good chemical fit between plant and insect. For adapted insects, it could regulate time of colonization and abundance by a fine division of resources (Jones, 1983). Conversely, the ability to track variable host-plant resources in time and space is an important attribute for the specialized herbivore and is an important factor in shaping insect life-history characteristics, such as migration, diapause and reproduction that must be synchronized with resource availability in space and time (Denno, 1983). Denno illustrated this by a detailed study of the relationship between *Spartina alterniflora*, an intertidal marsh grass, and the monophagous planthopper *Prokelisia marginata*.

3.4.2 Habitat characteristics

In addition to the characteristics of individual host plants, the vegetational 'texture' of the habitat influences the invertebrate community. Texture encompasses a range of features, such as the density and patch size of the host plants, habitat heterogeneity and species diversity, that could influence herbivore movement, searching behaviour and host-finding success. Two aspects of texture are considered here: resource (host plant) concentration in relation to herbivore density, and habitat complexity in relation to invertebrate community diversity. Other features that affect diversity are considered under the general heading of colonization rates and species equilibrium.

(a) Resource concentration

The resource-concentration hypothesis (Root, 1973) suggests that specialized herbivores are more likely to find and remain in areas of concentrated resources (e.g. where host plants grow in pure stands) and under such conditions a few species are likely to become very abundant. By contrast, host plants in mixed stands are less easily detected by specialized herbivores and their reproductive success is likely to be lower because of the greater difficulty of finding food and because non-host plants may interfere with host finding, feeding and reproductive success (associational resistance). These ideas accord with the general observation that crop monocultures are more prone to pest outbreaks than are natural mixed vegetational stands, which tend to have more diverse faunas with lower herbivore load. However, there is a considerable amount of variation between species in the way they respond to vegetational texture. Data from over 200 studies reviewed by Andow (1991) generally support the resource-concentration hypothesis, but there are a significant number of species (especially polyphagous species) that have variable responses. Monophagous species generally tend to be more abundant in monoculture, but some can be more abundant in polyculture.

It has also been suggested that higher populations of natural enemies may contribute to reduced herbivore load in polyculture compared with monoculture. The ratio of predators to herbivores does tend to be lower in crop monoculture than in old field polyculture (Root, 1973; Allan, Alexander and Greenberg, 1975), but there is not much evidence to show that regulation of natural enemies is more effective in polyculture than in monoculture (Karieva, 1983; Andow, 1991).

(b) Habitat complexity and productivity

Complex habitats offering a wide range of resources and structures generally support richer and more complex invertebrate communities than do simplified habitats, but the mechanisms involved and the relative importance of the various aspects of habitat complexity in maintaining faunal diversity are not well understood. In seminatural grassland habitats (e.g. the old field studied by Murdoch, Evans and Peterson, 1972), features of habitat complexity such as the structural diversity and species diversity of plants are often positively correlated and it is not possible to say which is of greatest significance for invertebrate diversity. A further complication in interpreting the results of comparative field studies is that the communities being studied may differ considerably in productivity. Allan, Alexander and Greenberg (1975) compared the arthropod communities on the foliage of two crop fields (alfalfa and *Dactylis glomerata*) with those found on a bluegrass (*Poa*) field ungrazed for 5 years and an old field that had been ungrazed for 12 years and had many shrubs and trees. Plant species richness and complexity (foliage height diversity) were considerably lower in the crop fields than in the fallow fields. The fallow fields generally had greater numbers of arthropod species than did the crop fields, but not strikingly so (fallow : crop species ratio 1 : 1.6); evenness was lowest and population density was highest in the alfalfa field (Table 3.3). The authors concluded that the main difference between the crop and the fallow fields was the greater unevenness in abundance due to a high proportion of a few pest species. They suggested that habitat complexity and plant productivity both enhance species richness and in crop fields low complexity may be counterbalanced by greater productivity.

Table 3.3 Species richness, diversity, evenness and density of arthropods in crop and fallow fields (from Allan, Alexander and Greenberg, 1975)

	Orchard grass	Alfalfa	Alfalfa	Fallow: blue grass		Fallow: old field	
	(June)	(June)	(June)	(June)	(July)	(June)	(July)
Number of morphospecies (*S*)*	130	92	148	165	233	195	195
Number of families	66	52	74	70	87	86	81
Number of orders	9	8	11	11	11	11	12
Diversity *H'*	3.21	1.61	2.39	2.65	3.62	3.65	3.32
$_eH'$	25	5	11	14	37	39	28
Evenness	0.684	0.357	0.477	0.519	0.664	0.693	0.629
Number of arthropods/100 sweeps	401	943	966	809	555	262	355

*Defined as taxa keyed at least to family, and numbered as distinct types.

(c) Colonization rates

Comparisons between communities of different ages are complicated by time-related phenomena such as colonization rates. Current thinking about colonization of new habitats has been strongly influenced by ideas on island biogeography (McArthur and Wilson, 1963) which have been concerned with the process of colonization of islands and the factors involved in determining the equilibrium numbers of species present. Briefly, the equilibrium number of species in a given habitat is a function of the rate at which new species arrive (immigration rate) and the rate at which species leave (extinction rate) (Figure 3.7). The equilibrium number of species when the habitat is saturated will be determined by the size, structure and range of resources available, but the time taken to reach this equilibrium will be influenced by factors such as the size of the species pool of potential colonizers and the distance from sources of colonizers. Colonization is an asymptotic process, proceeding most rapidly initially (Strong, Lawton and Southwood, 1984): for example, the rate of colonization of new grass leys in the first few years is very high (see Figure 3.6). Hence, the time available for colonization will have an important bearing on species richness and will inevitably complicate comparisons between young communities such as annual crops and older communities such as perennial grass swards. Time effects are considered further below under the heading of succession.

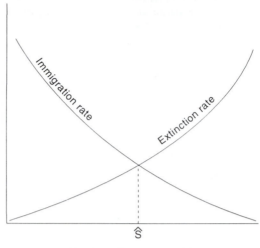

Number of species on island

Figure 3.7   Equilibrium model of the numbers of species on islands. The equilibrium number of species, $\hat{S}$, is determined by the intersection of the immigration and extinction curves. (After MacArthur and Wilson, 1963.)

(d) Species equilibrium (diversity)

The notion that particular habitats have a certain limit for species (equilibrium) is supported by the studies of Simberloff and Wilson (1969) on the recolonization by arthropods of defaunated islands of red mangrove (*Rhizophora mangle*) off the west coast of Florida. The upper limit for species richness in a given habitat may be considered to be determined by the 'divisibility' of resources (i.e. the extent to which they can be partitioned into workable niches; Putman and Wratten, 1984). However, this limit rarely appears to be reached (Lawton, 1982) and there is no generally accepted single theory to explain realized diversity in any given situation. Slobodkin and Sanders (1969) considered diversity to be a function of the severity, stability and predictability of the environment in which a community becomes established. Harsh environments support few species, but as severity decreases and predictability increases (e.g. along a latitudinal gradient from the tundra to the tropics) species diversity increases. In favourable environments, predictability is less important. Decreasing severity and increasing predictability will be accompanied by increasing productivity and spatial heterogeneity (i.e. greater ranges of resources for potential colonizers). Under favourable environmental conditions, where abiotic factors are not limiting, the role of biotic interactions increases markedly: these include invertebrate–host plant interactions and interactions between the invertebrate species within the community.

3.4.3 Species interactions

The difficulties of studying species interactions in even the simplest communities are highlighted by Usher *et al.* (1979) and Usher (1985), and the significance of these interactions in complex, multispecies communities can only be considered in a general way at present. Some inferences about the biotic processes likely to be of most significance at the community level can be drawn from the studies that have been conducted on the factors determining the abundance of individual species populations.

It is widely held that processes acting in a density-dependent manner tend to regulate population densities whereas factors operating in a density-independent manner tend to destabilize populations, and it has been suggested that these two types of processes determine the predictability or constancy over time of community structure. Strong, Lawton and Southwood, (1984) suggested that, depending on the form and relative magnitude of density-dependent and density-independent stages in the life cycles of component species, a series of theoretically ideal communities can be defined, ranging from essentially random assemblages in which density dependence is feeble or non-existent to highly deterministic systems structured by strong interspecific competition. Examining the amplitude of population fluctuations in a range of studies (maximum: minimum ratios) they concluded that most species fluctuate within one or two orders of magnitude, but some species vary little from year to year and a few are very unstable. However, the degree of stability in component populations and hence the predictability of community structure vary considerably among plant types and habitats. For example, the community of bracken-feeding insects at Skipwith Common in Yorkshire, England has a reasonably predictable but not absolutely constant structure with some fluctuation in numbers from year to year but with rare species tending to be rare in most years and common species always being common (Lawton, 1983; Strong, Lawton and Southwood, 1984). On the other hand, the leafhopper community of acidic grassland at Silwood Park in Berkshire, England appears to be highly labile in its population levels and in its species composition (Waloff and Thompson, 1980).

The relative importance of the different types of mortality factors and the extent to which density-dependent processes operate in the field will be considered in Chapter 4; for the purposes of this discussion of community structure and stability the conclusions of Strong, Lawton and Southwood (1984) appear appropriate. These authors concluded that the dominant density-dependent influences on phytophagous insect populations and communities usually operate vertically through the food chain and not horizontally with competitors, that density dependence is often too feeble to be detected in many populations and, at best, phytophagous insect communities will be only moderately deterministic in their structure.

3.5 SUCCESSIONAL CHANGES IN THE COMMUNITY

3.5.1 The process of succession

Communities are dynamic in nature, in a continuous state of flux. Short-term cyclical changes in species composition and relative abundance are associated with changes in weather, seasonal cycles of growth and decay and the phenology and life histories of the species in the community. More permanent changes in the composition of the community may also occur, reflecting colonization of new species and extinction of others; the pronounced changes in species composition and community structure associated with succession are most marked.

Broadly defined, succession is the process of change in the composition, structure and function of the community that develops when a new habitat becomes available for colonization (primary succession) or when a previous community has been disrupted or eliminated from an existing habitat (secondary succession). The classical view regarded succession as an orderly, directional and highly predictable developmental process, starting with a few species of early colonizers (pioneer stage) and proceeding through well-defined seral stages of increasing maturity and complexity until a stable, climax community appropriate to the habitat is reached. The species in each seral stage were considered to modify the physical environment to the extent that it was no longer suitable for them and they were replaced by species better adapted to the new conditions. This concept of succession as a highly deterministic process has been restated and elaborated by Odum (1969). He viewed succession as an orderly process of community development that results from the modifications of the physical environment by the community and culminates in a stabilized ecosystem, in which maximum biomass (or high information content) and symbiotic function between organisms are maintained per unit of available energy flow. Odum listed 24 attributes of ecological succession, the main features being a change from production (P)/respiration (R) > 1 in the early stages to $P/R \approx 1$; increasing diversity and stability as succession proceeds; a change from predominantly small, short-lived organisms with simple life cycles to large, long-lived organisms with complex life cycles; and change from rapid, open cycling of nutrients to slower, closed cycles with mineralization of dead organic matter making an important contribution.

An alternative view is that succession is essentially a random process that results from the characteristics of individual species and proceeds not just to a single predictable climax community but to any one of many possible climaxes. Drury and Nisbet (1973) consider that most of the phenomena of succession can be understood as consequences of differential growth, survival and colonizing abilities of species adapted to different points on environmental gradients.

Several alternative mechanisms might determine which species replace the opportunistic species with broad dispersal powers and rapid growth characteristics that are the early colonizers of new sites. Connell and Slatyer (1977) consider three models: (1) The facilitation model holds that the entry and growth of later species is dependent on earlier species preparing the ground; this is the classical view of succession. (2) The tolerance model suggests that a predictable sequence is produced by the existence of species that have evolved different strategies for exploiting resources: later species will be those that can tolerate lower levels of resources than earlier ones and can invade and grow to maturity in the presence of those species that preceded them. (3) The inhibition model suggests that all species resist the invasion of competitors. The first occupants pre-empt the space and will continue to exclude or inhibit later colonists until the former die or are damaged, thus releasing resources. Only then can later colonists reach maturity. Connell and Slatyer consider that the species composition shifts gradually and inexorably towards species that live longer, not because these species are more likely to colonize, but because once a long-lived species becomes established it persists by definition. Thus, ecological succession need not necessarily be 'directional', for simply by virtue of their life-history characteristics long-lived species will eventually dominate.

Most works on succession have been primarily concerned with plants, and insects have generally been ignored or assigned a passive role. The overall effects of insect herbivores on plant succession and plant community structure in grassland are probably relatively minor compared with the profound effects of vertebrate grazers (Crawley, 1983); however, selective herbivory can alter botanical composition while saprophagous invertebrates can influence plant succession by altering habitat conditions and the rate of decomposition and mineralization of organic residues (Chapter 6).

Several successional processes may be in progress simultaneously over different timescales in any given community. Longer-term changes in grassland might be associated with maturation of a new grass ley, while short-term changes might be related to cycles of plant growth and decay and to habitat alterations brought about by management practices such as grazing, cutting and burning. Some examples of studies that have documented changes in the invertebrate community associated with longer-term primary succession in land recovering from severe disturbance and secondary succession in agricultural land, and with rapid succession in decaying organic residues, are considered below. Ideally, successional studies should be carried out in a given site over an extended period of years, but in practice this has rarely been accomplished and most studies are based on examination of contemporary communities of known or presumed successional age.

3.5.2 Primary succession in severely disturbed land

Soils that have been extensively disturbed by mining generally present very hostile environments for invertebrates. Features likely to inhibit faunal establishment include lack of organic matter and suitable food, unfavourable moisture conditions, excessive fluctuation in surface temperatures, and extreme acidity and metal toxicity due, for example, to the weathering of pyritic mine spoil. However, faunal colonization can proceed rapidly following site restoration. Hutson (1980) recorded a peak population density of 131×10^3 mites m^{-2} in mine spoil in England once the grass and clover seed had germinated, while Curry and Momen (1988) recorded 174 species and 74×10^3 arthropods m^{-2} within 2 years of reclaiming and seeding cutover peat in Ireland.

The rate and extent of faunal establishment will depend on factors such as the extent of site rehabilitation, the size and shape of the reclaimed area, and the proximity of suitable colonizers (Majer, 1989). The main requirements in terms of rehabilitation are liming of very acid soils, stabilization of surface moisture and temperature regimes, and the provision of food in the form of organic waste and plant litter for the decomposer community. Generally, the degree and extent of faunal development reflects that of the vegetation, but, as will be seen below, the processes of floral and faunal development are not always in complete parallel.

The early colonizers of restored land tend to have the characteristics of 'r-selected' species (*sensu* Southwood, 1977). They are generally small and short-lived, with the capacity for rapid multiplication and rapid dispersal. Several species of Acari and Collembola dispersed phoretically or by wind are usually among the early colonizers, and where there is sufficient organic matter large numbers of dipterous larvae are soon found (Dunger, 1969a,b; Curry and Momen, 1988). As site conditions become more favourable larger sized and more long-lived species with poorer powers of dispersal, lower reproductive rates and longer generation times ('K-selected' species) are able to become established. These two extreme forms of life-history adaptation tend to predominate in the early and late stages of succession; many of the species in maturing sites will tend to be intermediate in their characteristics between these two extremes. A third group characterized by their ability to tolerate adverse conditions ('*A*-selected' species) may be recognized in sites where conditions remain harsh (Greenslade, 1983).

While faunal succession in restored land can vary considerably from site to site, some of the main features of the process are illustrated by the two examples considered below.

(a) A disused chalk quarry

Parr (1978) sampled the soil microarthropod fauna of plant communities of varying successional age in a disused chalk quarry in Yorkshire, England. The data, when analysed using the method of principal coordinate analysis (Gower, 1966), revealed two clearly separated plant successional groups – an 'early' successional group dominated by dicotyledonous plants of prostrate growth form and a second, 'late' successional group representing a grassland community dominated by tall grass species and small shrubs. Early-successional and late-succes-

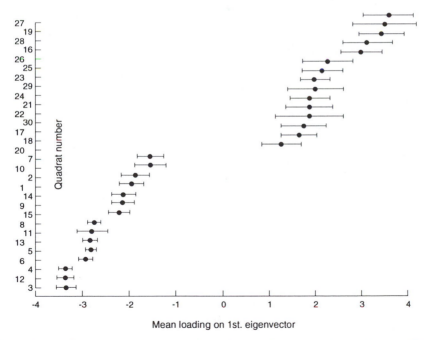

Figure 3.8 Ordination of quadrats based on their microarthropod fauna. The quadrats are ranked according to successional age (time since abandonment of limestone quarrying). The bars represent standard errors based on eight sampling occasions. (After Parr, 1978.)

sional microarthropod communities could also be recognized (Figure 3.8). The late-successional microarthropod community had a higher population density, a higher total number of species, higher numbers of species per sample, and higher species diversity than the early-successional community (Figure 3.9). The close correspondence between plant and microarthropod communities suggests related patterns of community development. The increase in microarthropod population density and species diversity was probably due to the increase in soil organic matter and to the general amelioration of the soil environment associated with increased biomass of surface vegetation. One difference between successional trends in the plant and microarthropod communities was that, while arthropod diversity continued to increase with successional age (distance from the quarry) in line with the predictions of Odum (1969) and Regier and Cowell (1972), plant diversity decreased at the point of transition from the dicotyledonous stage to the grassland stage. The

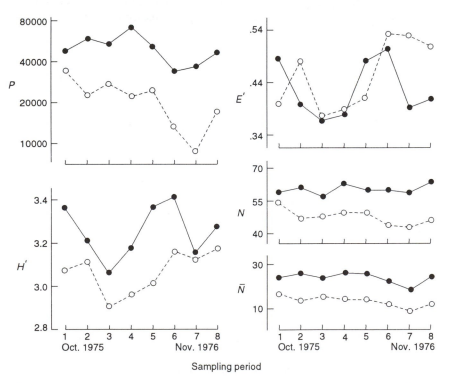

Figure 3.9 Seasonal variation in community parameters for 'early' (○) and 'late' (●) successional microarthropod groups. P = log density of microarthropods (numbers per square metre); H' = diversity; E' = equitability; N = total number of species in group. $\overline{N}$ = mean number of species per sample. (Parr, 1978.)

establishment of an *Arrhenatherum* grassland vegetation may have pre-cluded many of the prostrate rosette-forming dicotyledons in this site, thus depressing plant species diversity although allowing other species in for the first time (Usher, 1979). Thus, while diversity tended to increase with succession, this trend was reversed with the advent of a newer vegetative structure. By contrast, increase in soil arthropod community diversity could continue uninterrupted as soil organic matter accumulated and as soil development continued.

(b) Reclaimed mine spoil

Data on faunal succession in rehabilitated open-cast coal-mining dumps collected over a period of 25 years are summarized by Dunger (1989). Forty dumps in southeastern Germany were studied; most had been afforested with *Populus*, *Alnus* and *Robinia* and seeded with clover and lupin. The following are some of the main conclusions:

1. The most important requirement for faunal rehabilitation of coal dumps were: (a) stabilization of the water regime; (b) nutrient build-up facilitated by rapid production of organic matter by N-fixing plants; (c) detoxification (e.g. neutralization of mineral acids by liming).
2. Succession proceeded at different rates above ground and in the soil; the process was generally slower below ground.
3. Succession did not proceed uniformly across all groups with a steady increase in density and diversity. For some groups, a 'pioneer peak' was followed by decreasing density and diversity.
4. Succession processes were modified by individual site factors and rehabilitation proceedures. Moisture was the most important factor; the amount of dead organic input was of secondary importance.
5. Five stages of succession were recognized. The first, pioneer, stage was characterised by hot dry conditions, the absence of organic matter, and colonization by *r*-strategists, notably the collembolans *Entomobrya lanuginosa* above ground and *Proisotoma minuta* below ground. This stage could last for up to 30 years in harsh sites but as little as 2 years under favourable conditions. The second stage was influenced by the method of rehabilitation and the colonization rates of the various organisms. Increasing litter accumulation from the herb layer and high densities of collembolans, dipterous larvae, predatory carabid beetles and spiders occurred. This stage could end at the fifth year or it could extend until the ninth year or longer on acidic dumps. The third, fourth and fifth stages reflected vegetational development from shrub with a well-developed litter layer colonized by epigeic earthworms – *Dendrobaena* spp. initially and later *Lumbricus rubellus* – to fully developed woodland where *K*-strategists including the anecique

earthworm *Lumbricus terrestris* was predominant. Under optimum conditions the woodland stage may commence 20–25 years after rehabilitation.

3.5.3 Secondary succession in agricultural land

Trends in the exopterygote insect fauna of three sites at different stages in secondary succession in England (an early-successional site consisting of a harrowed field, a mid-successional grass–forb old field and a late-successional, predominantly link-woodland site) conformed with the predictions of classical succession theory (Brown and Southwood, 1983). The taxonomic diversity and the trophic diversity (major types of feeding habits) of the group as a whole increased with successional age. Niche breadth of sap-feeding herbivores as indicated by the records of their host plants was inversely related to the successional age of the habitat, and herbivore species colonizing the early-successional site had, on average, shorter generation times than those of later stages.

Marked increases in the species richness and abundance of arthropods were recorded in a maturing grass ley in Ireland over a period of 3 years (Figure 3.6, Table 3.4). The trend of increasing species richness and evenness in the arable land–grass ley–old pasture sequence is well illustrated by the rank/relative abundance curves already discussed (Figure 3.5). However, colonization and population development in maturing leys can sometimes occur much more rapidly. Arthropod species richness and population densities had reached similar levels within 2 years of sward establishment on reclaimed cutover peat to those in more mature (5–6-year-old) neighbouring leys (Curry and Momen, 1988).

3.5.4 Succession in organic residues

Several studies have been concerned with the role of invertebrates in the decomposition of organic residues and the succession of species associated with residue decay. Distinctive aspects of this kind of succession are that the timescale is rapid and the resource is progressively depleted so that a characteristic equilibrium community never becomes established.

(a) Dung and carrion

Some of the main features of invertebrate succession in dung and carrion were reviewed by Doube (1987). The succession process is strongly influenced by spatial and temporal patterns of distribution: the temporal supply of corpses, for example, is normally highly irregular and unpredictable, while the daily supply of dung from domestic animals and its

Table 3.4 Numbers of arthropod taxa (S) and mean population densities (N) ($\times$ 10)m^{-2} in a maturing grass ley (after Purvis and Curry, 1980)

Soil cores

Age of leys (years)	1		2		3	
	S	N	S	N	S	N
Acari	21	1.3	54	22.4	75	61.9
Collembola	11	0.9	15	11.4	19	41.3
Insecta	5	0.08	13	1.8	18	9.8
Other arthropods	4	0.06	6	0.4	6	1.5
Total	41		88		118	

Small vegetation samples (funnel extracted)

Age of leys (years)	1		2		3	
	S	N	S	N	S	N
Acari	21	25.5	32	27.7	36	29.0
Collembola	2	0.1	7	0.3	9	0.3
Insecta	10	2.6	15	1.3	13	1.3
Other arthropods			2	0.03	4	0.06
Total	33		56		62	

Vegetation ('D vac' suction samples)

Age of layers (years)	2		3	
	S	N	S	N
Acari	50	0.2	65	1.3
Collembola	19	1.2	25	1.4
Insecta	119	0.3	166	0.5
Other arthropods	12	0.04	20	0.07
Total	200		276	

quality can be relatively predictable. Size is important, especially in the case of carrion. Most large corpses are consumed by vertebrate scavengers primarily, while the corpses of small animals are often consumed by necrophagous beetles and fly larvae. Dung of herbivores, omnivores and carnivores attracts different arrays of colonizing species. The dominant dung and carrion species vary with latitude and between continents. For example, aphodid dung beetles and flies are the dominant coprophages in northern Europe; geotrupid and scarabaeid dung beetles become important in southern Europe; and scarabaeids dominate in many tropical regions.

Despite marked differences related to seasonal and local environmental factors, there are some general features of arthropod succession in dung that appear to apply widely in temperate regions (Laurence, 1954; Mohr, 1943; Valiela, 1974; Breymeyer, 1974; Olechowicz, 1976b). The early colonizers of fresh dung are mainly coprophagous Diptera of the families Muscidae, Scathophagidae, Sepsidae and Borboridae, and scarabaeid dung beetles. Succession begins with a small number of species, and the number of taxa present and the complexity of the food web increase as the interior of the dung becomes aerated and accessible to air-breathing fly larvae and their predators. In later stages of decomposition an increasingly complex community develops as general litter-dwelling species become established (Chapter 5).

(b) Grass residues

Trends in the invertebrate populations associated with grass residues confined in litter bags decaying in the soil and on the soil surface in *Agrostis–Festuca* grassland in Ireland over a period of 10 months are shown in Figure 3.10. Population trends for the major invertebrate taxa were different for litter on the surface and in the soil. There were relatively high populations of nematodes throughout in both cases. Enchytraeid worms were fairly abundant throughout while epigeic earthworms were fairly abundant in surface litter initially. Among the arthropod groups other than Acari and Collembola, the most pronounced trends were exhibited by Diptera, with high initial numbers of nematoceran larvae in surface litter diminishing rapidly as decomposition proceeded, and by Pauropoda and Protura, which increased considerably in numbers in buried litter during later stages. Collembolan populations increased rapidly initially in the surface litter, particularly in bags from which macroinvertebrates were excluded, and later declined in abundance, while in buried litter the population buildup was much more gradual. Acarine populations increased gradually throughout the study. Densities of smaller invertebrates such as Collembola, Acari, Nematoda and to some extent Enchytraeidae were lower in coarse-mesh litter bags to

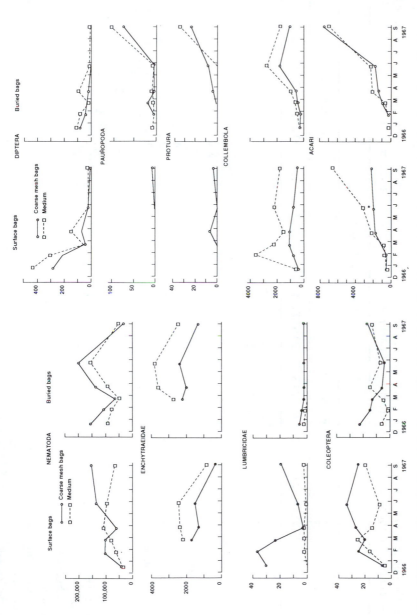

Figure 3.10 Changes over time in populations of the main invertebrate groups in decaying grassland herbage (total numbers from four litter bags). (From Curry, 1969c.)

Table 3.5 The dominant species/genera of Acari and Collembola in decaying sward remains confined in litter bags (after Curry, 1969c); 1–7 are relative abundance classes (percentage of total population) over the range 5% to 70%

	Recovery date																	
	7-mm mesh bags						Surface bags						0.5-mm mesh bags					
	19.12.66	23.1.67	20.2.67	20.3.67	23.5.67	4.9.67	19.12.66	23.1.67	20.2.67	20.3.67	23.5.67	4.9.67	19.12.66	23.1.67	20.2.67	20.3.67	23.5.67	4.9.67
Hypogastrura denticulata	4	4	2	1	1		6	7	3	3	2						2	
Olodiscus minima	2																	
Achipteria coleoptrata	2		1															
Chamobates borealis							2											
Isotoma olivacea	1	2	4	3				2	4		1			2	4	4	1	
Isotoma notabilis		2	1		1		1	1	1		1			1	2		2	2
Pergamasus lapponicus		1	1				1								1			
Bdellidae sp.			1				1											
Minunthozetes semirufus				3	2	2				3	2	2				3	3	2
Liebstadia similis				1		1				1		1				1	1	
Eupodes sp.					1					1		1						
Folsomia quadrioculata												1						
Schwiebia spp.												1						
Tyrophagus putrescentiae												1						
Pygmephorus spp.												1						2
Variatipes quadrangularis												1						

Species								
Hypogastrura denticulata	2						2	3
Isotoma notabilis	2	2	3	2	2	3	1	3
Onychiurus spp.	2	1	3	1	2	3	1	1
Iphidozercon minutus	1	3	2	2	2	3	3	2
Folsomia quadrioculata	1	1	2	1	1	1	1	2
Folsomia candida	1							1
Alliphis halleri	2				2			
Olodiscus minima						1		
Eupodes sp.							1	
Isotoma olivacea							1	
Schwiebia spp.	2							1	2
Coccotydeus spp.	3							2
Pygmephorus spp.	1							3
Variatipes quadrangularis	1							1

which larger invertebrates had access, suggesting inhibitory effects by such groups as earthworms on other soil fauna.

A total of 119 species or species groups of Acari and Collembola were recorded from the decaying herbage, and numbers of species increased as decomposition progressed. Changes in the relative abundance of the dominant species over time indicated successional trends in the case of surface litter, with three fairly distinct phases being discernible (Table 3.5). The first phase was dominated by the collembolan *Hypogastrura denticulata* and extended over the first 2 months; the second phase dominated by the collembolan *Isotoma olivacea* extended over the third and fourth months; and the third phase dominated by the oribatid mite *Minunthozetes semirufus* extended from the fourth to the tenth month. The succession of dominant species in buried litter was less clearly defined.

3.6 COMMUNITY STABILITY

Complex natural communities are generally perceived as being less prone to pest outbreaks and more stable than highly simplified monocultures. This greater stability of natural communities is frequently linked with their greater species diversity although stability is not necessarily a consequence of diversity (Murdoch, Evans and Peterson, 1972; Murdoch, 1976; Horn, 1974; Hurd and Wolf, 1974). Indeed, some authors have suggested that diversity may be the result of stability rather than a cause. This argument has been developed by May (1973, 1976), who concluded from theoretical studies that stability is not an automatic mathematical consequence of species interactions; on the contrary, May argued that as a system becomes more complex it becomes more fragile and any correlation between species diversity and stability arises from the fact that complex communities can be supported only by very stable (predictable) environments. Unstable (unpredictable) environments can support only structurally simple, robust communities. Much of the controversy surrounding this topic may be at least partly due to the use of the term stability in different senses and recent authors have recommended greater precision in the use of the many related but often quite different concepts involved. For example, it is possible to distinguish between local (neighbourhood) and global stability. Thus, the local 'component' communities (plant–invertebrate associations) in an old grassland may show considerable fluctuations in species composition and abundance while the overall 'compound' (global) community may be quite stable. Furthermore, it is useful to distinguish between two major aspects of stability, constancy and resilience, where constancy refers to stability in the sense of species composition and where population size and resilience refers to the ability of the community to survive and function after perturbation arising from weather or other abiotic environmental factors.

Putman and Wratten (1984) consider that constancy is likely to be less in a complex system comprising large numbers of species and interactions, while complex communities are more likely to be resilient because damage to one part can be compensated for by adjustments in other parts. Constancy and resilience need not necessarily be correlated: a community may be very resilient but still fluctuate greatly (i.e. have low constancy). However, it is clear that resilience need not necessarily be associated only with diverse communities. Complex systems may have large numbers of relatively weak feeding interactions that are easily disrupted, while the few links in the food webs of simple systems may be robust and not easily disrupted.

Odum (1975) concluded that quite stable systems can have either low or high diversity depending on the energy base of the community. Where a strong external source of energy and nutrients is available he considered that it can be exploited more efficiently by a community with a concentrated, specialized structure of low diversity than by a community with more dispersed, diverse structure. On the other hand, higher diversity may be optimal in systems which are dependent on internal production and transfer of energy; that is, high diversity tends to be associated with self-contained systems with limited energy input while systems heavily subsidized by energy input tend to have low diversity. This accords with the general experience that fertilized grasslands tend to have lower species diversity than unfertilized sites, although the interpretation of such situations is often complicated by the effects of differences in management intensity (Chapter 5).

3.7 COEVOLUTION

It has been suggested that reciprocal interactions between plants and their herbivores and between herbivores and their enemies have strongly influenced the evolution of natural communities, and some authors consider that natural systems owe their stability to the coevolution of their interacting species rather than to species diversity (e.g. Murdoch, 1976). Certainly, there is ample evidence to indicate the importance of herbivory as a selection pressure on host plants. Grazing by large herbivores is a major factor in the maintenance of much of the earth's grasslands, and the evolution of Gramineae and Bovidae are inextricably linked (McNaughton, 1979). The vast array of physical and chemical defensive adaptations found in plants, the many examples of adaptations in insects to counter plant defenses and the utilization of volatile plant chemicals as cues in host finding by specialized herbivores have been interpreted as responses to strong reciprocal selection pressure between grazers and plants (e.g. Ehrlich and Raven, 1964).

However, Strong, Lawton and Southwood (1984) argue that the sustained reciprocal and intense interactions that are necessary for coevolution in the strict ('focused') sense are rare and that, more often, diffuse coevolution occurs in response to a variety of conflicting demands. While acknowledging that diffuse coevolution could have been involved in the rise of the angiosperms and the subsequent diversification of phytophagous insects they conclude that coevolution does not provide a general mechanism to explain the contemporary structure of phytophagous insect communities. Rather, they consider that 'the insect fauna of plants is a mixture of the coevolved, the pre-adapted and the opportunistic in varied and unpredictable proportions'.

3.8 DEFINING AND COMPARING COMMUNITIES

Several approaches have been used to characterize, delimit and compare invertebrate communities, to examine their relationships with soil, vegetation and other habitat features and to evaluate their responses to stresses of various kinds. Criteria used include the relative abundance, frequency of occurrence and fidelity of dominant species, various measures of interspecific association, and the degree of similarity between assemblages in terms of co-occurring species. The application of these criteria has been greatly enhanced by the advent of a range of computer-based multivariate approaches that can explore patterns and relationships in very large data sets. Some examples of these approaches and their applications to the study of grassland invertebrate communities are considered below.

3.8.1 Dominance, frequency and fidelity

A common approach has been to identify characteristic species defined by such criteria as dominance (percentage contribution to the population), frequency (percentage of samples in which the species occurs) and fidelity (degree of restriction to a particular habitat) adapted from the Braun-Blanquet school of phytosociology. Weis-Fogh (1948) applied this approach to the analysis of the microarthropod fauna sampled at seven locations along a 12-m sampling line on a sandy permanent pasture in Denmark. The sampling line represented a moisture gradient from a dry end covered with a *Gnaphalium arenarium–Hieracium pilosella* plant sociation to a moist end with a *Carex demissa–Hieracium auricula* sociation. Successive changes in the composition of the fauna corresponded with changes in ecological factors such as soil moisture and structure and the nature of the organic remains, and there were distinct differences between the faunas in the wet and dry parts of the site. Based on dominance and constancy a dry-soil community dominated by the collembolan *Folsomia* and the prostigmatic mite *Variatipes* and a moist-soil community characterized by the oribatid mites *Tectocepheus* and *Schelor-ibates*, were erected. Similar approaches were used by Haarløv (1960), Davis (1963) and others to delimit microarthropod communities in grassland soil, while Wood (1967c) defined nine groupings of species from moorland soils in Yorkshire, England based on differential species of high frequency and fidelity.

3.8.2 Interspecific associations

Another approach is to formally establish interspecific associations based on species co-occurrence to determine what species should be grouped together as a community and where community boundaries should be drawn. A simple method is to construct 2 × 2 contingency tables based on the presence or absence of the pairs of species being compared. Where significant associations are found based on χ^2 analysis, coefficients of association may be calculated (e.g. Cole, 1949; Debauche, 1962) and species with significant coefficients may then be grouped following procedures such as those of Fager (1957) to define 'recurrent groups' or associations of species that tend to occur together. Davis (1963) studied the microarthropod fauna (mites and collembolans) of established

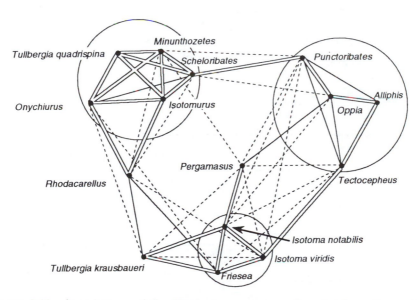

Figure 3.11 Associations of the 15 most common species of microarthropods based on Pearson's index. Double lines indicate that the index was significant at the 1% level, single lines indicate significance at the 5% level and broken lines indicate all other positive but non-significant indices. (From Davis, 1963.)

grassland and of grasslands on sites recently reclaimed from open-cast mining in Northamptonshire, England. Interspecific associations between 15 of the commonest species were calculated using the Pearson Index (Debauche, 1962) and cluster analysis was used to combine species showing positive association. The results indicate three strong species associations (Figure 3.11). One comprising the collembolans *Tullbergia quadrispina, Isotomurus palustris*, etc. was characteristic of old grassland; the second group (the collembolans *Isotoma viridis, I. notabilis and Friesea mirabilis*) was associated with the reclaimed mine sites; and the third (the mites *Punctoribates punctum, Alliphis halleri*, etc.) was a more loosely knit group not showing any strong site preference.

3.8.3 Indices of similarity

Indices of similarity are widely used for making comparisons between sites based on the extent to which they have species in common. Some indices such as that of Sørensen (1948) consider only presence or absence while others also take abundance into account (e.g. Bray and Curtis, 1957; Whittaker, 1972). Once pair-wise comparisons have been made sites may be classified in a way that reflects their overall similarity. A simple approach is to construct a dendrogram that reflects highest, lowest or mean similarities between sites or groups of sites depending on the method of calculation. An example is provided by Curry and Tuohy (1978), who compared the acarine faunas of grass leys and old pasture sampled at three locations in Ireland (Figure 3.12). The resulting classification displays a recurring feature of such studies (e.g. Wood, 1967c;

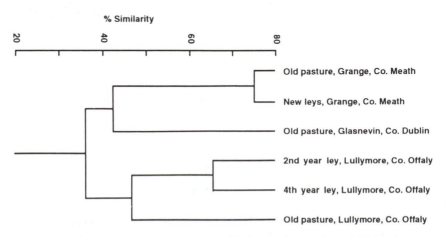

Figure 3.12   A comparison of the acarine faunas of grass leys and old pastures at three locations in Ireland. (After Curry and Tuohy, 1978.)

Haarløv, 1960) – namely, that even fairly dissimilar sites that are located close together are likely to have more similar faunas than sites that are widely separated from one another. Differences between adjacent sites were most pronounced at Lullymore where the old pasture was largely unmanaged and had a mature grassland fauna. At Grange, by contrast, differences between leys and old pasture were obscured by the overriding influence of management, notably cutting, which had the effect of repeatedly causing the community to revert to an early successional stage.

3.8.4 Leafhopper communities

Relationships between leafhopper communities and their habitats have been studied by several workers. Kontkanen (1950) used a range of approaches, including criteria of constancy and fidelity (Braun-Blanquet, 1932) and similarity to distinguish four community types named after the dominant species of leafhoppers (Auchenorrhyncha) in a range of biotopes in Finland:

1. *Neophilaenus lineatus–Tettigella virida* community on the wettest type of peat soil meadows, swamps and fens.
2. *Philaenus spumarius–Arthaldeus pascuellus–Elymana sulphurella* community on moist and fresh biotopes (peat soil meadows).
3. *Psammotettix confinis–Doratura stylata–Empoasca paolii* community, rich in species, occurring on dry biotopes–grass meadows of different kinds.
4. *Macrosteles sexnotatus–Notus flavipennis* community on poor *Carex–Scheuchzeria* bogs.

While there was a broad equivalence between leafhopper communities and higher-level community groupings in plant sociological classification, the leafhopper community types did not coincide with lower-level plant community groupings. Most meadow and swamp leafhoppers were fairly polyphagous and could spread according to their moisture requirements, regardless of the detailed composition of the vegetation. When classified according to their moisture preferences, six groups of species could be recognized depending on their preference for dry or wet biotopes or their ubiquitousness. Only three species with narrow tolerance ranges showed marked habitat restriction.

Müller (1978) studied the spatial and temporal distribution of 69 breeding leafhopper species in a catena of four grassland associations on a south-facing slope of the Leutra Valley near Jena, Germany. There was considerable overlap between the plant associations in regard to their leafhopper fauna, but local accumulations of species based on their percentage representation in the different associations could be recognized and were considered to represent leafhopper associations in the plant communities. The author considered that these were coupled with the plants not by host plant relations so much as by particular structural and microclimatic qualities of the plant. When the distribution of all 1001 arthropod species sampled by sweep netting was considered most were found to be widely distributed but distinct groupings (based on percentage representation) could be recognized, which reflected the four plant associations.

3.8.5 Synusiae and isovalent groups

Alternative species groupings may be obtained based on the relationships between species and environmental factors rather than on their relationships with one another. Strenzke (1952) grouped oribatid mites associated with various plant associations in the North German coastal plain into synusiae, according to their dependence on five environmental factors – moisture, organic matter, acidity, degree of ground cover and salinity. Each community comprised a few typical species, all sharply restricted to certain habitats. However, Knülle (1957), likewise studying the oribatid mites in a range of districts in the coastal plain of northern Germany, noted that the species belonging to particular synusiae were often not restricted to the corresponding habitat type. Analysis of species distribution in relation to soil moisture led to the definition of groups of species termed isovalents whose distribution reflected the tolerance of their members for various soil moisture conditions. Often these groups crossed vegetational boundaries and might contain members from more than one synusia (i.e. the isovalent groups linked various synusiae together and established the extent of their faunal similarities). This approach shows how understanding of communities can be extended by taking into account responses to a range of environmental variables but is limited by the fact that species distributions are frequently influenced by many factors, to which the responses of the various species are often not well understood.

Regardless of the approach used, attempts to define and classify invertebrate communities and to relate them to habitat features have been at best only moderately successful for several reasons. Many of the commoner species are ubiquitous in distribution and occur in a wide range of different habitats. In many cases where synusiae or associations have been defined these are only of local validity and are not replicated in similar conditions elsewhere. Frequently, invertebrate associations do not correspond with macrohabitat features such as soil or vegetational types; they may often reflect environmental factors that do not vary in parallel with macrohabitat features. However, changes in parameters such as frequency of species occurrence can be very useful in evaluating community response to environmental stress. The advent of computer-based multivariate techniques has greatly facilitated developments in this area.

3.8.6 Multivariate methods

Limitations of the techniques described so far are that in most cases only a small proportion of the total number of species is taken into account in characterizing the community, and in the case of the similarity dendrograms only a small proportion of the total number of indices is used in arriving at the groupings. There is thus a considerable amount of information loss, which may obscure important relationships. Computer-based ordination techniques that can handle large data matrices offer considerable potential for examining patterns of variation in communities (see Gauch, 1982 for a review). A few examples of techniques that have been successfully used in grassland invertebrate studies are briefly con-sidered below.

Principal component analysis (PCA) has been widely used in ecological studies. It is considered to be an efficient summarization technique; that is, it has the ability to reduce the relationships between large numbers of samples each containing large numbers of species to a summary form while retaining most of the information about intersample relationships. However, as with most ordination methods, the ecological interpretation of the axes of ordination is often difficult (Gauch and Whittaker, 1972; Gauch, Whittaker and Wentworth, 1977). This is a limitation when the object is to interpret relationships between samples in terms of underlying environmental variables, but if the main object is to detect clusters and the relationships between clusters, PCA can give useful results. The Bray–Curtis or polar ordination technique (Bray and Curtis, 1957) allows the ecologist to choose the axes, and this approach has been used successfully in studies where community response to an environmental gradient or applied stress in being assessed. Dindal, Newell and Moreau (1979) used this approach to show the responses of earthworm communities to municipal wastewater irrigation in Pennsylvania, USA (Figure 3.13).

Detrended correspondence analysis (DCA) is a versatile technique that has proved useful in a wide range of applications. Figure 3.14 shows an ordination of 17 grassland sites in the south of Ireland based on their staphylinid beetle fauna. The ordination with respect to axis 1 clearly reflects the influence of management: fields that were intensively managed and cut for silage are separated from less-disturbed sites that were lightly grazed or cut for hay. The two glacial esker grassland sites labelled 'E' are of conservation interest. One of these, a site in a heritage zone designated for landscape conservation, was located outside the old

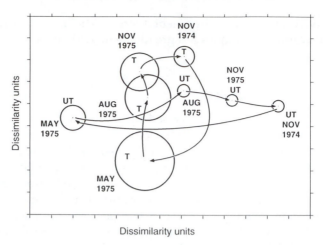

Figure 3.13 Bray–Curtis ordination of earthworm communities from sites in Pennsylvania, USA, treated (T) and untreated (UT) with wastewater effluent. Each circle represents a community in time; the size denotes density and frequency of earthworms within a community; the proximity of circles indicates degree of similarity between sites; the arrows indicate patterns of seasonal change. (After Dindal, Newell and Moreau, 1979.)

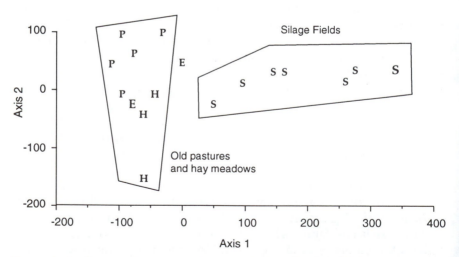

Figure 3.14 Detrended correspondence analysis ordination (Hill, 1979a) of staphylinid beetle presence/absence data from Irish grassland sites. S = silage fields; H = hay meadows; P = permanent pasture; E = glacial esker grassland. (From Curry and Good, 1992.)

pasture and hay meadow cluster. Its location suggests a degree of faunal impoverishment that could have been due to overgrazing by cattle.

A second example illustrates the use of DCA to evaluate the response of the soil microarthropod fauna to a gradient of pig slurry contamination downhill from a storage lagoon subject to periodic overflow after heavy rainfall (Figure 3.15). The analysis was based on the frequencies of occurrence and abundance of 61 acarine and collembolan species in soil samples taken at intervals along the gradient. The ordination divided the samples into two broad groups – those located near the source of contamination and those farther away. The major features separating the two groups were that the dominant species occurred more frequently and were more abundant in heavily contaminated soil where a surface organic layer had developed, while some species such as *Isotoma viridis* and *Sminthurinus aureus* occurred more frequently and were more abundant in less-contaminated soil (Bolger and Curry, 1984).

Relationships between invertebrate communities and environmental variables can be explored more fully by using a range of complementary

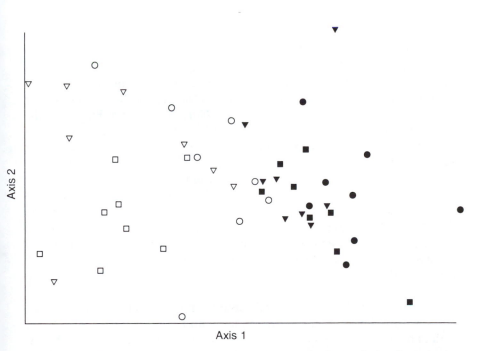

Figure 3.15 Soil microarthropod samples along a gradient of pig slurry contamination ordinated by detrended correspondence analysis. Distance from source: ▽, 5 m; □, 15 m; ○, 35 m; ▼, 55 m; ■, 75 m; ●, 95 m.

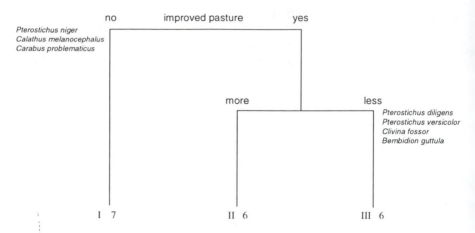

Figure 3.16 Classification of upland grassland sites based on their ground beetle (Carabidae) communities. Indicator species for each division and the number of sites in each group are indicated. (After Rushton, Luff and Eyre, 1989.)

multivariate techniques. For example, Rushton, Luff and Eyre (1989) used both ordination (DECORANA; Hill, 1979a) and classification (TWINSPAN; Hill, 1979b) techniques in their study of the effects of pasture improvement and management on the ground beetle and spider communities of upland grassland sites in Northumberland, England. The classification technique uses two-way indicator species analysis to produce a hierarchical classification of sites, grouping those with similar faunas and identifying 'indicator species' that differentiate the groups of sites. The 19 sites studied were classified into two major groups, improved and unimproved (Figure 3.16). Improved sites were subdivided into two further groups that reflected the intensity of pesticide use for leatherjacket control. Stepwise linear discriminant analysis (James, 1985) was then used in an attempt to identify the environmental variables (site management, soil and vegetation characteristics) that determined the differences between site groups indicated by the TWINSPAN analysis.

In a study of the ground-beetle fauna of intensively managed agricultural grasslands in northern England and southern Scotland, Eyre, Luff and Rushton (1990) used ordination and classification approaches to identify site groupings and then used canonical correspondence analysis (Ter Braak, 1986) to relate the community data to measured environmental variables.

3.9 SUMMARY

While the classical notion of the biological community as an integrated, highly organized and structured superorganism has little relevance in the real world, the community concept has nevertheless proved useful in considering the ways in which populations interact with one another and with their environment. In its broadest sense, the concept has been applied to the assemblage of species populations occupying a designated area or habitat while it has also been applied to the populations associated with discrete habitat units, such as plant species or even individual plants or other microhabitats. Community structure is defined in terms of parameters such as biomass and trophic structure, species composition and distribution of individuals among species, species diversity and stability.

The nature of the invertebrate community is very much influenced by the general structure of the habitat. Grassland is simpler in its architecture than mixed deciduous woodland and supports a correspondingly less varied fauna in the vegetation and litter layers but a richer soil fauna, reflecting more evenly distributed soil organic matter.

The ratio of faunal to plant biomass tends to be highest ($\propto 1:100$) in mesic grasslands and lowest in arid land ($< 1:1000$). The vast bulk of invertebrate biomass (70–98%) occurs below ground, and decomposers comprise the dominant trophic group in most habitats although their dominance is less marked under arid conditions.

Unutilized grasslands resemble most other natural ecosystems in that only a small proportion of net primary production is used by herbivores; the vast bulk ($> 90\%$) is returned to the soil where it supports highly complex decomposer food webs. Conceptual models of these food webs with computer simulation of energy and matter circulation offer what is possibly the best approach to enhancing understanding of the functioning of grassland soil communities; a good example of the application of this approach to short-grass prairie is provided by Hunt *et al.* (1987).

Analysis of species–abundance and species–area relationships in invertebrate communities reveals the existence of fairly consistent patterns that suggest a degree of community 'organization'. Factors implicated in maintaining community structure may include herbivore–host plant relationships and the influence on these of host-plant physical and chemical characteristics; habitat features such as complexity and the degree of resource concentration; and interspecific interactions with particular reference to the role of natural enemies.

The classical view of succession as a highly deterministic process is not now widely held, but the process does tend to have certain characteristic features. These include reduced net production, increasing diversity and stability, a change from predominantly *r*-selected to predominantly *K*-

selected species, and a change from rapid, open nutrient cycling to slower, relatively closed cycles as succession proceeds.

Successional processes involving grassland invertebrates occur over various time and spatial scales, ranging from rapid local changes over periods of months associated with decaying dung and plant residues to the widespread changes over periods of several years associated with community development and maturation in disturbed sites.

While community diversity and stability tend to be correlated, these two attributes are not necessarily causally related. Stability involves aspects such as constancy and resilience that may be influenced differently by different sets of factors. These factors could include the predictability and severity of the environment, the nature of the resource supply for the community and the degree to which the species in the community are adapted to one another through coevolution.

Criteria such as dominance, constancy and fidelity, coefficients of interspecific association and indices of similarity have been used to define local communities or synusiae, but a comprehensive classification of invertebrate communities paralleling that of vegetation and soil is unrealistic. However, the advent of computer-based multivariate techniques for classification and ordination has provided a valuable set of tools for evaluating the responses of invertebrate communities to different management practices and environmental stresses.

4

Factors affecting the abundance and composition of the grassland fauna

Climate and weather, the structure and composition of the sward and its nutritional quality, the quantity and quality of litter returned to the soil, and physical and chemical characteristics of soil are the main habitat factors influencing the composition and abundance of the grassland fauna. The biotic interactions considered to be most important are natural enemies, disease and competition. Habitat features that influence the fauna have already been considered in a general way in Chapter 1, while habitat features and biological interactions that influence the structure and functioning of the invertebrate community have been considered in Chapter 3. In this chapter, the ways in which selected factors influence population density are considered, with particular reference to food quality, natural enemies, competition and disease. Some case studies, in which the relative importance of the various factors has been analysed, are then reviewed. The effects of management practices, which can be very pronounced, will be considered separately in Chapter 5.

4.1 FOOD QUALITY AND PLANT DEFENCES

Grassland invertebrates are unlikely to be limited by the quantity of food available, but there is considerable evidence that food quality may often impose constraints on population growth and development. In the context of herbivorous species food quality may be determined by a range of plant characteristics that influence host finding and host preference and suitability, or which, conversely, may render plants resistant to attack by particular herbivores. Characteristics that can have a role in plant defence against herbivores may be either physical or chemical. Physical features that may make it difficult for herbivores to feed on plant tissues include the presence of spines or hairs, tough plant epidermis, fibrous tissues in leaves and stems, etc., while chemical factors may include the content and availability of primary nutrients and water and the concentration of secondary plant substances or allelochemicals. These may act as cues for host finding or as phagostimulants for adapted herbivores, and as repellants or toxins for non-adapted species.

4.1.1 Physical factors

Physical attributes that influence the suitability of grasses for herbivores include the extent and nature of leaf pubescence – notably, the presence of leaf trichomes, which may be impregnated with silica or calcium carbonate; the hardness of the plant epidermis; the presence and extent of lignified or silicified fibres in leaves and stems; and leaf shape. Shape seems to be an important visual cue in host selection by grasshoppers (Mulkern, 1967), while grass-harvesting termites (*Trinervitermes geminatus*) appear to select narrow-leaved grasses in preference to species with broader leaves (Ohiagu and Wood, 1976). Young frit fly larvae (*Oscinella frit*) have difficulty in penetrating mature grass stems and larval mortality in grassland is regularly around 40%, compared with about 25% in oat tillers of more uniform age (van Emden and Way, 1973). Only young tillers are susceptible; once oat plants reach the four-leaf stage they are resistant to attack.

4.1.2 Chemical nutrients

Insect growth and development can be impaired by defficiencies in any of the range of nutrient substances that they require in their diet. Water content of leaves is often below 50–60% for long periods, for example, and this can be limiting under some circumstances (Edwards and Wratten, 1980). Most attention has been given to plant mitrogen as an indicator of nutritional quality because of its fundamental importance for amino acid synthesis. The relatively low N levels in most plant tissues suggest that N may be a limiting nutrient for many herbivorous and detritivorous invertebrates. Data from several grassland studies indicate that the level and availability of plant N has an important influence on rates of insect growth, reproduction and population density.

(a) Plant nitrogen

Animal tissues typically comprise 50% or more protein, about 7–14% N by weight, and animals have a high dietary requirement for N, especially during periods of active growth and reproduction. The N content of plant tissues can range from less than 1% to about 7% of dry mass (Mattson, 1980), with highest concentrations (3–7%) occurring in young actively

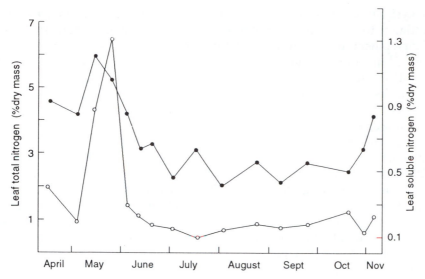

Figure 4.1   Changes in total nitrogen (●) and soluble nitrogen (○) in leaves of *Lolium perenne* throughout the growing season at Silwood Park, Berkshire, England. (From Prestidge and McNeill, 1983a.)

growing tissues or in storage tissues such as found in seeds; concentrations decline sharply during the growing season and are lowest in plant litter. Figure 4.1 shows typical changes in grass leaf N throughout the growing season at Silwood Park, Berkshire, England (Prestidge and McNeill, 1983a). Total leaf N was highest in spring during the phase of rapid grass growth; it declined rapidly in early summer, remained low throughout the summer and increased sharply in late autumn at the time of the autumn flush of growth.

For convenience N can be classified as either protein N (PN) or non-protein N (NPN) (Mattson, 1980). Non-protein N comprises a huge variety of compounds including inorganics such as nitrates and ammonia, organics such as amines, amides, amino acids, chlorophyll and other pigments, and a wide array of secondary compounds. Pronounced qualitative variations in N content relate to season and plant phenology. Young, moisture-rich tissues have much higher levels of NPN and soluble PN than mature tissues, except for senescent tissues when proteins are being hydrolyzed. Young tissues are particularly rich in amino acids, soluble proteins and N-based compounds whereas N in older tissues is mainly in the form of structural or insoluble protein. Levels of soluble N in grass leaves at Silwood Park increased rapidly in spring (by six- to ten-fold) for a period of 2–3 weeks as leaf proteins were degraded and metabolites were translocated to new growing points (Figure 4.1). In autumn, there was a small increase in soluble N associated with proteolysis in senescent tissues and translocation to new leaf material. In addition to variations in N status related to seasonal changes in plant development, factors such as moisture and temperature stress can influence N quantity and quality, while fertilization can also have substantial effects (Mattson, 1980, etc.).

Studies at Silwood Park have given valuable insights into relationships between food quality as measured by plant N and grass-feeding Hemiptera. McNeill (1973) identified the availability of high-N feeding sites at a time when the general level of N in grass is low as an important factor influencing larval mortality and fecundity in the grass bug *Leptopterna dolabrata*. In a laboratory experiment where larvae were fed on leaves alone, few reached adulthood and those that did were very weak, survived only for a day or so, and had underdeveloped gonads. When flowerheads were added there was a six-fold increase in numbers reaching normal adulthood. The degree of synchronization between the life cycle of the leafhopper *Dicranotropis hamata* and plant N level was found to have a profound influence on its reproductive success (Table 4.1). When the spring peak of soluble N in leaves of creeping soft-grass (*Holcus mollis*) coincided with most of the population becoming adult a much higher natality was observed, resulting in a much higher population of nymphs in the following generation.

Preferential feeding by adult leafhoppers on developing grass flowers and seeds and a relationship between high fecundity levels and high N levels in leaves during leaf senescence were reported for five leafhopper species (Hill, 1976). Prestidge (1982a) examined the effects of food quality on feeding, growth and oviposition rates of four leafhopper species reared in the laboratory on leaves of Yorkshire fog (*Holcus lanatus*) collected from field plots fertilized with different rates of ammonium nitrate applied in the spring. Instar duration, adult consumption rate, nitrogen utilization efficiency and oviposition rate were influenced by leaf N level, but the exact nature of the relationships between these parameters varied with the different species, which had different strategies of N utilization.

Table 4.1 Timing of the spring peak in soluble N and reproduction in the leafhopper *Dicranotropis hamata* (after McNeill and Southwood, 1978)

Year	Mid-date of N peak	Date of adult peak	Nymphs per female
1970	8 June	25 May	3
1971	1 June	20 June	5
1972	8 May	4 May	30
1973	21 May	24 May	16
1974	13 June	30 May	1

Relationships between total plant N and herbivore performance are often not clear-cut, since total N may not always be a good indicator of available N. Factors that influence N availability to herbivores include moisture levels and allelochemicals (Mattson, 1980). Changes in plant N induced by fluctuating moisture regimes can be critical in determining the survival of newly emerged herbivorous insects and the course of outbreaks of pests such as locusts (White, 1976, 1978), while utilization efficiency appears to be directly correlated with plant moisture for some lepidopterous caterpillars at least (Scriber, 1977). Most of the plant's defences are located in non-vascular tissues and are thus largely avoided by sap-sucking invertebrates (Raven, 1983); consequently, total plant N is a more useful indicator of food quality for sap suckers than for chewers (Raven, 1983; Wint, 1983). Wint (1983) utilized an enzyme-inhibition assay technique to estimate the levels of leaf-protein complexing agents and by combining these estimates with values of total leaf protein he was able to estimate the amount of protein actually available to a feeding insect. He demonstrated relationships between available protein and performance of larval winter moth (*Operophtera brumata*) fed on leaves from a range of plant species, whereas no correlations occurred between total N and larval growth rates.

The amino acid component of plant N is particularly important, and especially the composition of the amino acid fraction (Prestidge and McNeill, 1983b). Hill (1976) found a close link between the life histories of five leafhopper species and periods of high concentrations of amino acids in the plant. There was also a good relationship between the abundance of the five dominant species and the quality of the amino acids present.

Ratios of carbon to nutrients, and particularly to nitrogen are good indicators of food quality, which is often low for saprophagous invertebrates feeding on plant litter and soil organic matter (Swift, Heal and Anderson, 1979). Earthworms such as *Lumbricus terrestris* prefer litter with high N and sugar content (Satchell, 1967) and grow well in soil amended with dung (Meinhardt, 1974; Lofs-Holmin, 1983b). *Millsonia anomala*, the dominant geophagous earthworm in humid savannah at Lamto, Ivory Coast, responded to improved substrate quality by growing faster and reducing its rate of soil ingestion (Lavelle, Schaefer and Zadi, 1989). Conversely, its capacity to increase its soil-ingestion rate enables it to exploit soil containing little organic matter (1.2%). Litter quality also had a marked effect on the growth and fecundity of the woodlouse *Armadillidium vulgare* and could thereby have a major influence on its population dynamics in grass heathland (Rushton and Hassall, 1983a, 1987). Many saprophagous invertebrates obtain at least part of their food requirements from microbial biomass, but the quality of this is also influenced by the nature of the organic substrate. Booth and Anderson (1979) reported effects of fungal food quality on the growth rate and fecundity of the collembolan *Folsomia candida* fed on *Coriolus* and *Hypholoma* that had been grown in liquid media containing various concentrations of N. Microbivory and altered ingestion rates are among the feeding strategies that enable invertebrates to cope with poor food quality; these and other adaptations are considered more fully in the following section.

(b) Adaptations to unfavourable food quality

Adaptations that enable herbivores to cope with low and variable plant N content include (1) variable feeding rates, digestive efficiencies, duration of feeding and developmental times; (2) switching between plants or between sites on the same plant or utilization of alternative sources of N; (3) modification of plant physiology; (4) evolution of larger body size; and (5) life cycle adaptation to N availability.

Variable feeding rates and utilization efficiencies
An increased food-ingestion rate appears to be a common response to low levels of food N, and because this involves a more rapid throughput of food with a shorter residence time in the gut there is a decrease in the efficiency of assimilation. Figure 4.2 shows a positive relationship between assimilation efficiency and increasing food N content in the case

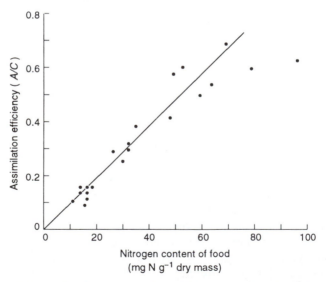

Figure 4.2 Relationship between total food N levels and assimilation efficiency (*A/C*) in energy terms of food ingested by the grass bug *Leptopterna dolabrata*. (After McNeill and Southwood, 1978.)

of the grass bug *Leptopterna dolabrata* (McNeill and Southwood, 1978), although this simple relationship does not always hold (cf. Prestidge, 1982a). Hill (1976) reported that nymphs of the Leafhopper *Dicranotropis hamata*, which hatch in June–July at a time when leaf N levels are low, have a high ingestion rate but a low efficiency of N utilization (24%), while older nymphs and adults feeding on good-quality food the following spring have a low ingestion rate but a high efficiency of N utilization (43%). Other species that breed throughout the summer have high feeding rates and low N utilization efficiencies throughout their life cycle.

Different food utilization strategies were reported among four species of leafhoppers feeding on *Holcus lanatus* leaves with different levels of N (Prestidge, 1982a). *Zyginidia scutellaris*, a mesophyll feeder, consumed 26–72% (dry mass basis) of its body weight each day but only assimilated 30–52% of its N intake, suggesting a strategy of rapid throughput of food and assimilation of the more easily digestible components. *Eucelis incisus*, a xylem sap feeder ingesting very watery food low in carbohydrates and nutrients, consumed only 3–12% of its body weight equivalent a day but had a relatively high N assimilation efficiency (53–63%). It combined a high-volume food intake with a high N-extraction efficiency. Two phloem-feeding species, *Dicranotropis hamata* and *Elymana sulphurella*, consumed 9–25% of their body weight a day and assimilated 25–59% of their daily N intake. Both consumption rates and N-assimilation efficiency were influenced by the N status of the leaves being fed on, but the nature of the response to changing N levels differed from species to species. Interspecific differences in food utilization are indicated by Figure 4.3. In each case, N-utilization efficiency peaked over a very narrow range of food N concentration, which in no case corresponded with the highest level. The maximum utilization efficiency was shown by *Eucelis incisus*, the xylem-feeding species that ingested food with the lowest N level. The fact that the four species reached their maximum utilization efficiencies at different levels of plant N led Prestidge to hypothesize that leafhopper species may be associated with particular N concentrations in grasses. This hypothesis was partially supported by field data that showed that some oligophagous leafhopper species are associated with particular N levels in grasses and occur on grasses that have N concentrations within narrow ranges; however, other species show little association with plant N level (Prestidge and McNeill, 1983a). Gibson (1976) also found evidence that some grass-feeding Miridae tend to be associated with characteristic bands of N in their host plants.

Protracted feeding periods and life cycles may also be adaptations to low food quality, an extreme case being cicadas (Magicicadae spp.) feeding on root xylem sap of deciduous trees that have 13–17-year cycles (Mattson, 1980). The slow developmental times of soil macroarthropods

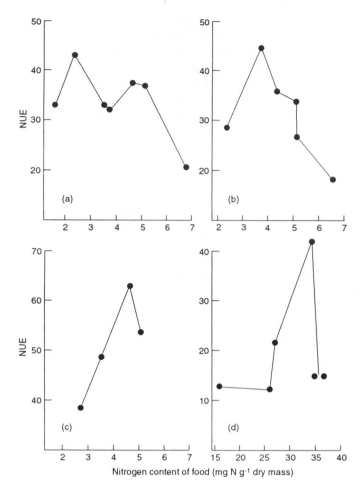

Figure 4.3 Nitrogen utilization efficiency (NUE) for four leafhopper species feeding on Yorkshire fog (*Holcus lanatus*) leaves with different levels of nitrogen. (a) *Dicranotropis hamata*; (b) *Elymana sulphurella*; (c) *Eucelis incisus*; (d) *Zyginidia scutellaris*. (After Prestidge, 1982a.)

such as isopods, millipedes and some scarabaeid beetles may also reflect the low-nutrient status of the organic matter on which they feed.

The quality of the food consumed by some invertebrates may be improved through the activities of microorganisms. Animal–microbial symbiosis is highly developed in the case of termites where the digestion of plant structural polysaccharides is greatly enhanced by the activities of the microflora in their hind gut, and, in the case of fungal-growing termites and ants, by deliberate fungal cultivation on faeces in fungal gardens. Coprophagy is common in isopods, millipedes and termites, and this has the effect of greatly enhancing nutrient assimilation from poor substrates (Hassall and Rushton, 1982, 1985). Few invertebrates are able to utilize plant structural compounds to any degree, and most saprophagous species derive a significant amount of their nutrients from microbial biomass or products of microbial activity associated with decaying residues.

Altering food source
Some insects feed on different parts of their host plants at different seasons and these changes of location appear to be correlated with N availability. In the case of grasses, leaves provide an adequate source of N early in the season, but in midseason, when leaf N levels are low and flowers are available, the latter are preferred by leafhoppers (Hill, 1976). Shifts in locations of the grass bug *Leptoterna dolabrata* also appear to be linked with the quest for maximal available N (McNeill, 1971, 1973). Young nymphs feed on the leaves of *Holcus mollis*, but in early June as N levels in the leaves drop they move to flowerheads as they become available. In the limestone grassland site studied by McNeill, the earliest available flowerheads were those of *Holcus lanatus*, and McNeill concluded that competition for those scarce high-N feeding sites was an important density-dependent mortality factor influencing late instars and the fecundity of surviving females.

The frit fly (*Oscinella frit*) is an example of a species in which different generations utilize different plant parts. First-generation larvae in May–June are associated with the central shoot, while second-generation larvae feed on florets of oats in July–August.

Occasional carnivory appears to be relatively common among detritivorous and some herbivorous invertebrates as a mechanism for supplementing dietary nutrients. Many detritivores eat exuviae and dead or injured invertebrates, while partial predation is common among herbivorous mirid bugs and other groups. As seen in Chapter 2, nematophagy appears to be common among 'detritivorous' soil microarthropods.

Modification of plant physiology
Ungulate grazing is widely acknowledged to be a major factor in the maintenance of much of the earth's grasslands and moderate levels of grazing increase primary production and prolong the vegetative phase of grass growth (Chapter 6). The impact of invertebrate herbivores on grass is probably relatively low compared with that of large ungulates, but many of the former appear to be able to modify plant physiology and growth processes to their own advantage to some extent. Salivary juices injected into plants by sap-sucking insects such as aphids modify plant metabolism, while leaf miners, gall formers and plant parasitic nematodes also bring about local changes that enhance the quality of food

available to them. Saliva of grasshoppers has been shown to increase plant growth (Dyer and Bokhari, 1976).

Evolution of larger body size

Mattson (1980) argues that it may be advantageous for consumers of poor food material to evolve larger body size on the basis that larger animals with lower surface-to-volume ratios have lower respiratory rates and higher production efficiencies than smaller ones and that larger animals with larger heads and stronger mouthparts may be better able to macerate tough plant tissues than smaller ones. Also, large size may allow the development of highly elaborate digestive systems necessary to digest poor, tough food. However, evidence relating body size to food quality is circumstantial.

Life-cycle adaptation to host plant N

The life-history patterns of some herbivores appear to be closely correlated with the availability of N in their host plants. McNeill and Southwood (1978) examined the relationships between the life histories and population dynamics of a range of phloem-, mesophyll- and xylem-feeding sap-sucking insects (Homoptera and Heteroptera) and nitrogen availability in *Holcus mollis* over a period of 7 years. The peaks of population density corresponded very well with the peaks of N availability, while the peaks of total numbers of species lagged behind those of population density as groups of species replaced one another (Figure 4.4). The first peak of density is formed by species that mainly feed on leaf-mesophyll cell sap, while the second and highest peak is formed by phloem-feeding species. The midseason peak corresponds to species feeding either on flowers and developing seeds or on phloem, while the autumn peak corresponds to both leaf and phloem feeders.

Despite these adaptations to low and changing N status, however, it is clear that food quality, and N availability in particular, are important factors limiting the abundance of grassland invertebrates.

4.1.3 Allelochemicals

Many of the chemical substances in plants without any apparent function in plant metabolism are known to play a role in plant–herbivore interactions, acting in some cases as olfactory cues in host finding, or as phagostimulants for specialized herbivores; more often, these allelochemicals have an anti-herbivore defensive role. From a functional point of view these defensive chemicals may be grouped into those that act as toxins (e.g. alkaloids, cyanide, glucosinolates) and those that are non-toxic but reduce the digestibility of plant material and the availability of plant nutrients to herbivores (digestibility-reducing substances such as

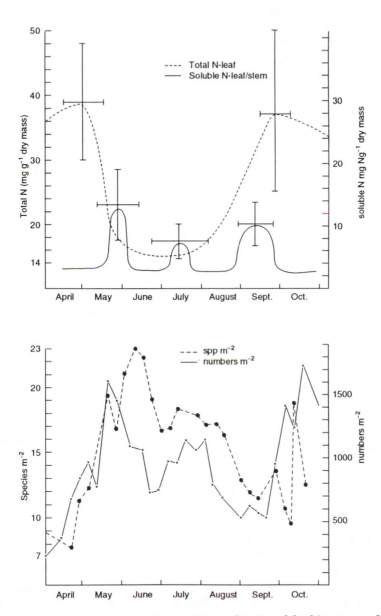

Figure 4.4 Species richness and population density of herbivores on *Holcus mollis* in relation to the nitrogen status of the host plant. The data are means over a period of 7 years; the bars on the N level diagram indicate standard errors. (From McNeill and Southwood, 1978.)

tannins and other polyphenols). It has been suggested that toxic allelo-chemicals are associated with ephemeral plants or plant parts that are unpredictable in occurrence and thus difficult to find by herbivores ('unapparent') whereas long-lived ('apparent') plants and mature plant tissues that are at high risk from herbivores tend to have higher con-centrations of digestibility-reducing substances (Rhoades and Cates, 1976; Feeny, 1976). Toxins are relatively 'cheap' to produce and are effective in relatively small concentrations against non-adapted enemies, but they confer no protection against adapted enemies, which may utilize them as attractants or phagostimulants (e.g. mustard oils in crucifers). Toxic allelochemicals are mostly associated with N-rich plant tissue (Mattson, 1980). Digestibility-reducing allelochemicals are said to be metabolically more costly to produce and are required in relatively high concentrations to be effective, but they are effective against all enemies to a degree by reducing the food available to them and hence their growth rate and fitness. These types of defensive chemicals (tannins and other phenols) are generally associated with tough plant tissues of high fibre content and low nutritive value. The characterization of allelochemicals as being either toxins or digestibility-reducing substances is, however, an over-simplification. Thus, tannins can be phagostimulants or feeding deter-rents for different insect species; or they may be stimulatory at lower concentrations and either have no effect or be deterrents at high con-centrations (Bernays, Cooper Driver and Bilgener, 1989). There is no direct evidence that tannins affect digestion in insects, and they can cause toxic effects (gut lesions) in non-adapted insects.

Considerable variation in allelochemistry and in physical plant defen-ces may occur between leaves within plants, between seasons and between plants (Krischik and Denno, 1983; Jones, 1983). This hetero-geneity may have evolved at least partially in response to herbivore pressure and may be an effective deterrent to the build-up of large her-bivore concentrations.

The relative proportions of toxins and digestibility-reducing substances change with plant age. Young grass leaves of high nutritional value produce toxins that inhibit at least some herbivores. Bernays *et al.* (1974) fed five locust and grasshopper species (Acrididae) on a range of grasses and found that fewer seedling grasses were eaten than were maturer grasses. Time-lapse film studies with *Locusta migratoria* showed that less time was spent feeding on seedlings, and that locomotor activity and mortality levels were higher. Distastefulness appears to be associated with organic inhibitors such as alkaloids and steroids (Bernays and Chapman, 1974). Mature grasses, on the other hand, rely for their defence on digestibility-reducing substances, such as increased fibre content allied with tough, silicified tissues and low nutrient content.

During artificial selection for agronomically desirable traits many of the characteristics that confer natural resistance against pests – high fibre content, leaf toughness, low nutrient content, etc. – are eliminated or drastically reduced, so that modern high-yielding cultivars are often more prone to pest attack than their wild ancestors. One of the results of introducing European grass cultivars into Australia and New Zealand is that the indigenous grass grub *Costelytra zealandica* and a variety of scarabaeid beetles have become major pests of sown pastures (Davidson, Hilditch *et al.*, 1979; East, King and Watson, 1981). Larval growth rates of the scarabs *Rhopea*, *Anaplognathus* and *Sericesthis* were two to three times higher than usual when feeding on the roots of temperate grasses in a pot experiment (Davidson and Roberts, 1968), while improved pastures produce larger pupae and adults and increased fecundity compared with native grasslands.

4.2 NATURAL ENEMIES

Undisturbed grasslands support a wide range of predatory and parasitic species that are generally considered to have an important role in the natural regulation of invertebrate populations. However, there is a lack of critical information on this subject; much of the available information on the effectiveness of natural enemies relates to pest species in highly simplified crop monocultures and is of doubtful relevance for more complex natural systems. Nevertheless, some work on cereal aphid enemies does appear to offer useful pointers in relation to grassland and will be included in the following account. For present purposes, it is convenient to group natural enemies into oligophagous (quite specialized) and polyphagous (attacking a wide range of invertebrates). Vertebrate predators will be considered under a separate heading.

4.2.1 Specialized enemies

The most abundant and widespread of the specialized enemies are the parasitic Hymenoptera, and, in the case of aphids, syrphid larvae (Diptera), lacewing larvae (Chrysopidae, Neuroptera) and larval and adult ladybirds (Coccinellidae, Coleoptera). It is known that these enemies can have a considerable effect on cereal aphid populations in years when they are abundant (Vickerman and Wratten, 1979). For example, Sunderland and Vickerman (cited by Vickerman and Wratten, 1979) compared aphid populations in field cages designed to exclude predators such as ladybirds and syrphids with those in the open field. Aphid densities reached a peak of 15 per wheat ear on 17 June in the open field and then declined rapidly. In the cages, aphid populations reached a peak of 120 per ear on 24 June, remained at a high level until 1 July and then declined. They calculated that 84% of the apparent loss in the population in the field could be attributed to aphid-specific predators in that year (1976) when those predators were particularly abundant. However, populations of specialized predators in cereal fields are low in most years (Potts and Vickerman, 1974) or their effects may not be felt until too late in the season to avoid economic damage.

Some authors consider parasites to be of little importance in reducing aphid numbers in cereal fields. Latteur (1973) and Rautapää (1976) reported levels of parasitism usually less than 5% but Dean (1974) recorded peak levels of 49% and 39%, respectively, for *Metopolophium dirhodum* and *Sitobion avenae* in Britain in 1971. Dean (1974) considered parasites to be important in keeping aphid numbers low in 1971 and suggested that parasites are more likely to synchronize with and help to control aphids than are predators. It is likely that there is a greater degree of synchronization between aphids and their specialized enemies in the less-disturbed grassland habitats than in arable crops, and hence the role of such specialized enemies in natural population control may be correspondingly greater.

The impact of specialized enemies on other grassland herbivores is not clear. Waloff (1975) reported that some of the commoner leafhopper species in acid grassland in the UK are heavily parasitized. Thirty-six parasitic species were bred out of leafhopper hosts in captivity and a further 17 were collected in the field. *Psammotettix confinis* (Cicadellidae), the commonest leafhopper at Silwood Park, was the most heavily parasitized and nine species of Pipunculidae (Hymenoptera) and six of Dryinidae (Cyclorrhapha, Diptera) were bred from it. The greatest mortality occurred during the egg stage, and this was the key factor affecting population size. The overall percentage parasitism of cicadellids by Pipunculidae was 10–25% and this could reach 40% in some generations, while the percentage parasitism by Dryinidae rarely exceeded 20%. On the other hand, Delphacidae were attacked by only three species of parasitoids. Andrzejewska (1979a) likewise drew attention to the role of specialized parasites in causing leafhopper mortality during the early stages of the life cycle.

Three parasites – a mermithid worm and a tachinid fly internally and a second tachinid fly externally – were recorded from the garden chafer *Phyllopertha horticola* in the English Lake District (Milne, 1984). Up to 5% of larvae in culture were found to be infested with the mermithid, but Milne concluded that parasitism in the field is rare. On the other hand, Moore (1983) reported that over 50% of stem-boring dipterous larvae could be attacked by parasitic Hymenoptera in established ryegrass swards. The main parasite was the wingless ground-dwelling braconid wasp *Chasmodon apterus*, which was responsible for over 90% of larval parasitism.

4.2.2 Polyphagous predators

A wide range of polyphagous predators occurs in grassland, including predatory mites, spiders, harvestmen, ants, rove beetles and ground beetles, while in rough grassland earwigs may be abundant. Polyphagous predators are thought to be particularly important in the natural control of cereal aphids because, unlike specialized enemies, they persist in the crop during periods of low pest density and can prevent early season build-up of aphid numbers (Edwards, Sunderland and George, 1979; Sunderland and Vickerman, 1980; Sunderland, Fraser and Dixon, 1986; Chiverton, 1986).

Most grasslands support a rich and varied predatory mite fauna: some species such as the nematophagous *Alliphis halleri* have specialized feeding habits, but most feed on a variety of suitably sized prey. Mites of the family Bdellidae appear to exert a significant degree of natural control of the collembolan *Sminthurus viridis* (lucerne flea). The lucerne flea was first recorded in South Australia in 1884 and since then it has spread over much of the agricultural land in that area and has become an important pest of improved pasture and legume crops (Wallace, 1967). Its major predator, *Thoribdella* (*Bdellodes*) *lapidaria*, is now well established in southern Australia and appears to have a substantial impact on the density of the lucerne flea over much of its range. This impact is apparent in pastures where bdellid mite numbers were greatly reduced by DDT applications that had no apparent effect on the lucerne flea (Wallace, 1954). Flea numbers subsequently increased substantially in DDT-treated areas due presumably to predator destruction: no such increase occurred in areas where bdellid mites were absent (Wallace, 1967). Analysis of data over 8 years showed a highly significant negative relationship between *T. lapidaria* numbers 8–9 weeks previously and flea numbers (Figure 4.5). When predator density early in the season was greater than 20 m^{-2} this was sufficient to prevent outbreaks of *S. viridis* later in the season, but if predator density was less than 10 m^{-2} *S. viridis* numbers could increase to outbreak proportions.

Predatory ants are highly mobile and very active, with low production efficiency and high food requirements, and they can be among the most important of all invertebrate predators in undisturbed habitats when numbers are high (Petal, 1967). The impact of ants on invertebrate populations in an unmanaged *Stellario–Deschampisetum* meadow community near Warsaw, Poland, has been studied in some detail (Petal, 1972; Petal *et al.* 1971; Petal, 1980). *Myrmica* spp. were dominant, *Myrmica laevinodis* and *M. ruginodis* being most abundant. The average ant density was 225–300 m^{-2}, and by observing prey being brought to selected nests in 5-min time intervals it was calculated that 40–80 g fresh mass prey m^{-2} yr^{-1} was consumed. Predation was most intense in early summer

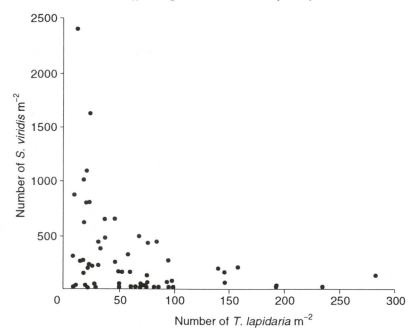

Figure 4.5 Relationship between the population density of the lucerne flea *Sminthurus viridis* and that of the mite *Thoribdella lapidaria* 8–9 weeks previously on pasture near Perth, South Australia. (From Wallace, 1967.)

from mid-May until mid-July, when there was a high requirement for dietary protein to feed young. Spiders (mainly migrating juveniles of the family Lycosidae), aphids and plant bugs, planthoppers and flies accounted for 78–88% of prey consumed by *M. laevinodis*. Spiders and flies were preyed upon more or less in proportion to their occurrence in the field, but consumption of planthoppers was always high even when their density was low. Kajak *et al.* (1972) calculated that ants killed (eliminated) 43% of the planthoppers, 32% of adult flies and 49% of spiders produced in the Polish meadow referred to above. Because they prey mainly on juvenile arthropods and on newly emerged adult flies, ants can have an important role in limiting the abundance levels of these invertebrates in meadows.

Spiders also have a major role as predators in natural habitats. Kajak (1971) estimated prey consumption by spiders on the basis of the numbers of prey caught in marked webs during measured time intervals and calculated that a spider population of mean density 52 m^{-2} in an unmanaged Polish meadow was capable of consuming the total annual production of Diptera and Auchenorrhyncha. Spider predation was most intense in August and September in contrast to ant predation, which was most intense in May–June: while ants eliminated 10–15% of flies in spring declining to 2% in July, spiders eliminated 30–50% of emerging flies in July–September (Kajak, Olechowicz and Petal, 1972). Flies are much more important prey for spiders than for ants, forming 70–80% of the diet of web spiders in Polish meadows compared with 5–12% for ants. The intensity of spider hunting for flies was correlated with the intensity of dipteran emergence, with peaks in spring and late summer. Spiders were the most abundant predators of leafhoppers in grassland at Silwood Park, England, greatly increasing in numbers and in importance as predators in late summer and autumn (Waloff, 1980). In all, 35 species of spiders were recorded as feeding on leafhoppers in the field.

Spider predation was considered to be insignificant in intensively managed grassland in Poland (Kajak, 1971); however, small species of the family Linyphiidae can be abundant in grass leys and arable crops and recent work has shown that these may have a sizeable impact on aphid populations (Sunderland, Fraser and Dixon, 1986). The inability of spiders to share a limited space with members of the same or other species probably makes them ineffective in controlling extreme prey densities associated with pest outbreaks, but they can have an important stabilizing effect in preventing insects attaining high densities (Riechert, 1974). Spiders are poor 'model' predators in the classical biological-control sense. They lack prey specificity, strong functional and numerical responses to increasing prey density, and the ability to 'track' population changes in specific prey species. Self-limiting competition maintains their density well below that dictated by the food supply. However, while individual taxa cannot effect biological control, Riechert and Lockey (1984) concluded that the diverse spider communities that are characteristic of most natural systems apparently can.

Among the 71 species of predators feeding on Auchenorrhyncha at Silwood Park were 9 species of Heteroptera (mainly Nabidae) and 10 species of Carabidae (Waloff, 1980), but the impact of these on prey populations was not estimated. Carabid and staphylinid beetles are believed to have an important role in the natural control of aphids and other pests in arable land (Sunderland and Vickerman, 1980; Luff, 1987); they have received less attention in grassland. As in the case of spiders, they are not good 'model' predators and can probably effectively regulate prey populations only within certain density limits. However, they undoubtedly contribute to the overall effectiveness of the polyphagous predator community in limiting species abundance and maintaining overall community diversity and stability in unmanaged grasslands. Under conditions of intensive management the role of polyphagous predators may be less important and the significance of specialized enemies may be relatively greater than in undisturbed habitats (Kajak, 1980).

4.2.3 Vertebrate predators

Insects constitute a major part of the diet of many species of birds. They are particularly important in the chick diet of the grey partridge (*Perdix perdix*), and the decline in numbers of this species over the past 30 years has been linked with declining insect food in English cereal farmland (O'Connor and Shrubb, 1986). Many bird species consume larvae of the garden chafer (*Phyllopertha horticola*), including rooks, jackdaws, starlings, oyster-catchers, green plovers, curlews and gulls (Milne, 1984). Above a certain density, voraciously feeding third-instar larvae of garden chafers cause wholesale severing of grass roots, resulting in yellowing and fading of the turf in affected patches. Birds, apparently attracted by the faded colour, turn over the loose turf and eat the larvae. However, Milne concluded from his observations in the English Lake District that only a very small fraction (0.01–3.4%) of the larval population was actually consumed. Predation by starlings (*Sturnus vulgaris*) has been reported to reduce high densities of grass grub (*Costelytra zealandica*) by up to 50% (East, 1972; East and Pottinger, 1975) in New Zealand in situations where starlings were abundant and where the sward was closely grazed, the soil was moist and larvae were plentiful enough to keep starlings in the area.

Bird predation can have a very significant impact on grasshopper numbers: Bock, Bock and Grant (1992) reported 2.2 times more adults and 3 times more nymphs in plots from which birds had been excluded for 3 years in semiarid grassland in Arizona. The rook (*Corvus frugilegus*) feeds on a wide range of grassland invertebrates including earthworms, leatherjackets and aerial insects (Murton and Westwood, 1977). Earthworms form an important component of the diet of many vertebrate species such as shrews (Soricidae), the European hedgehog (*Erinaceous europaeus*), the mole (*Talpa europaea*), which is a specialized worm feeder, the red fox (*Vulpes vulpes*), the European badger (*Meles meles*) and a wide variety of birds (MacDonald, 1983). Any predation will affect invertebrate numbers to some degree, but the effects of vertebrate predation on long-term population fluctuations are probably not great.

4.3 DISEASE

Insects are subject to various diseases caused by bacteria, fungi, protozoans, viruses and other organisms, but little is known about their influence on invertebrate populations in grassland. There are many pathogens associated with tipulid larvae that do not generally appear to influence population trends greatly (Carter, 1976; Pritchard, 1983). Tipula iridescent virus (TIV) has sometimes caused heavy larval mortality in France (Ricou, 1967) but it is sporadic in occurrence and Carter (1978) was not able to introduce high enough levels of infection in field trials to control populations in northern England. The Japanese beetle, *Popillia japonica*, has been controlled successfully by *Bacillus popilliae* in the USA, (Splittstoesser, 1981).

Three species of the fungus *Entomophthora* are commonly found attacking cereal aphids in northern Europe, and were considered to have killed more aphids than any other organism in 1970–71 (Dean, 1975). However, the influence of fungal pathogens is limited by the need for high atmospheric humidity for development and levels of infestation are not usually high until late in the season.

King and Mercer (1979) noted the effects of coccidian protozoans (*Adelina* sp.) on female black beetles (*Heteronychus arator*; *Scarabaeidae*) in New Zealand. Infected females had reduced fat-body levels with the result that few overwintering females survived to oviposit in the spring. They reported a decline of 66% in field populations of black beetle over three generations, during which time the level of *Adelina* infection in females in spring increased from 8% to 50%. Natural epizootics of nuclear polyhedrosis virus (NPV) occur in *porina* (*Wiseana* spp.; Hepialidae, Lepidoptera) in New Zealand (Kalmakoff, 1979), soil being the major reservoir of the virus. Older pastures appear to accumulate the virus and it is suggested that enzootic control of *Wiseana* could occur under those conditions. Pathogens such as milky disease (*Bacillus* sp.) and protozoans (*Nosema* spp., *Mattesia* sp.) can be significant causes of death in third-instar grass grub larvae in New Zealand and may influence populations in subsequent generations (Miln, 1979).

4.4 COMPETITION

Interspecific competition in field populations is difficult to demonstrate. Gibson (1976) studied the distribution and abundance of five species of Stenodemini (Heteroptera) and one species of Rhopalidae (Heteroptera) in relation to grass species in an area of limestone grassland near Oxford, England. The spectrum of food plants utilized by different bugs over-lapped considerably, but ecological separation was usually achieved by emergence timing and/or feeding on different parts of the same grass species. One pair of species (*Notostira elongata* and *Megaloceraea recticornis*) had almost complete overlap, and one species suffered very high mortality at the time of overlap, suggesting that the two species were competitors. Subsequent investigations (Gibson, 1980; Gibson and Visser, 1982) provided evidence that competition capable of causing mortality in the field did occur, but only rarely and under special circumstances, and involving only one of 15 possible pairs of species.

Another example of weak interspecific competition in the field involving garden chafer (*Phyllopertha horticola*) and the beetle *Dascillus cervinus* in English grassland was provided by Milne (1984) (see section

4.5.5(a)). Available data tend to suggest that interspecific competition is rare in nature, and that when it does occur it is usually transitory and its effects on population processes are relatively minor. However, this may not always be the case. Interspecific competition does not have to be continuous to be significant: competition for scarce resources at critical times of the year can have important effects on life-history parameters such as size and fecundity that can subsequently affect population dynamics. Hassall and Dangerfield (1989) concluded from their studies of isopod populations in heath grassland in East Anglia that interspecific competition for food between *Armadillidium vulgare* and *Porcellio scaber* juveniles could occur at a time when the young are going through a very important stage of growth.

Large numbers of arthropod species, and high population densities, are often found in dung pads in tropical and subtropical regions, suggesting intense competition (Doube, 1987). While both interspecific and intraspecific competition have been experimentally demonstrated in some situations (e.g. Giller and Doube, 1989), potentially competing species can often be separated spatially and temporally to varying degrees and the significance of competition in determining population abundance in any given situation can be difficult to assess. Competition among dung arthropods in northern temperate latitudes is probably of little importance. Interspecific competition between imported dung beetles and dung-breeding flies has been exploited for the biological control of *Musca vetustissima* and other nuisance flies in Australia (Waterhouse, 1974; Ridsdill-Smith and Matthiessen, 1988) (see Chapter 6 for further details).

Intraspecific competition in the form of larval combat has been shown to be a major cause of mortality among scarabaeid larvae at high density in pasture soils. This has been found to be the case for *Aphodius howitti* (Carne, 1956) and for grass grub (East, 1979a; East, King and Watson, 1981) in New Zealand, and for garden chafers in England (Milne, 1984) (see section 4.5.5).

4.5 NATURAL CONTROL OF POPULATIONS

The relative importance of various factors in determining the distribution and abundance of invertebrates has been the subject of considerable debate in the past, with much controversy surrounding the opposing views of Nicholson (1957, 1958) and Andrewartha and Birch (1954) concerning the role of density-related processes in stabilizing population density. Weather is a major factor in determining resource availability and its action tends to have a destabilizing effect on populations; however, Strong, Lawton and Southwood (1984) concluded from their review of published studies that most invertebrate herbivores are at least relatively deterministic, with fluctuations in abundance from year to year being within one to two orders of magnitude. This relative stability would suggest some degree of Nicholsonian regulation, although true density-dependent regulation may often be feeble or non-existent in field populations (Strong, Lawton and Southwood, 1984; Stiling, 1988).

Key-factor analysis based on the methods developed by Varley and Gradwell (1960) provides a simple technique that has become widely used for evaluating the contribution of different mortality factors to population change. Key factors are defined as those factors that cause greatest mortality and that largely determine population trends from year to year; namely, those factors that contribute most to perturbation of densities away from population equilibria (Stiling, 1988). Density-dependent factors, on the other hand, are those factors that tend to operate most heavily at higher densities and as population stabilizing mechanisms.

The relative importance of different factors in determining the abundance of selected grassland invertebrates that have been studied in sufficient detail are discussed below.

4.5.1 Leafhoppers (Auchenorrhyncha, Hemiptera) in acidic grassland

The population complex of leafhoppers associated with *Holcus* spp. in acidic grassland in England was studied for 5 years (Waloff and Thompson, 1980). Life tables were compiled for six species and key-factor analysis was carried out. In five cases, mortality during the egg stage was identified as the key factor (e.g. Figure 4.6). Weather, state of the food plants, natural enemies and dispersal were the factors influencing population level and there was no evidence for density-dependent processes. The authors suggest that the absence of any obvious density-dependent processes may be associated with the lability of the populations and their powers of dispersal.

4.5.2 Grass bug (*Leptopterna dolabrata*)

This was the commonest grass-feeding bug in acidic grassland at Silwood Park, Berkshire, in the south of England (McNeill, 1973). Populations were normally relatively low, reflecting the generally poor nutritional quality of the vegetation. Key-factor analysis revealed two mortality factors operating in a density-dependent manner that appeared to regulate population densities. These were late-instar nymphal mortality (instar III–IV) and variations in the potential fecundity of females from year to year. These factors appeared to be mediated by the availability of high N feeding sites at a time when the general level of N in the grass was low; in this case, therefore, intraspecific competition for high N food

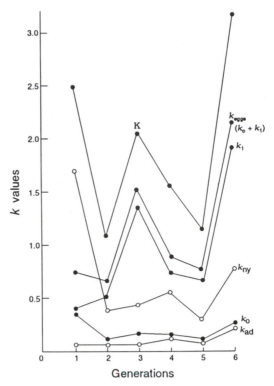

Figure 4.6 Key-factor analysis of mortality in six generations of *Errastumus ocellaris* (Hemiptera, Cicadellidae) in acidic grassland in England. K = total generation mortality; k_0 = reduction in fecundity from average maximum; k_1 = mortality of eggs; $k_{eggs} = k_0 + k_1$; k_{ny} = nymphal mortality; k_{ad} = mortality of adults due to parasitism. (From Waloff and Thompson, 1980.)

appeared to be the main factor regulating population abundance. The effect of natural enemies was not found to be important.

4.5.3 Cinnabar moth (*Tyria jacobaeae*)

The caterpillars of this species are commonly found feeding on ragwort (*Senecio jacobaea*), a widely distributed biennial weed of grassland that is particularly common in sand dunes and dry sandy soils with an open sward. Under such conditions, Dempster (1975) and Dempster and Lakhani (1979) reported density-dependent caterpillar mortality which they attributed to starvation (competition for food). Delayed density-dependent reduction in adult fecundity resulting from scarcity of food for larvae was also recorded. Pupal mortality and larval mortality due to starvation were the key factors in determining trends in numbers from

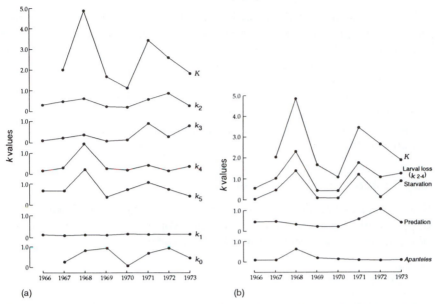

(a)   (b)

Figure 4.7 Key-factor analysis of mortality in the cinnabar moth (*Tyria jacobaeae*). (a) Mortality in all stages; (b) impact of starvation, predation and parasitism by *Apanteles* on larval mortality. K = total mortality; k_0 = failure to reach maximum fecundity; k_1 = egg mortality; k_2 = early larval mortality; k_3 = mid-larval mortality; k_4 = late larval mortality; k_5 = pupal mortality. (From Dempster, 1975.)

one year to the next (Figure 4.7). Intraspecific competition for food occurred only above a density of about 1000 larvae kg^{-1} ragwort. A more recent study by Crawley and Gillman (1989) in a dense, mesic grassland sward at Silwood Park, Berkshire, revealed no important differences in the population dynamics of cinnabar moth compared with the earlier studies. Populations fluctuated less at Silwood Park, but that was attributed to the reduced abundance of the host plant.

4.5.4 Grasshoppers and locusts (Acrididae)

Weather appears to play an extremely important part in determining the population dynamics of Acrididae (Dempster, 1963; Chapman, 1976). Moisture is required for egg development and green-food production, but wet weather is detrimental to nymphs and adults, which need warm dry sunny conditions. Fecundity is reduced by cool damp weather. Mosaic patterns of bare ground and sparse and dense vegetation providing the range of microclimates required by the different stages in the life cycle are optimum for the development of high-density populations. There are many diseases, parasites and predators associated with this group but the effects of these are usually insufficient to bring about density-dependent population regulation. Dempster (1963) concluded that numbers are kept generally low by unfavourable weather, while natural enemies and pathogens probably dampen peaks in population fluctuations. Where exceptionally favourable weather allows populations to rise emigration occurs, possibly due to intraspecific competition, bringing numbers down again. Particularly severe weather may cause local extinctions of populations at the edges of the species range and populations may become re-established through immigration in favourable years.

4.5.5 Pasture scarabs (Scarabaeidae)

Several species of scarabaeid beetles may cause a greater or lesser degree of pasture damage in various parts of the world. The population ecology of a few species has been studied in some detail.

(a) Garden chafer in England

One of the most long-term ecological studies ever conducted was that carried out by Milne (1984), who studied the garden chafer *Phyllopertha horticola* in the English Lake District over 29 years. During this period, population density was generally low, except for two occasions when populations reached outbreak levels (Figure 4.8). The main factors influencing population density were as follows:

Weather
Excessive rainfall and subzero temperatures can cause heavy mortality to soil stages. Late autumn frosts can catch third-instar larvae still feeding close to the soil surface before going deeper in the soil to hibernate. When other environmental factors (oxygen concentration, soil moisture and food) were adequate, larval growth rate depended directly on soil temperature during the 110-day feeding period and this determined pupal weight and fecundity. Excessive rain could reduce oxygen level enough to slow down development and fecundity in the next generation.

Migration
Female migration out of the area resulted in a net loss of eggs in the order of 4–5% or less each year, a relatively unimportant effect.

Disease
Two pathogens were of some significance: black disease, a bacterial disorder caused by *Micrococcus nigrofaciens* which causes affected larvae to turn black, and white disease caused by a granulosis virus which results in the body of the affected larva or pupa becoming encased in a relatively

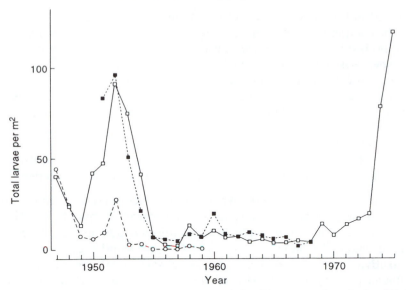

Figure 4.8 October population densities of third-instar garden chafer (*Phyllopertha horticola*) larvae in three grassland sites in the English Lake District. (After Milne, 1984.)

hard, chalk-white coating 1–2 mm thick. The incidence of lethal disease was far more common in wet habitats than in dry. Disease appeared to affect the population in an imperfectly density-dependent manner.

Parasites
Three parasites of the garden chafer were recorded – an unidentified mermithid worm and a tachinid fly (*Dexia rustica*) internally, and one external parasite, the larva of another tachinid fly (*Dexia vacua*). Only the mermithid worm and *Dexia vacua* were deemed to be of any significance; both were scarce and contributed little to population fluctuations and both were very imperfectly density-dependent. Parasitism was less important than disease in its influence on population trends.

Predators
The garden chafer was subject to a variety of predators, both invertebrate and vertebrate, including a range of insect-eating bird species, although no really substantial predation by any organism was found. Predation acted in an imperfectly density-dependent manner.

Interspecific competition
Three other species of chafer (the cockchafer *Melolontha melolontha*, the Welsh chafer *Hoplia philanthus* and the brown chafer *Serica brunnea*)

occurred in the same habitat but in low numbers, suggesting that interspecific competition between chafer species was likely to be negligible. A
dascillid beetle, *Dascillus cervinus*, was common in Lake District pastures.
This species is similar in size, life history and feeding habits to the garden
chafer. Only it or the chafer tended to be present in individual turf
samples, suggesting interspecific competition. However, interspecific
competition was considered to be weak and density-independent.

Intraspecific competition
Intraspecific competition at high population density manifested itself in
three ways, all of which were perfectly density-dependent in action:

1. Increasing larval mortality. When larvae touch there is a strong like-
   lihood of puncture wounds caused by snapping mandibles. These
   punctures almost invariably become infected and eventually prove
   fatal. Figure 4.9 shows a very close ($r = 0.97$) and highly significant
   positive correlation between mortality and initial egg density in a
   flower-pot experiment; 94% of the variation in mortality (r^2) was
   associated with density.
2. Decreased fecundity. The numbers of mature eggs formed depends on
   the amount of fat-body stored by the female third-instar larva before

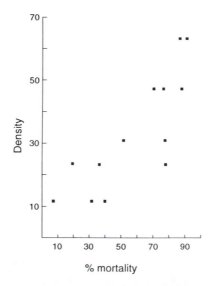

Figure 4.9 Relationship between garden chafer density (number of live eggs
introduced per pot) and mortality rate in a flowerpot experiment with *Agrostis–
Festuca* turf samples. Mortality was calculated from numbers of surviving third-
instar larvae. (From Milne, 1984.)

hibernation and is directly related to its weight. A strong negative correlation was found ($r = -0.86$, $r^2 = 0.74$) between average weight per female pupa and population density in the field, indicating reduced fecundity at high densities.

3. Decreased proportion of females. Field data showed a highly significant negative correlation ($r = -0.88$) between pupal density and the percentage occurrence of females in samples.

Milne concluded that for most of the time (when populations were low) natural control was due to the combined action of density-independent factors (weather and interspecific competition) and imperfectly density-dependent factors (pathogens, parasites and predators) but whenever this combined action fails, increase to the point of 'collective suicide' is prevented by the one factor that acts in a perfectly density-dependent manner – namely, intraspecific competition. On the other hand, decrease to extinction is prevented ultimately by density-independent factors alone.

(b) Pasture scarabs in Australia and New Zealand

The native grass grub (*Costelytra zealandica*) has become a major pest of sown grass-clover pastures in New Zealand and its population ecology has been studied by several authors (e.g. East, 1979a; East, King and Watson, 1981). Life-table studies indicate that population fluctuations are largely determined by density-independent mortality of first- and second-instar larvae in December–February. This is inversely related to summer soil moisture when below wilting point (Kain, 1975), probably reflecting desiccation, high temperature and starvation. Adult dispersal and activity of natural enemies were generally unimportant in their effects but protozoan disease may be significant (King and Mercer, 1979). When the weather is favourable and the population has increased, the upper limit of population growth is set by larval mortality arising from larval combat. This effect is strongly density-dependent above a certain threshold density, which is linearly related to pasture growth in autumn and winter (i.e. to food supply).

The black beetle (*Heteronychus arator*) is an African species that is confined to areas of New Zealand where the mean annual temperature equals or exceeds 12.8 °C (Watson, 1979). Outbreaks occur sporadically and are associated with above-average spring temperatures. Flight, oviposition, feeding, egg and larval development cease when air temperatures decline below 15–17 °C (King, Mercer and Meekings, 1981a). Life-table studies (Watson, 1979; Watson and Wrenn, 1980; King, Mercer and Meekings, 1981b) revealed a close association with grasses that are adapted to warm climates, notably paspalum (*Paspalum dilatatum*) and kikuyu (*Pinnesetum clandestinum*), which has a more restricted distribution. Pastures where either of these grasses was dominant provided favourable habitats throughout the year whereas pastures based on temperate species (e.g. ryegrass) were favourable only during spring and summer when larval development occurs. In paspalum-dominant pastures population fluctuations were largely determined by larval mortality between December and February and by variations in natality. Summer larval mortality was directly related to temperature in the previous spring, with earlier oviposition and more rapid development in warm springs resulting in greater survival over the summer. Unlike the grass grub, the adult black beetle is highly mobile and density-dependent dispersal of adult beetles in spring prior to oviposition could be a major cause of variation in natality. Populations and flight activity are high in warm springs, resulting in widespread colonization of temporary habitats such as ryegrass swards leading to pasture damage, while in cooler years populations tend to be confined to the more favourable habitats afforded by paspalum and kikuyu. The role of predators was not considered to be significant, but diseases caused by protozoans (Microsporidia and Coccidia) and viruses may be.

East, King and Watson (1981) successfully used simple models based on k values derived from life tables to predict the patterns of population change in grass grub and black beetle populations (Figure 4.10). In these models, population fluctuations were determined by physical factors (variations in soil moisture and temperature), with the upper limits of

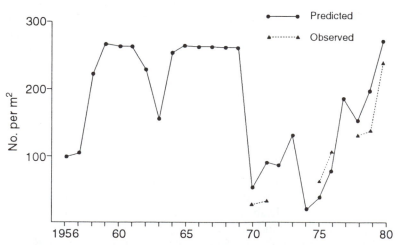

Figure 4.10 Predicted and observed autumn populations of grass grub (*Costelytra zealandica*) larvae in sandy soil, the Waikato, New Zealand. (From East, King and Watson, 1981.)

population growth being set by density-dependent larval mortality in the case of grass grub and variation in natality in the case of black beetle. Parasites and predators appeared to be unimportant and it would be interesting to know. to what extent this might have been due to suppression of natural enemies by the widespread use of DDT and other organochlorine insecticides in the 1950s and 1960s.

Under the more extreme climatic regimes found over much of Australia, soil moisture and temperature may override density effects in determining the pattern of scarab population fluctuations (Davidson, Wiseman and Wolfe, 1970, 1972a,b).

Disease may have a relatively important effect in determining scarab population trends in some situations. Milne (1984) considers that pathogens such as milky disease (*Bacillus* sp.) and a few species of microsporidian protozoans (e.g. *Nosema* spp. and *Mattesia* sp.) that infest grass grubs can have a considerable influence on the slope and intercept of the density-dependent (larval combat) relationship, resulting in more effective compensation for density changes and a greater stability of the population at lower levels. Pathogens were a significant cause of death in late third-instar larvae and may be involved in determining fluctuations in populations of grass grubs in subsequent generations.

4.5.6 Porina (*Wiseana* spp.)

The *Wiseana* (Hepialidae, Lepidoptera) species complex contains some important pests of pasture in New Zealand. The population ecology of *Wiseana cervinata* was reviewed by Barlow, French and Pearson (1986). Graphical methods of key-factor analysis did not reveal any obvious key factors that influence the abundance of autumn larval populations; however, further analysis by the Manly (1977, 1979) method suggested that variation in survival of surface-dwelling eggs and young larval stages determines the densities of damaging autumn larvae in any one year. Survival of surface-dwelling eggs and young larvae was inversely related to summer temperature in summer-dry areas. Larval mortality varied widely depending on the incidence of the fungal pathogen *Metarhizium anisopliae* and a nucleopolyhedrosis virus. Viral infection could cause up to 90% mortality but did not appear to regulate populations from year to year. The authors concluded that populations of *W. cervinata* are regulated by almost perfect density-dependent variations in recruitment, possibly due to moth dispersal and immigration.

4.6 SUMMARY

The mechanisms that determine insect population densities, and particularly the relative importance of density-dependent and density-independent processes, have long been central and sometimes controversial topics in ecology. Weather, notably temperature and moisture, is undoubtedly of major importance in determining population density (1) by its direct effect on metabolic processes and growth rates of individuals and populations and (2) by its indirect effect on the food supply and the general conditions of the habitat. The effects of weather are most apparent under extreme climatic conditions and in areas with strongly seasonal climates. Weather acts in a density-independent manner, tending to have a destabilizing effect on populations.

Food supply, and especially food quality, is a major limiting factor for many invertebrates. The range of plants that can be exploited by herbivorous insects is severely limited by plant physical and chemical characteristics. Most insects are able to exploit only a limited range of plant species and may be restricted to particular parts of their host plants that may provide food of suitable quality only for very restricted periods of time.

Interspecific competition, when it occurs in grassland invertebrates, tends to be of the diffuse kind and does not generally appear to have a major impact on population trends. By contrast, intraspecific competition at high population density is the factor that will ultimately control population increase when all other factors fail to do so.

Natural enemies (predators and parasites) and disease often cause high levels of mortality but the degree to which they are responsible for regulating invertebrate population densities in grassland is open to question. The indications are that enemies can exert a significant degree of natural control only within particular population ranges. Interaction effects between food quality and natural enemies may be important in this context (McNeill and Southwood, 1978). Relatively modest changes in birth and survival rates or in generation time induced by small nutritional differences in host plants could lead to changes in the rate of increase in herbivore populations, which could carry them into or out of range of control by their predators or parasites; thus small nutritional differences may cause large changes in equilibrium population levels.

The relative importance of different factors in influencing insect abundance was assessed by Stiling (1988) based on published data from 63 studies involving 58 insect populations. He examined the frequency of population processes acting as key factors (i.e. factors that contribute most to perturbation of densities away from population equilibria) or as density-dependent factors (i.e. factors that operate as negative feedback mechanisms, bringing the population back towards an equilibrium). Overall, no single process could be regarded as a key factor of overriding importance, different sources of mortality being important in different systems. The most common key factors were those acting from the next trophic level up – predators, parasites and viral and fungal infections.

Dispersal, migration (especially by adults) and reduced fecundity also figured prominently as density-perturbing factors. In about half the studies there were no obvious density-dependent factors operating, suggesting that many populations commonly fluctuate between an upper ceiling and a lower floor rather than being maintained around a more fixed equilibrium. When density dependence was detected it was more likely to be due to a source of mortality acting from the trophic level below (the host plant) than from the trophic level above. Any given type of mortality or population process was more likely to perturb densities away from a mean level than to be density dependent, acting to return populations to equilibrium. However, Stiling's conclusions in regard to the incidence of density dependence have been disputed by Hassell, Latto and May (1989). On re-examining the same data they found that the proportion of studies in which density dependence was detected increased markedly with the number of generations available for analysis, suggesting that many studies may be too short to allow reliable conclusions to be drawn about the occurrence of density dependence.

The concept of 'spreading of risk' appears to be relevant to the interpretation of the population dynamics of at least some invertebrate groups (den Boer, 1981). Den Boer (1985, 1986) concluded from a study of carabid beetles over a long period in the Netherlands that populations are subject to continuous local extinctions, and population persistence results from dispersal and refounding of these local populations. Survival of the overall population depends on the extent to which local populations fluctuate out of phase, leading to spreading of risk of extinction.

5

Influence of management on the grassland fauna

5.1 GENERAL RESPONSES TO MANAGEMENT

The aim of grassland management is usually to increase economic productivity, so increasing the intensity of management is normally accompanied by a dramatic increase in net primary production (NPP) – up to a point – with a corresponding increase in the efficiency of utilization. Under extensive grazing management in semiarid short-grass prairie, for example, cattle may assimilate less than 5% of NPP and as much as 75% may return to the soil without being utilized (Coleman *et al.*, 1976). At the other extreme, under conditions of intensive grazing or mowing as much as 90% of shoot production may be ingested by grazing animals or harvested (Andrzejewska and Gyllenberg, 1980; Hutchinson and King, 1980). However, an important difference between the latter two methods of utilization from the point of view of the organic matter cycle and the invertebrate decomposer fauna is that in the case of grazing systems there is a high return of ingested plant material (up to 60%) to the soil in the form of dung (Figure 5.1). This does not occur in the case of conserved forage removed for animal feed, unless the organic matter cycle is deliberately closed by returning dung from animal feeding yards to the pasture.

Changes in NPP and in the magnitude and rates of energy and nutrient fluxes in the managed grassland ecosystem undoubtedly have major effects on the biotic community. Some of these effects such as changes in floristic composition have received considerable attention; others are less well understood. Management may affect the invertebrate fauna indirectly by altering the extent and quality of the food supply and directly by drastically altering the habitat.

Certain general responses of invertebrate communities to increasing the intensity of management can be identified. While the invertebrate communities of extensively managed grasslands tend to be complex, reflecting the varied habitat provided by a complex vegetational structure and a

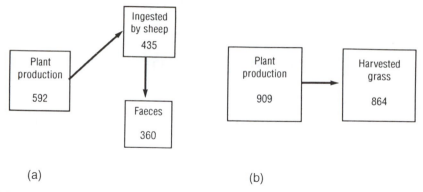

(a)   (b)

Figure 5.1 Production and utilization of plant biomass (g dry mass m^{-2}) in Polish grasslands: (a) sheep pasture; (b) fertilized hay meadow. (After Andrzejewska and Gyllenberg, 1980.)

well-developed surface mat of decaying plant material, the communities of intensively managed grasslands tend to be simpler in structure and to be dominated by species that can tolerate disturbance and are adapted to exploit the greater productivity of managed swards. These trends are illustrated by an inverse relationship between leafhopper species diversity and the intensity of management in Polish meadows (Figure 5.2); some species of leafhoppers and grasshoppers were favoured by man-

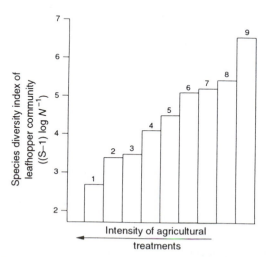

Figure 5.2 Relationship between leafhopper species diversity and the intensity of management in Polish grasslands. 1–5, Managed meadows; 6–9, unmanaged meadows. (From Andrzejewska, 1979a.)

agement (Andrzejewska, 1979a). Parallel trends may be noted in decomposer communities: for example, microarthropods tend to attain greatest density and species diversity in old pastures with a well-developed litter layer. Curry (1969a) recorded 220 000 microarthropods m^{-2} and 200 species in *Festuca–Agrostis* grassland lightly grazed by cattle and deer while a ryegrass sward intensively managed for silage production on a similar soil type had only 16 000 microarthropods m^{-2} and 100 species (Curry *et al.*, 1980). On the other hand, earthworms, and particularly larger species such as *Lumbricus terrestris*, appear to benefit from the greater fertility and productivity associated with intensive management. Mean earthworm biomass was 162 g m^{-2} in the *Lolium* grassland referred to above, compared with less than 90 g m^{-2} in the *Festuca–Agrostis* site. However, a moderate level of mangement can promote biological diversity and indeed is essential to maintain semi-natural grassland. For example, grazing or judiciously timed cutting is necessary to preserve the characteristic fauna of chalk grassland (Morris, 1971a).

The management practices that most influence the grassland community include reseeding, grazing, mowing, fertilizer application, burning, soil water control and pesticide use; the influence of these practices on the fauna are considered below.

5.2 GRAZING

Grazing can influence grassland invertebrates by altering the botanical composition and structure of the sward and by altering the nature and rate of organic matter return to the soil. Effects on populations occupying the vegetation and litter are likely to be most marked, but soil populations may also be affected by changed environmental conditions resulting from trampling and removal of vegetation and litter.

5.2.1 Floristic changes

The importance of grazing in grassland evolution and in maintaining seminatural, plagioclimax grasslands has already been referred to in Chapter 1. Reversion to coarse grasses and scrub was noted in chalk grassland in England when grazing pressure was reduced after the rabbit population declined because of an outbreak of myxomatosis in 1954 (Wells, 1971). Moderately grazed old grasslands frequently have high floral diversity that may decline when grazing pressure is reduced, while intensive management usually, but not invariably, tends to bring about a reduction in species diversity (Harper, 1969). Floristic changes depend on such factors as the degree of resistance of plants to grazing, the degree of selectivity by grazing animals, the frequency of defoliation, the level of plant nutrients and climate. Floristic change associated with grazing is a fairly long-term process and from the point of view of invertebrates the effects are less apparent than those associated with the physical disruption of the habitat caused by defoliation and trampling and the deposition of dung. In contrast to permanent pasture, short-term grass–legume leys have low floral diversity, comprising a small number of cultivars capable of responding to high levels of fertilizer input and frequent defoliation by grazing or cutting; such leys have an impoverished fauna compared with old grassland.

Grazing effects on invertebrates mediated through changes in floral composition are more likely to be seen in the cases of oligophagous herbivores. This is well illustrated by the example of the cinnabar moth (*Tyria jacobaeae*) (Dempster, 1971). The host plant of this insect is usually ragwort (*Senecio jacobaea*) and occasionally groundsel (*Senecio vulgaris*). Ragwort cannot become established in a closed grass sward; it is favoured by overgrazing where bare patches in the sward allow seedlings to become established. When ragwort is abundant, high densities of the moth occur but the population tends to fluctuate violently and the moth may fail to survive periodic crashes due to starvation. Undergrazing, on the other hand, reduces the availability of food and creates conditions that are favourable for natural enemies, thus greatly increasing mortality; under such conditions the moth may also fail to persist.

5.2.2 Changes in habitat structure

Selective grazing of more palatable species can result in a structurally heterogenous sward, with heavily grazed areas interspersed with clumps of ungrazed vegetation. An extreme form of this is seen in tussock formation, with consequences for the distribution and abundance of surface fauna (Chapter 1). This situation is most likely to develop under conditions of uncontrolled, low-intensity grazing of rough pasture; where grazing is controlled, with periods of intensive grazing followed by recovery periods, the result is likely to be a greater degree of uniformity within the sward.

An important effect of grazing is to reduce the size and complexity of the habitat available to above-ground invertebrates. Most grassland invertebrate species are associated with the inflorescences of flowering plants (Evans and Murdoch, 1968) and specialist endophagous species in particular are likely to be severely depleted in closely grazed swards. Morris (1967) reported marked increases in populations of the seed weevils *Apion loti* and *Miarus campanulae* in ungrazed chalk grassland plots compared with grazed plots in Britain. The increase was in response to increased numbers of flowers and fruits of bird's foot trefoil (*Lotus corniculatus*) and harebell (*Campanula rotundifolia*), their respective larval foodplants. Also, knapweed (*Centaura nigra*) on chalk supports a number of more or less specialized invertebrate feeders, but these are virtually absent from the rosettes in intensively grazed grassland. Reduction in sward complexity associated with grazing has a pronounced effect on numbers of spiders, which require a complex architecture for web attachment: Cherrett (1964) recorded twice the spider numbers from moorland grassland at Moor House, England, when sheep were excluded.

There appears to be a general relationship between plant biomass and invertebrate biomass in grasslands, especially those that are managed similarly and are in the same climatic zone (Andrzejewska, 1979a), and many studies have reported reductions in above-ground and surface invertebrates in heavily grazed swards (Morris, 1967, 1969, 1971b, 1973; Andrzejewska, 1979a; Purvis and Curry, 1981). Figure 5.3 compares the log abundance of the range of taxa (206 species in all) trapped by suction sampling from grazed ryegrass/clover ley plots compared with similar conserved plots in Ireland (Purvis and Curry, 1981). The two sets of plots had been treated similarly early in the season, both being conserved for silage until the end of June when the herbage was cut; thereafter, the treatments diverged. Figure 5.4 (h) shows the degree of dissimilarity between these two management types throughout the year based on the dissimilarity index (D)

$$D = \frac{\Sigma \mid a-b \mid}{n},$$

where $\mid a-b \mid$ is the difference in log abundance of each taxon between the two treatments and n is the total number of taxa. The two sets of plots were very similar early in the year while they were receiving similar management, but were markedly dissimilar later in the year under contrasting management regimes.

In a series of papers, Morris (1967, 1968, 1969, 1971b, 1973) reported on the responses of the invertebrate fauna of chalk grassland on the Barton Hills, Bedfordshire in southern England to cessation of grazing by comparing the faunas of intensively grazed plots with those of plots fenced off to exclude rabbits and sheep. A general increase in numbers of most invertebrate groups living in the vegetation layer and in the turf (superficial soil and surface mat) was recorded in the enclosures. Some specialized seed weevils responded very dramatically to increased abundance of flowers of their host plants, as already noted. Among the herbivorous species, only one, *Chorthippus brunneus*, a grasshopper that prefers short grass, did not benefit from the cessation of grazing. Orthoptera appear to oviposit in compacted soil and are favoured by stock trampling (Richards and Waloff, 1954; Dempster, 1963). The

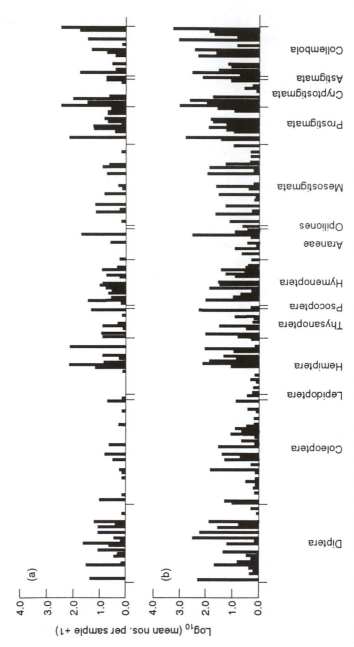

Figure 5.3 Mean log abundances of arthropod species in ryegrass/clover plots (a) conserved for silage until June, regrowth grazed by sheep; (b) conserved for silage, cut in June and August or September. (From Purvis and Curry, 1981.)

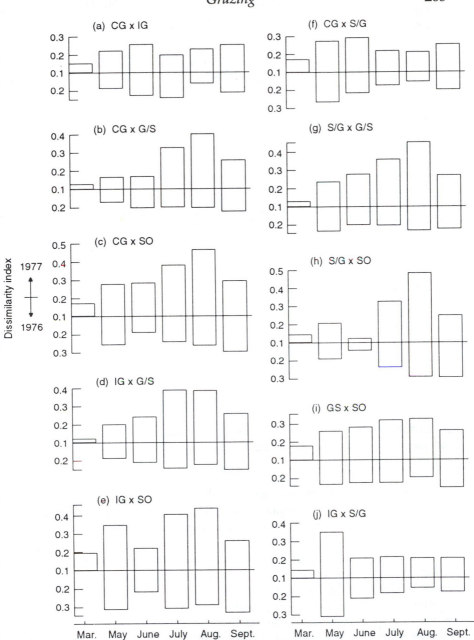

Figure 5.4 Indices of faunal dissimilarity (see section 5.2.2) between plots receiving different management treatments. CG = continuous grazing; IG = intermittent heavy grazing; S/G = conserved for silage cut in June, regrowth grazed; G/S = grazed early in the season and conserved for silage from June; SO = conserved for two silage cuts. (From Purvis and Curry, 1981.)

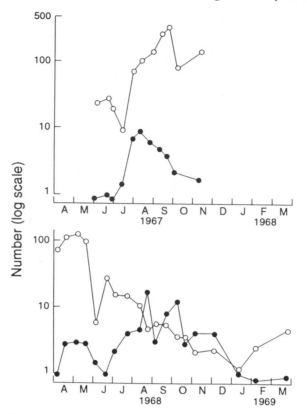

Figure 5.5   Mean numbers of adult lacebugs (*Agramma laeta*) per 2 m^{-2} in grazed (●) and ungrazed (○) chalk grassland. (From Morris, 1969.)

intensively grazed plots had low numbers of Heteroptera and most species showed a positive response to the exclusion of grazing animals. A few species, notably the lacebug *Agramma laeta* (Tingidae), the most abundant species recorded, increased markedly in numbers for the first few years of exclosure, but then declined (Figure 5.5). This species was observed to feed on a sedge, *Carex flacca*, which apparently undergoes changes in the absence of grazing that render it unsuitable as food for *Agramma laeta*. Another lacebug, *Acalypta parvula*, responded in a similar manner. It appears to be a moss feeder, and its later decline in numbers may be related to deterioration of the habitat caused by increased growth of grasses and build-up of dead plant material and litter. Most leafhopper species (Auchenorrhyncha)were also more abundant in ungrazed grass-land, with the exception of *Macrosteles laevis* and, to a lesser extent, *Psammotettix cephalotes*, which were more abundant in the grazed areas. *Macrosteles laevis* is a characteristic species of disturbed grassland habitats, with the ability to recolonize mown meadows very rapidly (Andrzejewska, 1962).

Adverse effects of heavy grazing on invertebrate populations may not always be manifested in terms of reduced population density. Al Dabbagh and Block (1981) actually recorded higher densities of the isopod *Armadillidium vulgare* in grass heath heavily grazed by rabbits (mean 500 m^{-2}, maximum 900 m^{-2}) compared with those in a lightly grazed, tussocky site (mean *c.* 400 m^{-2}). However, there were differences between the two sites in terms of age structure, generation distribution and cohort composition, which led the authors to conclude that the population of the heavily grazed site was unstable, with annual changes in density and age structure, while that of the lightly grazed site was more stable, with similar recruitment from year to year.

5.2.3 Herbage utilization

The ways in which grazing intensity can alter herbage utilization within sheep pasture are illustrated by data from Armidale, New South Wales, Australia, where the effects of sheep stocking levels on the soil-invertebrate community and on the partitioning of energy from shoot production between sheep, invertebrate herbivores and decomposers were studied (Hutchinson and King, 1980) (see Figure 6.4). The proportion of shoot production ingested by sheep at the lowest stocking density was only 59% and this increased to 94% at the highest stocking density and at the same time the proportion ingested by invertebrate herbivores declined from 14.5% to 4.2%. The proportion of shoot production unutilized by herbivores and consumed by invertebrate decomposers declined from 27% at 10 sheep ha^{-1} to less than 2% at 30 sheep ha^{-1}, and there was a concomitant increase from 22% to 35% in the proportion of plant material returned to the soil via sheep faeces. Thus, with increasing grazing pressure, there was a marked decrease in energy consumption by invertebrate herbivores, a less-marked decline in the overall quantity of matter entering the decomposer food chain but a pronounced decline in the relative importance of plant-litter decomposition with a parallel increase in the vertebrate faecal pathway for litter return. Dung on the surface provides an important, if transient, habitat and food source for invertebrates, and invertebrate activity makes an important contribution to dung decomposition (Chapter 6).

5.2.4 Effects on soil invertebrates

Although the soil-invertebrate community does not experience the same degree of disruption associated with defoliation and trampling as that experienced by the more exposed herbage and surface fauna, nevertheless grazing intensity alters some important parameters of the soil-litter environment that affect invertebrates. Reduction in surface litter and vegetation cover, for example, can drastically alter the soil microclimate. Thus, Davidson, Hilditch *et al.*, (1979) recorded a maximum soil temperature of 28 °C in lightly grazed pasture in New South Wales compared with 42 °C in heavily grazed areas. This greater temperature variation in heavily grazed pasture was responsible for high mortality in scarab larvae and, presumably, in other soil invertebrates also.

Adverse effects of heavy grazing on soil fauna are likely to be experienced most acutely by hemiedaphic invertebrates living in the litter and superficial soil layer; this was generally the case for macroarthropods in chalk grassland in England (Morris, 1968) and in sheep pasture at Armidale, New South Wales (Hutchinson and King, 1980). Exceptions include certain species of Homoptera (Hutchinson and King, 1980) and Acrididae (Knutson and Campbell, 1976), which appear to prefer sparsely covered ground conditions. Mesofauna (Acari, Collembola and Enchytraeidae) showed general trends of declining density with increasing sheep-stocking rates at Armidale (King and Hutchinson, 1976). In the case of collembolans, the greatest change in abundance occurred in surface-dwelling species; stocking density had virtually no effect on species living at a greater depth in the profile. Hutchinson and King (1980) noted that the two main groups of euedaphic soil invertebrates at Armidale – root-feeding scarabaeid beetles and larger oligochaetes – reached maximum densities at an intermediate stocking rate (20 sheep ha^{-1}), the stocking level at which primary productivity of the pasture was also at a maximum. Other studies confirm that earthworms are favoured by high pasture fertility and productivity and rapid organic matter turnover (Waters, 1955; Cotton and Curry, 1980a,b), whereas many other grassland invertebrate groups such as collembolans and oribatid mites reach greatest abundance under conditions of slow turnover of organic matter and litter accumulation.

5.2.5 Effects on grassland pests

The relationships between grazing intensity and the population densities of certain soil-invertebrate pest species have been reviewed by East and Pottinger (1983), who recognized three types of responses. They concluded that increasing the stocking rate of grazing animals reduces the populations of most pasture invertebrates (type I response), although populations of some species may reach a peak at intermediate stocking levels (type II response) or increase up to high stocking rates (type III response). Most of these responses could probably be attributed to direct mortality of foliage and surface-dwelling invertebrates resulting from trampling, or to the indirect effects of grazing animals on living space, microclimate and food supply for the invertebrates living in the soil and litter. Insect pest species that show a type I response include porina caterpillars (*Wiseana* spp.), the Australian soldier fly (*Inopus rubriceps*) and grass grub (*Costelytra zealandica*) in New Zealand. Studies in the Waikato region of New Zealand showed that grass grub populations declined from high to low levels as stocking rate increased over the range 9 to 20–25 stock units per ha (Kain and Atkinson, 1970; Dixon and Campbell, 1978; East, 1979b).

Some scarabaeid larvae such as *Sericesthis geminata* in Australia have type II responses. Roberts and Morton (1985) reported a significant quadratic relationship between total scarabaeid biomass (in logarithmic units) and grazing pressure from sheep in temperate sown pasture at Armidale, New South Wales. Biomass peaked at about 12 ewes ha^{-1} but showed little change over the range 10–15 ewes ha^{-1}. The dominant genera, *Anaplognathus* and *Sericesthis*, both showed this response (Figure 5.6). Seastedt, Ramundo and Hayes (1988) offer additional evidence from laboratory and field studies that moderate grazing (or clipping) of foliage

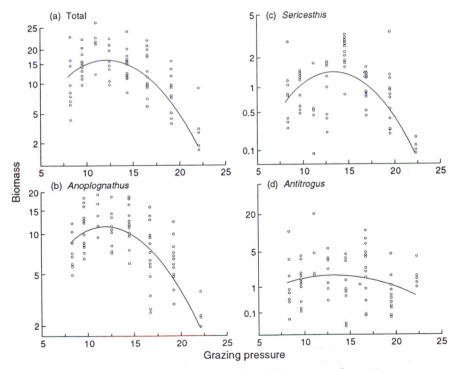

Figure 5.6 Relationships between scarabaeid biomass and grazing pressure (sheep ha^{-1}) at Armidale, New South Wales, Australia. (After Roberts and Morton, 1985.)

often increases densities or biomass of below-ground herbivores and detritivores in prairie soils in the USA. This response occurs in spite of a neutral or reduced growth response of roots to foliage removed, and the authors suggest causal mechanisms involving increased root quality (N content) and changes in assimilation efficiency.

Examples of invertebrate herbivores with a type III response include frit fly, *Oscinella frit*, which prefers short swards (Southwood and Jepson, 1962; Purvis and Curry, 1981), grasshoppers (Richards and Waloff, 1954; Dempster, 1963) and some ant species (Hutchinson and King, 1980).

There is considerable interest in the potential for controlling soil pests by manipulating stock density, particularly in Australia and New Zealand, and this topic will be dealt with at greater length in Chapter 6. Prospects would appear to be most favourable for the control of species with type I and, to a lesser extent, type II responses by increasing grazing intensity and/or by 'mob stocking' at critical periods in the life cycle. However, in practice, the use of grazing animals to control pests may be limited by such factors as possible damage to the sward and the weather.

5.3 MOWING

The effects of mowing are similar to those of grazing in that defoliation drastically reduces the size and complexity of the above-ground habitat, prolongs the vegetative phase of grass growth and increases primary production, up to a point. Mowing differs in several important ways. It is non-selective and therefore the sward heterogeneity associated with grazing is absent and the physical disruption associated with trampling does not occur. Under agricultural management most of the shoot production is removed from the site, with little return of organic matter to the soil by way of plant litter or dung (Figure 5.1), but the intervals between successive defoliations are normally longer than those experienced under intensive grazing and significant invertebrate populations can develop during these recovery periods (Curry and Tuohy, 1978; Curry and O'Neill, 1979; Purvis and Curry, 1981).

The response of the invertebrate fauna to mowing will vary depending on the intensity of grassland utilization and on the frequency and timing of mowing. Seminatural grasslands require some minimal level of management for their maintenance, and judiciously timed annual cutting can be an effective alternative to grazing for maintaining floral and faunal diversity in areas of conservation interest (Morris, 1971a, 1979; Wells, 1971). However, most invertebrate species are adversely affected if mowing occurs during their most susceptible stages of development. Grasslands that are intensively managed for hay and silage production normally have impoverished faunas relative to those of seminatural habitats, but the species present are better adapted to disturbance and populations recover rapidly after cutting. In view of those different responses, the effects of mowing on the fauna of seminatural grasslands and of intensively managed agricultural grasslands will be considered separately.

5.3.1 Seminatural grasslands

Abundance, species richness and diversity of leafhoppers (Auchenor-rhyncha) and Heteroptera in chalk grassland in England were consistently reduced by cutting in July (Morris and Lakhani 1979), while the effects of cutting in May were much less, leading the authors to conclude that cutting in May was a good method of maintaining high richness and diversity of Hemiptera in chalk grassland. The leafhopper community tends to be stratified vertically and tall grasslands tend to support most species (Andrzejewska, 1965); this group could therefore be expected to be particularly adversely affected by cutting. Of 71 leafhopper species recorded from the study plots on chalk grassland, 23 were adversely affected by cutting, while 14 of the 42 heteroptera species were reduced in numbers by the same treatment (Morris, 1979, 1981a). Most species were not reduced by cutting in May, with the exception of a few whose adults emerge early in summer, and the effects of the May cut were generally short-lived. Cutting in July had more severe effects that often persisted into the next spring. Most leafhoppers become adult in late summer (July–September) and are particularly affected by cutting in July, but some are characteristic of short grassland and eight species were found to increase in abundance in cut plots of chalk grassland (Morris, 1981b); these included *Adarrus ocellaris*, which was scarce in unmanaged grassland and increased progressively in cut plots over three successive years. *Macrosteles laevis* also responded positively to cutting. Again, the timing of cutting was of significance: for example, *Adarrus ocellaris* responded more positively to cutting in May and July than to cutting in July alone, while several species increased more in plots cut only in July than in plots cut in both May and July.

When limestone grassland plots that had been cut during three successive seasons (1973–75) were allowed to remain uncut there was a progressive increase in mean number of individuals (N), species richness (S), and diversity (D) for both Heteroptera and Auchenorrhyncha during the following three seasons (1976–78) (Figure 5.7). The mean rates of increase from 1975 to 1978 were much greater in plots cut in July and in both May and July than in those cut in May only. Also, S and D, but not N, were higher on previously cut plots than on unmanaged control plots in the case of leafhoppers, suggesting a positive effect of rejuvenation on this group after cutting.

Effects of cutting on the coleopteran fauna of limestone grassland were

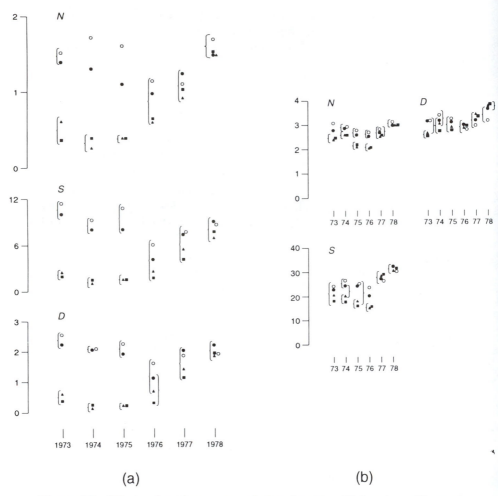

Figure 5.7 Effects of cutting on population densities (N, log ($n + 1$)), species richness (S) and diversity (D, Brillouin index) of (a) Heteroptera and (b) Auchenorrhyncha in limestone grassland. Values not significantly different are bracketed. ●, Cut in May; ▲, cut in July; ■, cut in both May and July; ○, uncut control. (From Morris and Plant, 1983.)

less marked than those on Hemiptera, probably because of the lack of marked vertical stratification of the beetle fauna in grassland (Morris and Rispin, 1987). No effects of cutting on overall population densities were apparent although significant treatment-related differences in species richness, diversity and evenness were noted. Significantly more species and individuals of predacious and saprophagous families were recorded on uncut plots, but no significant differences were recorded for phytophagous species.

5.3.2 Managed grassland

The response of the leafhopper population to repeated mowing of a fertilized Polish meadow is illustrated in Figure 5.8. Mowing was always followed by a decrease in population density, followed by a rapid increase in numbers with vegetation regrowth. The recovery in numbers was attributed to colonization by immigrants, by local animals that survived mowing, and by newly hatched animals (Andrzejewska, 1979a). Generally, the average density of phytophagous insects is higher in mown than in unexploited meadows, and this can be attributed to the higher nutritive value of plants in the mown meadow over the growing season, resulting from the high growth rate and the slowing down of the process of maturation and plant death. The dominance structure of leafhopper communities in managed meadows differs from that in natural meadows in that managed meadows have a much higher proportion of invasive *r*-selected species – eurytopic species with high fecundity and the ability to cover large distances – such as *Macrosteles laevis, Javesella pellucida and Streptanus aemulans*. Generally, agricultural management for grazing or cutting reduces the biomass of polyphagous predators including spiders (Kajak, 1980) and ants (Petal *et al.*, 1971).

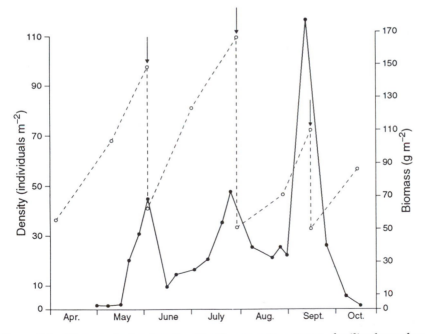

Figure 5.8 Changes in the density of leafhoppers in a mown, fertilized meadow in Poland. Arrows indicate dates of mowing; - - -, plant biomass; ————, leafhopper density. (From Andrzejewska, 1979a.)

Studies on the fauna of grassland swards cut two or three times per year for silage at Grange in Co. Meath, Ireland, confirmed that the surface-layer fauna are markedly depleted by cutting and removal of the herbage, the overall numerical reduction being in the order of 75% (Table 5.1). However, populations recovered rapidly between cuts, the dominant taxa–prostigmatic and acarid mites, aphids, thrips, isotomid and sminthurid collembolans and flies – being well adapted to disturbed conditions. In a comparative study at Celbridge, Co. Kildare, of the effects of various management systems – continuous sheep grazing, heavy intermittent sheep grazing, early-season grazing followed by late-season conservation for silage, and early-season conservation followed by late-season grazing – the factor apparently influencing the fauna most was the length of the sward. Populations tended to be low when the grass was short and to increase when the grass was allowed to grow. Comparisons between treatments throughout the season revealed greatest faunal similarity when treatments were most similar and greatest dissimilarity when treatments differed most in terms of their effects on the sward (Figure 5.4). Shoot-mining Chloropidae (Cyclorrhapha, Diptera) were the only group not to decline in numbers following cutting (Table 5.1). These insects (mainly *Oscinella frit*) were

Table 5.1 Immediate effects of cutting and removal of herbage on the fauna of old pasture at Grange, Co. Meath (after O'Neill, 1981 and Curry and Tuohy, 1978)

	Numbers of individuals (m^{-2})	
	Before cutting	After cutting
Acari		
Tyrophagus longior	15 704	3 504
Tarsonemidae	10 956	2 312
Pyemotidae	2 248	1 408
Araneae	43	2
Collembola	1 012	204
Coleoptera	24	4
Diptera		
Chloropidae	7	30
Opomyzidae	12	2
Others	108	16
Hemiptera		
Aphididae	116	3
Cicadellidae	7	1
Hymenoptera	33	8
Thysanoptera	240	24

more abundant in the stubble within hours of cutting than in the uncut sward at Grange, confirming the preference of frit fly for short swards.

A noteworthy feature of the study at Grange is how similar the faunas of four different grassland swards are in terms of species occurrence (Table 5.2) and population densities of the major groups, species richness and species diversity (Table 5.3), despite the marked botanical differences between the swards. Management appears to have had an overriding influence on the nature and composition of the fauna, favouring eurytopic, mobile species with short life cycles and high reproductive rates and adversely affecting less mobile species with longer life cycles and lower reproductive rates characteristic of undisturbed habitats.

5.3.3 Mowing and soil invertebrates

Soil invertebrates may respond differently to cutting in some respects than to grazing, but such differences are not well documented. On the positive side the absence of grazing animals would eliminate the direct mortality due to trampling that might occur under heavy stocking regimes, while the presence of long grass for significant periods of time might create more favourable microclimatic conditions in the soil than would occur under closely grazed swards. On the other hand, the absence of dung and the low level of shoot-litter return associated with repeated herbage cutting and removal could lead to a depletion of surface organic matter and litter-dwelling invertebrates. The scarcity of soil micro-

Table 5.2 Percentage similarity between grassland plots at Grange, Co. Meath (after O'Neill, 1981); A and B are replicate plots

		Red clover		Ryegrass		Ryegrass/ red clover		Old pasture	
		A	B	A	B	A	B	A	B
Red clover	A		68	63	76	57	73	61	60
	B	70		52	72	47	64	65	64
Ryegrass	A	73	70		68	81	76	62	62
	B	73	69	77		59	80	59	61
Ryegrass/red	A	76	71	80	78		63	60	63
clover	B	78	76	75	76	77		58	61
Old pasture	A	71	73	79	76	78	80		77
	B	72	75	80	75	76	80	83	

Values above the diagonal are based on the index of Whittaker (1975); those below the diagonal are based on the Sørensen (1948) coefficient.

Table 5.3 Population densities of major groups, species richness and diversity of the above-ground arthropod fauna in grassland plots at Grange, Co. Meath (after Curry and O'Neill, 1979); A and B are replicate plots

	Mean, numbers per m² (and numbers of species present)							
	Red clover		Ryegrass		Ryegrass/red clover		Old pasture	
	A	B	A	B	A	B	A	B
Acari*	183 (28)	142 (19)	237 (23)	237 (22)	275 (24)	168 (23)	451 (27)	296 (27)
Araneae	35 (6)	47 (5)	90 (9)	121 (6)	78 (9)	68 (10)	52 (9)	92 (9)
Collembola	1893 (18)	1248 (14)	1130 (16)	2547 (15)	1009 (13)	1708 (16)	1332 (15)	1090 (15)
Coleoptera	43 (24)	18 (16)	31 (17)	44 (20)	51 (16)	37 (19)	76 (19)	60 (19)
Diptera	178 (29)	150 (28)	131 (23)	120 (25)	181 (27)	151 (26)	218 (26)	260 (26)
Hemiptera	47 (10)	79 (6)	93 (9)	60 (11)	65 (10)	37 (9)	108 (10)	74 (10)
Hymenoptera	57 (10)	48 (12)	56 (13)	46 (12)	62 (13)	56 (12)	52 (13)	54 (13)
Thysanoptera	35 (6)	48 (5)	94 (5)	130 (5)	28 (4)	48 (5)	50 (5)	85 (5)
Diversity (D)†	0.85	0.88	0.86	0.86	0.85	0.89	0.89	0.91

*Only the larger species caught by suction sampling are included.
†Simpson, 1949.

arthropods in grass silage fields at Johnstown Castle, Co. Wexford, Ireland, would tend to support this view (section 5.1). However, there was no indication of adverse effects on earthworms that may benefit from the high level of soil fertility and of organic-matter turnover associated with intensive management.

5.4 FERTILIZERS

The effects of fertilizers on grassland invertebrates are largely a consequence of their effects on the vegetation. These include increased net primary production and changes in the sward composition, with an overall decrease in species diversity, a decline in the importance of many dicotyledon species and an increase in the proportion of some grass species in the sward (Rorison, 1971; van der Maarel, 1971; Traczyk, Traczyk and Pasternak, 1976). Increased nutrient content of the vegetation following fertilizer application can also significantly affect its quality as a food source for invertebrate herbivores (Andrzejewska, 1976a,b; Prestidge, 1982b).

The effects of organic and inorganic fertilizers in terms of nutrient enrichment may be comparable, but these two types of fertilizers differ in that organic forms provide additional food material for the decomposer community. For this reason, it is convenient to treat them separately when considering their influence on the fauna.

5.4.1 Mineral fertilizers

Increase in the nutrient content and food quality of vegetation is frequently reflected in greater fecundity, faster development and increased production and turnover of invertebrate herbivores (see Chapter 4). Andrzejewska (1976a) studied the effects of mineral fertilizers on the phytophagous fauna of a Polish tall oat-grass (*Arrhenatheretum*) meadow in a field plot experiment. Mineral fertilizers were applied at the rate of 350 kg N, 120 kg P and 200 kg K ha^{-1} yr^{-1} to some plots and these were compared with unfertilized control plots. Mean population densities were only slightly greater in fertilized plots, but higher numbers of insects – notably flies, aphids and leafhoppers – emerged in them. Table 5.4 shows the relationships in biomass terms between unfertilized and fertilized plots. These relationships reflected the differences in biomass of turf (0–5-cm vegetation layer) and tillering nodes. The tillering nodes in the upper 3–5-cm layer of soil and litter supported the greatest density of phytophagous invertebrates. This was the main hatching area and the initial dwelling place of the earlier larval stages of grasshoppers, leafhoppers and dipteran larvae, while a relatively high concentration of older larvae and imagines occurred in the turf zone and some species permanently

Table 5.4 Biomass of phytophagous invertebrates in unfertilized and fertilized grassland plots (after Andrzejewska, 1976a)

Group	Mean biomass (mg dry mass m^{-2})		Biomass of phytophagous insects that hatched and emerged (mg dry mass m^{-2})	
	−NPK	+NPK	−NPK	+NPK
Orthoptera	14.5	17.9	30.0	72.9
Homoptera				
Auchenorrhyncha				
(adults)	12.4	12.6	62.8	74.1
Auchenorrhyncha (larvae)	9.0	10.6	36.0	42.3
Aphidoidea	0.02	0.5	8.22	95.73
Lepidoptera (above-ground larvae)	1.2	0.5	27.6	63.6
Diptera (larvae)	116.8	190.4	584.0	952.0
Coleoptera (larvae)*	446.0	209.0	–	–
Lepidoptera (larvae in soil)	1 389.0	991.0	–	–
Nematoda	88.5	89.5	–	–

*Curculionidae, Scarabaeidae and Elateridae.

inhabited this stratum. On the other hand, mineral fertilizers caused a reduction in root biomass and there was a significant decrease in phytophagous animals including coleopterous and lepidopterous larvae in this stratum. Leafhoppers comprised 12–72% of the insect fauna in the vegetation layer, and their response to fertilizers was examined in some detail (Andrzejewska, 1976b).

Mean population densities were not significantly different in the two sets of plots but, overall, about 15% more larvae hatched out and more adults emerged in the fertilized plots than in the unfertilized. There was about a 12% difference in biomass of green plant parts (5-cm turf layer and tillering nodes) between the treatments, and the author concluded that the response in leafhopper production to fertilizers was correlated with the increase in biomass of green plant parts available for food, for hatching sites and for refuges. Leafhopper species richness was slightly lower in the fertilized plots, but the most marked difference between the two sets of plots was in terms of community structure. There was a marked increase in the relative abundance of species characteristic of simplified habitats, such as *Javesella pellucida*, *Macrosteles laevis* and *Streptanus sordidus* – species able to exploit the greater productivity and enhanced food quality associated with the fertilized plots. Under the same conditions, Olechowicz (1976a) found 11–40 times more aphids and 2–3 times more Diptera and Hymenoptera emerging in the fertilized plots than in the unfertilized, while emergence of Lepidoptera and Heteroptera decreased after fertilizer application.

In contrast to the Polish meadow, where increased larval hatching and adult emergence was not reflected in increased mean population density, increased population levels of Auchenorrhyncha in fertilized grassland have been recorded in other studies such as that of 4-year-old grassland plots in Ohio, USA, which had received mineral fertilizers or sewage sludge (Sedlacek, Barrett and Shaw, 1988). Total numbers of leafhoppers also increased markedly in grass plots receiving N fertilizers compared with unfertilized plots in England (Prestidge, 1982b). The major effects of fertilizers were to increase food quality in terms of total and soluble leaf N content, and to increase grass biomass and hence living area. However, there was a marked difference in the way the two leafhopper families, Delphacidae and Cicadellidae, responded. Delphacids responded positively; these were considered to be predominantly *r*-selected species adapted to high levels of N, while Cicadellidae were more abundant on unfertilized grass and were apparently adapted to low N food. However, even within closely related species groups there can be marked differences in responses to food quality, and different species may be correlated with particular N levels in grasses (Prestidge and McNeill, 1983a; cf. section 4.1.2(a)).

For reasons already considered in Chapter 4, the responses of chewing insect herbivores to changes in food quality associated with fertilizer use may often be less marked than those sometimes seen in sap-sucking insects. Henderson and Clements (1977b) found no effects of N fertilizer levels on the density of shoot-fly adults or larvae, but Moore and Clements (1984) did find that larvae were more numerous in plots treated with high levels of N, especially in plots treated with ammonium sulphate. The fertilizer effect was attributed to increased numbers of shoots suitable for penetration by larvae, to greater density, creating more attractive conditions for oviposition, and to increased nutritive value of plants.

The responses of grassland-invertebrate decomposers to mineral fertilizers appear to be quite variable. Positive responses have usually been attributed to enhanced quantity and quality of plant litter, while population declines have been attributed to factors such as increased litter-decomposition rates and shorter residence time, changes in pH, and toxic concentrations of salts or volatile materials such as ammonia following high rates of fertilizer application. Siepel and van de Bund (1988) applied canonical correspondence analysis to microarthropod population data from a range of grassland sites in the Netherlands in an attempt to determine which environmental factors had most effect on the microarthropod community. They concluded that N fertilization had a major influence, with factors such as mowing or grazing being of minor importance by comparison. High levels of N fertilization were associated with a decrease in species richness and in numerical abundance of microarthropods, but since high levels of N fertilization were correlated with other environmental factors (e.g. vegetation cover, biomass production, C:N ratio and pH) it was not possible to say to what extent the faunistic changes noted were directly due to the fertilizers.

Olechowicz (1976a) reported increased emergence of saprophagous Diptera from heavily fertilized Polish meadow plots, while Apterygota increased by 36% in the same plots (Zyromska-Rudzka, 1976) (Figure 5.9). At the same time there was a decline in acarine abundance, which was most marked for Scutacaridae (96% reduction) and for Cryptostigmata (50%). The reduction of oribatid mite density was most marked in the case of smaller species such as *Brachychthonius* spp., whereas some larger species such as *Liebstadia similis* and *Scheloribates laevigatus* increased in numbers. There was evidence of increased fecundity and mortality among oribatids occurring simultaneously in fertilized plots, tending to accelerate population turnover while standing crop was maintained at a lower level. Moderate to severe reductions in populations of Acari, Collembola, Diptera, Coleoptera and Myriapoda were reported by Edwards and Lofty (1975a) in permanent pasture receiving 144 kg N ha^{-1} yr^{-1} in England compared with unfertilized pasture, but Andrén and Lagerlöf (1983) reported generally beneficial effects of inorganic fertilizers on soil invertebrates in Swedish arable crops and grass leys. Some negative effects were recorded in dry weather. Wasilewska (1976) did not record any response by nematodes to mineral fertilizers in Polish meadow plots, while Berger, Foissner and Adam (1986) reported sig-

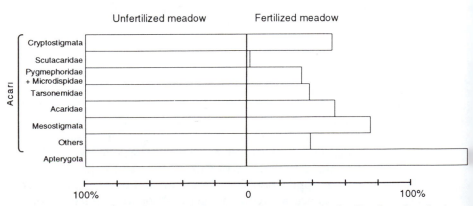

Figure 5.9 Relative density of mites and apterygotes in fertilized and unfertilized meadows. (After Zyromska-Rudzka, 1976.)

nificant reductions in the nematode populations of alpine meadows a few weeks after heavy applications of soluble mineral fertilizers, ammonium sulphate and lime.

Earthworm populations benefit from moderate applications of fertilizers, although the response in grassland is usually not as marked as that seen in arable land where the impact of fertilizers on plant production and return of organic residues to the soil may be more pronounced (Edwards and Lofty, 1982; Lofs-Holmin, 1983a). Earthworm numbers may be depressed at higher rates of application. Zajonc (1975) reported maximum population densities in meadows receiving 100 kg N ha^{-1}, with lower numbers at higher application rates. Numbers were depressed in meadow plots receiving 300 kg N ha^{-1} in Poland (Nowak, 1976), while Edwards and Lofty (1975a) recorded earthworm depression at much lower levels of N in English grassland. Large doses of sulphate of ammonia are particularly toxic to earthworms (Satchell, 1955; Edwards, 1977, 1983); this effect is likely to be more pronounced in acidic soils because of the increased acidification which sulphate of ammonia causes.

5.4.2 Organic manures

The responses of the grassland fauna to organic manures will depend *inter alia* on the rates and frequency of application. For example, the impact of a given annual loading will be very different, in the case of a grazed pasture with dung being added more or less continuously in small amounts, from the impact of periodic heavy applications of animal wastes from intensive livestock units. Grassland-invertebrate responses to organic manures will be considered firstly in relation to colonization and community development in the dung on the soil surface and secondly in relation to the impact of animal manure on the indigenous grassland fauna.

(a) The dung community

Herbivore dung, a rich source of energy and nutrients, is exploited initially by a few species of coprophagous dung flies and beetles (section 3.5.4) and, later, by an increasingly complex community comprising many general litter-dwelling species. Curry (1979) recorded a total of 144 arthropod species from cattle dung decaying on the soil surface; the dominant species are listed in Table 5.5.

The composition of the dung community varies considerably depending on factors such as age, location, climatic and other environmental conditions (Doube, 1987). Figure 5.10 indicates the seasonal occurrence of dipterous larvae and some other invertebrate groups in dung at Rothamsted Experimental Station, England (Laurence, 1954).

Table 5.5 Relative abundance values (percentage of numbers recorded) of the dominant arthropod species or higher taxa in cattle dung at Grange, Co. Meath. Slurry was applied at 550 m³ ha⁻¹ in June 1972 and August 1973. + indicates present with relative abundance < 1% (from Curry, 1979)

	Jan. 1973	March 1973	June 1973	June 1974
Collembola				
Hypogastrura denticulata	78	73	76	56
Isotoma olivacea	15	19	6	10
Isotoma viridis	1	3	3	11
Isotoma I notabilis	+	5	10	+
Isotomurus palustris	5	+	3	21
Diptera				
Dilophus	2	1		
Scatopse	11	7	79	
Psychoda		1		
Stratiomyidae	+	+	11	
Camptocladius	21	9	11	50
Smittia	59	67		1
Culicoides	3	8		
Sciara				43
Cecidomyidae	+	1		2
Acari				
Alliphis halleri	8	10	17	16
Arctoseius cetratus	13	7	31	8
Parasitus coleoptratorum	1	3	2	3
Parasitus fimetorum	1	2		2
Trachygamasus gracilis	4	12		6
Iphidozercon corticalis	3	2	+	+
Cocceupodes paradoxus	2	+	9	28
Stigmaeus antrodes	5	3	2	1
Schwiebia talpa	9	31	2	+
Tyrophagus longior	33	10	13	+
Pygmephorus gracilis			6	1
Uropoda orbicularis	+	3	5	+
Pygmephorus sellnicki				6
Ereynetes	+	+		22

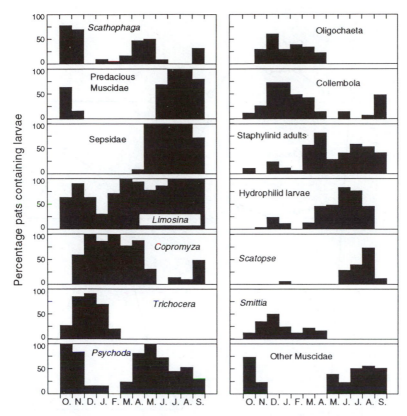

Figure 5.10 Seasonal occurrence of insect larvae, collembolans and oligochaetes in cowpats at Rothamsted, England. (From Laurence, 1954.)

Few genera were present throughout the year, one exception being *Limosina* (Sphaeroceridae), represented by several species, which was fairly widely distributed in cow pats in all months. Larvae of Muscidae and Sepsidae were absent in the coldest months, from December until April. Other genera, notably *Trichocera* (Trichoceridae), *Copromyza* (Sphaeroceridae) and *Smittia* (Chironomidae), occurred mainly in the winter and disappeared as the temperature increased. Another group, represented by *Scathophaga* (Scathophagidae), occurred abundantly in the spring, became less numerous in the summer, and reappeared in abundance in the autumn. *Scatopse* (Scatopsidae) and Sepsidae occurred mainly in the summer months when the temperature was high. Marked seasonal changes in the population densities of the most abundant larvae occurred (Figure 5.11).

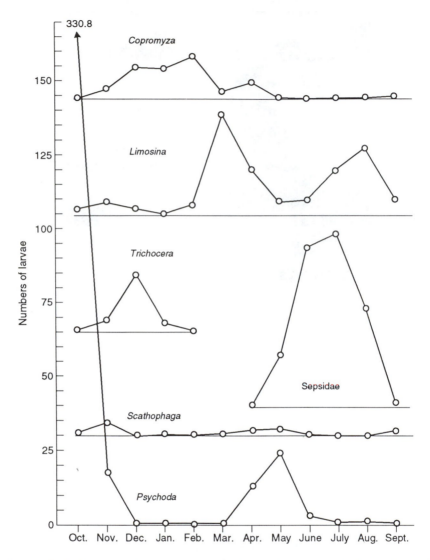

Figure 5.11 Seasonal changes in the population densities of dipterous larvae in cow pats. Numbers of larvae are geometric means per sample (4 × 2.5 cm cores). (From Laurence, 1954.)

(b) Effects on the soil fauna

Decomposer populations in soil are often food-limited and benefit from organic amendments, but adverse effects may arise following land-spreading of large quantities of animal manures or related organic wastes in the form of semiliquid slurry. Heavy applications of slurry can adversely affect soil aeration and the composition of the soil atmosphere (Stevens and Cornforth, 1974; Burford, 1976). Oxygen depletion, high levels of NH_3 and other volatiles, soluble salts of N and S and various organic decomposition products may be implicated in adverse effects of slurry on the soil community (Gisiger, 1961; Moursi, 1962; Curry, 1976b; Fortuner and Jacq, 1976; Huhta, Ikonen and Vilkamaa, 1979; Guiran Bonnel and Abirached 1980).

Earthworms
Fresh cattle and pig slurry are highly toxic to earthworms, causing 100% mortality to earthworms cultured in a soil/peat medium in the laboratory (Curry, 1976b). Ammonia and high salt concentrations are mainly responsible, with adverse effects occurring when the electrolytic conductivity exceeds about 3 mS cm^{-1} (Edwards, 1988; Hartenstein and Bisesi, 1989). Dead earthworms are commonly seen on the soil surface soon after slurry application, and further mortality is caused by birds such as gulls, crows and thrushes preying on worms temporarily expelled from the soil.

Heavy applications of slurry can drastically reduce earthworm numbers in the field. Curry (1976b) found no earthworms in grassland plots at Grange, Co. Meath, 2 months after treatment with cattle slurry at the rate of 550 m^3 ha^{-1} (Table 5.6). Within 9 months some recolonization was evident, and 14 months after slurry application population densities and

Table 5.6 Earthworm responses to cattle slurry applied to grassland plots at Grange, Co. Meath (after Curry, 1976b)

Treatment	Intervals since treatment (months)	Mean numbers (m^{-2})	Mean biomass (g m^{-2} fresh mass)
1. Control (average of six sampling dates)	–	311	97
2. Single application of 55 m^3 ha^{-1}	5	352	205*
3. Six applications of 110 m^3 ha^{-1} within 14 months	1–7	407*	188*
4. Two annual applications of 550 m^3 ha^{-1}	2	0*	0*
	9	10	7*
	14	453	148

*Differing significantly from control (P < 0.05).

Table 5.7 Composition of the earthworm fauna in untreated plots and in plots treated with cattle slurry at Grange, Co. Meath (number of adults m^{-2}) (after Curry, 1976b)

Species	Control plots	Six applications (110 m^3 ha^{-1}); last one 7 months before sampling	Two applications (550 m^{-3} ha^{-1}); last one 14 months before sampling
Allolobophora chlorotica	60	82	16
Aporrectodea caliginosa	35	46	22
Aporrectodea longa	2	1	2
Aporrectodea rosea	49	34	2
Dendrodrilus rubidus	1	6	49
Eiseniella tetraedra	0	0	4
Lumbricus castaneus	7	9	81
Lumbricus festivus	2	3	40
Lumbricus rubellus	3	10	15
Lumbricus terrestris	3	14	1

biomass in the treated plots were comparable with those in untreated controls although there was a marked difference in the composition of the population (Table 5.7). Mineral soil-dwelling *Allolobophora* and *Aporrectodea* species were scarcer and surface-dwelling, pigmented species characteristic of organic habitats *(Lumbricus castaneus, L. festivus, L. rubellus* and *Dendrodrilus rubidus)* were more abundant in the plots receiving heavy slurry treatment.

Any adverse effects of moderate levels of slurry application are transitory and the net effect is generally beneficial. Mean population densities and biomass were up to 60% higher in most instances in field plots 3–6 months after application of poultry, pig and cattle slurry at the rate of 55 m^3 ha^{-1} yr^{-1} at Grange (Curry, 1976b). In one instance (Table 5.6), the increase in biomass was over 100% but this reverted back to control levels within a few months. Other studies have shown that earthworms can tolerate, and respond positively to, quite heavy applications of slurry, provided the loading on any one occasion is not excessive. Earthworm numbers increased almost four-fold in one series of plots that had received over 5500 m^3 pig slurry in monthly instalments of up to 200 m^3 ha^{-1} over a period of 4 years in England (Unwin and Lewis, 1986).

Increases in earthworm population densities have been reported following irrigation with municipal waste (Dindal *et al.*, 1977) and with liquid manure (Zajonc, 1975). Farmyard manure has long been known to be a highly beneficial form of organic amendment for earthworms, especially in arable land where worms may often be food-limited. Edwards and Lofty (1982) reported a moderate increase in total earthworm numbers (+11%) in grassland plots treated with farmyard manure in England and a dramatic increase (+84%) in numbers of the surface-feeding *Lumbricus terrestris*. Nowak (1975) reported severe initial reduction in earthworm density in areas of Polish mountain pasture used for sheep folding where large amounts of dung had accumulated, but after about a year population densities were five times higher in such areas then in unaffected pasture.

Other invertebrates
The responses of other grassland invertebrates to organic manures are variable (Marshall, 1977). Suction sampling of grass fields at Johnstown Castle, Co. Wexford, Ireland, which had been treated with cattle or pig slurry for 4 years, revealed lower population densities of Collembola in slurry-treated plots than in plots receiving mineral fertilizers only (Table 5.8). Numbers of other surface-dwelling arthropods were not affected by slurry in any consistent way. The most abundant collembolan detritivorous species, *Sminthurinus aureus* and Isotomidae spp., were drastically reduced whereas the phytophagous *Sminthurus viridis* was less affected. Laboratory tests confirmed that volatile constituents (especially ammonia) given off by slurry cause heavy mortality in many species of Collembola.

Responses of soil arthropods to slurry recorded in various field experiments in Ireland ranged from drastic reductions in population densities of all species of Acari and Collembola in plots treated with 550 m^3 cattle slurry, to moderate increases in numbers of hemiedaphic collembolans such as *Hypogastrura denticulata*, *Isotoma notabilis* and *Pseudosinella alba* in plots treated with moderate levels of cattle and pig slurry (Bolger and Curry, 1980, 1984). Figure 5.12 illustrates the response of the microarthropod community as a whole to different levels of pig slurry applied to a sandy soil at Kilmore Quay, Co. Wexford. The ordination

Table 5.8 Mean numbers of collembolans (m^{-2}) with standard errors caught by suction sampling at Johnstown Castle, Co. Wexford. Slurry was applied in April and June at 80–100 m^3 ha^{-1} yr^{-1} (from Curry *et al.*, 1980)

	No slurry	Pig slurry	Cattle slurry
April 1977	658 ± 208	175 ± 54	207 ± 43
May 1977	1120 ± 302	371 ± 83	490 ± 73
June 1977	1600 ± 251	355 ± 87	511 ± 86
August 1977	987 ± 412	650 ± 239	618 ± 242

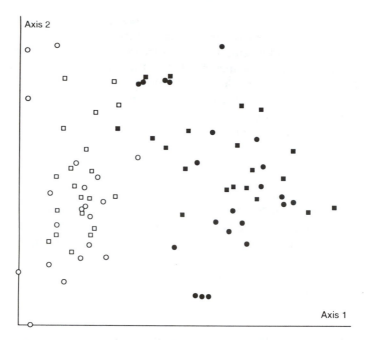

Figure 5.12 Ordination of soil samples taken from plots treated with increasing levels of pig slurry based on Detrended Correspondence analysis of the frequencies of occurrence of microarthropod species. Three applications per year: ○, control; □, 23 m³ ha⁻¹; ● 46 m³ ha⁻¹; ■, 115 m³ ha⁻¹ per application. (From Bolger and Curry, 1984.)

shows a broad separation of samples taken from untreated or lightly treated plots from those taken from more heavily treated plots, reflecting treatment-induced changes in the frequencies of occurrence of acarine and collembolan species.

Variable responses in nematode populations have been reported to animal manures. Microbivores tend to benefit from high input of dung or slurry and associated microbial activity (Wasilewska, 1974; Guiran, Bonnel and Abirached, 1980; Dmowska and Kozłowska, 1988); decreases in populations of omnivores have been reported in grassland receiving high inputs of sheep dung (Wasilewska, 1974), while Guiran, Bonnel and Abirached (1980) recorded a decrease in numbers of phytophagous nematodes in French grass leys heavily fertilized with pig slurry. Marked changes in species composition and a drastic reduction in species diversity were reported by Dmowska and Kozłowska (1988) after heavy applications of cattle slurry.

5.5 FIRE

Fire is a key element in the maintenance of tropical seral grassland, while in temperate latitudes it is sometimes used to improve the quality of rough pasture, particularly in upland areas. When used in the management of savannah grassland its effects include the elimination of woody plants that are less well adapted to resist fire than are grasses, the removal of inedible, dried vegetation, the stimulation of nutrient cycling through return of mineral nutrients to the soil in the form of ash, and the stimulation of tillering and grass growth (Spence and Angus, 1971; Singh and Joshi, 1979). A very significant proportion of net primary production in tropical grasslands may be consumed annually by fire; this may be as much as 53% in African tall-grass savannah (Sinclair, 1975).

The effect of fire on the fauna depends on the intensity and duration of the fire. During hot fires in coniferous forest the surface temperature may reach 1100 °C, consuming all the litter and litter-dwelling organisms (Daubenmire, 1968). Surface temperatures exceeding 700 °C have been recorded for very short intervals during fires in African savannah grasslands, but rarely does the surface temperature exceed 100 °C for more than a few minutes. Under these conditions the litter layer is only partially burned and the effects on the invertebrate community are not as great as those of hot fires.

Burning facilitates the regrowth of young shoots by removing dead plant remains and litter and by adding nutrients to the soil in plant ash. Immediate effects on the physicochemical properties of the habitat may include elevated pH, higher concentrations of available bases, elevated surface temperatures and reduced water availability in the surface horizons (Woodmansee and Wallach, 1981). Some loss of plant nutrients may occur through leaching in rainfall or through wind removal and volatilization. However, natural ecosystems appear to be able to retain nutrients after fire and to quickly replace those lost. Nitrogen losses, for example, may be replaced quickly by enhanced microbial mineralization and biological N fixation (Woodmansee and Wallach, 1981).

The direct effects of fire on the grassland fauna do not appear to be catastrophic, although there may be adverse consequences in the longer term resulting from the structural deterioration of soil (Lal, 1987). Burning is generally carried out when much of the vegetation is dry and unsuitable for many phytophagous groups. Many of the more active surface dwellers may be able to escape out of the habitat, while others are able to find refuge under stones and in sheltered situations (Gillon, D., 1971). The total biomass of litter-dwelling and hemiedaphic soil invertebrates is often significantly reduced by burning; this is partly a direct effect of burning *per se* and partly a consequence of the destruction of the litter layer and the changed microclimatic conditions after removal or depletion of the insulating litter layer. Faunal density and biomass were reduced by about 60% on average 1 month after fire in Ivory Coast savannah (Lamotte, 1975). Testate amoebae were virtually eliminated (Couteaux, 1980); these animals cannot migrate rapidly to refuges and were therefore largely confined to the roots of grass tussocks, which are little affected by fire. Soil microarthropods (mainly mites and collembolans) were considerably reduced by burning in the same site (Athias, 1976). Euedaphic soil animals are little affected by fire and may benefit indirectly from the greater productivity. For example, Lavelle (1974) found that small, litter-dwelling earthworm species were scarce in burned savannah but the total earthworm biomass was greater than in unburned areas.

Fire promotes earlier and more vigorous grass regrowth, and burned savannah supports a denser invertebrate phytophagous fauna during the growing season than does unburned grassland. Phytophagous insects are attracted into the area by the green young sward. Andrzejewska (1979a) reported higher population densities of Orthoptera, lepidopterous larvae, Homoptera and soil-dwelling scarabaeid beetle larvae in burned than in unburned savannah in Panama. Fire destroys a proportion of the acridid fauna (grasshoppers etc.) of Ivory Coast savannah and further mortality occurs due to bird predation on agitated grasshoppers fleeing from fire, but the rejuvenated habitat is beneficial for this group and greatest species diversity occurs in regrowth after fire (Gillon, Y., 1971)

5.6 GRASSLAND RECLAMATION AND IMPROVEMENT

Grassland productivity may often be severely limited by such factors as drought or excess moisture, low pH or low levels of soil fertility, and vegetation of low agronomic value. Grassland improvement measures that remove physicochemical limitations and radically change the composition of the vegetation can be expected to have marked effects on the invertebrate fauna. Some effects on the fauna of sward establishment, irrigation of drylands and reclamation of wetlands are considered below.

5.6.1 Sward establishment

Clearing native vegetation such as deciduous forest for grassland results in the virtual elimination of the vegetation-zone fauna and its rapid replacement by a fauna adapted to the new conditions. The disappearance of the litter layer also results in a marked decline in the diversity of soil and litter invertebrates, with litter-dwelling species being particularly affected. The effects of deforestation on the soil fauna appear to be less marked in temperate regions than in the tropics. Many temperate forest species adapt well to grassland; in the case of earthworms epigeic species decline in abundance with the disappearance of surface litter, but the anécique and endogeic species increase in importance as soil fertility and food quality improve. By contrast, epigeic species comprise most of the tropical earthworm fauna, and these together with other litter-dwelling macroinvertebrates are destroyed by clearing and cropping. Lavelle and Pashanasi (1989) reported that soil macrofaunal biomass and population density in cultivated plots in the Peruvian Amazonia are reduced to 6% and 17% respectively, of those in primary forest. The indigenous forest earthworm species have largely disappeared, but when adapted species are available for recolonization high earthworm population densities can be found under pasture. Macroinvertebrate biomass of up to 160 g m^{-2} was recorded from pasture where the endogeic peregrine species *Pontoscolex corethrurus* had become established (Figure 5.13).

Improvement of old grassland by ploughing, cultivation and reseeding reduces numbers of most invertebrate groups to some degree, although these effects are often short-lived. Most of the invertebrates in the ploughed-in old sward probably perish; however, some pest species (e.g. the frit fly *Oscinella frit* and the aphid *Rhopalosiphum padi*) are able to survive for some time in the inverted sward. Most accounts indicate fairly marked reductions in numbers of soil arthropods after cultivation of old pasture (Sheals, 1956; Edwards and Lofty, 1975a). In the latter case, overall arthropod density was depressed by about 50% 6 months after

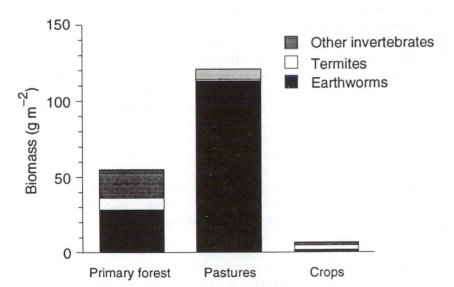

Figure 5.13 Macroinvertebrate biomass under different types of land use in the Peruvian Amazonia. (After Lavelle and Pashanasi, 1989.)

ploughing and reseeding, with hemiedaphic collembolans and crptostigmatic mites being most affected. Population densities of most arthropod groups and the dominant species in the top 7.5-cm soil layer of cultivated plots in Ireland were 20–50% lower than those in uncultivated control plots, but numbers of some groups (notably prostigmatic mites) were higher deeper in the cultivated soil (7.5–15 cm) than at a comparable depth in uncultivated plots (Figure 5.14). These were presumably associated with the decaying buried old sward. Once the new sward became established population densities in the surface soil recovered rapidly.

The initial effects of ploughing and mechanical cultivation on earthworm populations can be severe (Zicsi, 1969); however, once the initial stress has passed populations recover rapidly (Edwards and Lofty, 1975b). Endogeic species benefit from the increased food supply represented by the incorporated sward and litter, while in the longer term endogeic and anécique species benefit from the increased fertility and productivity of improved pasture. Epigeic species, which require a surface organic layer, are most adversely affected by pasture improvement.

Rotary cultivation can be particularly damaging to large invertebrates. Hunter (1968) reported reductions of up to 75% in slug populations following cultivation to a fine tilth in England, while similar cultivation reduced earthworm biomass (mainly *Aporrectodea caliginosa*) by 60–70% in a 4-year-old grass ley in Sweden (Boström, 1988a), but a year later the biomass had recovered to uncultivated control plot levels.

Most accounts suggest that minimal cultivation techniques have less drastic effects on the soil fauna than conventional cultivation (e.g. Edwards, 1975; Wilkinson, 1977), although Curry (1970) found that many soil-arthropod groups were equally affected when old pasture was surface-seeded after desiccation of the sward with the herbicides paraquat and dalapon as they were following conventional cultivation. Differences between conventional and minimal cultivation in terms of effects on the fauna are probably less important in grassland than in arable land where frequent cultivation results in an impoverished fauna. Anécique earthworms with relatively permanent burrows such as *Lumbricus terrestris* are particularly susceptible to damage by conventional cultivation, whereas some smaller endogeic species such as *Aporrectodea caliginosa* are better adapted to cultivation and are often the dominant species in disturbed habitats.

Grass leys established by minimal cultivation, with or without chemical desiccation, tend to be more prone to attack by pests such as frit fly and slugs than leys established by conventional methods. Chemical desiccation renders the sward unsuitable for herbivores, but some species appear to be able to survive for several days and may re-infest the new

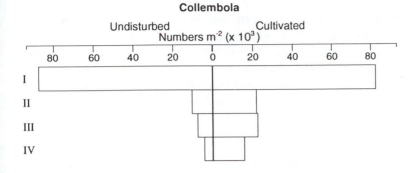

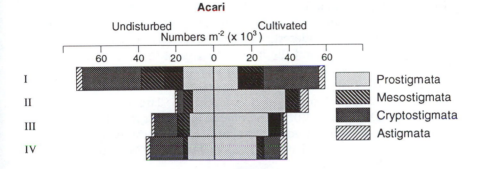

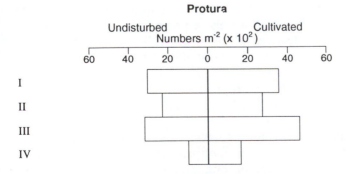

Figure 5.14 Vertical distribution of the most abundant microarthropod groups in undisturbed grassland and in plots that had been cultivated and seeded 12 months previously. I = 0–3.8 cm; II = 3.8–7.5 cm; III = 7.5–11.3 cm; IV = 11.3–15 cm.

sward unless a sufficient interval is left between desiccation and reseeding.

Cultivation may also affect population densities of pest species indirectly by influencing the abundance and activity of natural enemies. Moore, Clements and Ridout (1986) reported that ploughing greatly reduced parasitism of stem-boring larvae by the braconid wasp *Chasmodon apterus*, a wingless, ground-dwelling, slow-moving species, whereas direct drilling led to less-marked reduction.

5.6.2 Irrigation

As seen in Chapter 1, moisture has a major influence on the distribution and abundance of soil invertebrates. The dominance of macro-invertebrates in temperate soils decreases with decreasing soil moisture (Coleman and Sasson, 1980); in dry steppe soils, the contributions of macrofauna, mesofauna and microfauna to the total biomass are roughly similar, whereas in semidesert grasslands saprophagous macrofauna are almost entirely absent and the microfauna (protozoans and nematodes) living within the water film on soil particles are the dominant component of the soil-invertebrate biomass. In tropical soils with markedly seasonal rainfall patterns, the activity of decomposers is largely confined to the wet season (Dwivedi, 1979). Even in moist temperate grasslands, summer drought can cause reduced activity and high mortality in susceptible groups such as lumbricid (Gerard, 1967; Martin, 1978) and enchytraeid worms (Nielsen, 1955b).

The response of the phytophagous fauna to irrigation of savannah grassland in Panama was described by Andrzejewska (1979a). Small areas of savannah were managed, mown and irrigated to extend the growing season. The continuous supply of green vegetation allowed the phytophagous fauna to develop and feed throughout the year. By contrast, in non-irrigated savannah the vegetation withered rapidly at the end of the rainy season (mid-December) and only small quantities of green leaves remained. Most phytophagous invertebrates disappeared from the savannah surface during the dry season (Table 5.9) with only ants remaining abundant.

Invertebrates that can produce successive generations under such conditions (e.g. grasshoppers and leafhoppers) were especially favoured, as the high humidity and abundant food supply enabled such species to avoid their normal period of diapause so that they were able to produce several generations a year. A major part of the population comprised immigrants from adjoining habitats attracted by fresh, green vegetation. Soil-invertebrate herbivores can also benefit from irrigation: for example, irrigation may prevent summer temperatures from reaching lethal levels in temperate grasslands and so enhance the survival of soil pests such as

Table 5.9 Biomass (mg dry mass m^{-2}) of invertebrate herbivores in the sward of savannah grassland, Panama (after Andrzejewska, 1979a)

	Wet season		Dry season	
	May–June	Aug.–Sept.	Dec.–Jan.	Feb.
Unburned	199.2	29.3	100.8	2.4
Burned	1015.6	432.7	92.34	3.4
Irrigated	356.1	681.9		655.9

the grass grub *Costelytra zealandica* in New Zealand (East and Willoughby, 1980).

Removal of moisture restrictions by irrigation likewise promotes the activities of the decomposer community and accelerates decomposition processes. Protozoan activity increases dramatically when arid grasslands are watered (Elliott and Coleman, 1977; Dash and Guru, 1980). Increased microbial activity in wetted soils appears to be a major factor influencing the protozoan response. Clarholm (1981) observed a peak in bacteria 2 days after rainfall in field studies and after watering in pot studies; this was followed by a peak of protozoans 3 days later and a decline in bacterial numbers attributed to protozoan predation (Figure 5.15).

Dodd and Lauenroth (1979) reported the effects of watering semiarid grassland in Colorado on soil-arthropod populations. A four-fold increase was recorded in watered plots compared with unwatered controls, while above-ground macroarthropods also increased in watered plots (Kirchner, 1977). However, watering of arid soils only stimulates decomposer activity when levels of soil organic matter are adequate: Steinberger *et al.* (1984) reported that microarthropod and nematode densities in desert soil responded to watering only when litter levels were high. The distribution of lumbricid earthworms is often limited by soil moisture in drier regions, and substantial population densities may become rapidly established when dry lands are irrigated. Barley and Kleinig (1964) successfully introduced *Aporrectodea caliginosa* and the megascolecid species *Microscolex dubius* into sown, irrigated pasture on sandy loam soil in New South Wales, Australia. Within 8 years a population density of 300 m^{-2} was present. *Aporrectodea trapezoides* and related species were also successfully introduced into newly reclaimed irrigated land in Uzbekistan (Ghilarov and Mamajev, 1966). Peak population densities of *A. trapezoides* and *A. rosea* reached around 600 m^{-2} and 400 m^{-2}, respectively, in irrigated lucerne fields in South Africa (Reinecke and Visser, 1980). Salinity in the root zone can be a major problem in irrigation agriculture, restricting plant growth and soil decomposition

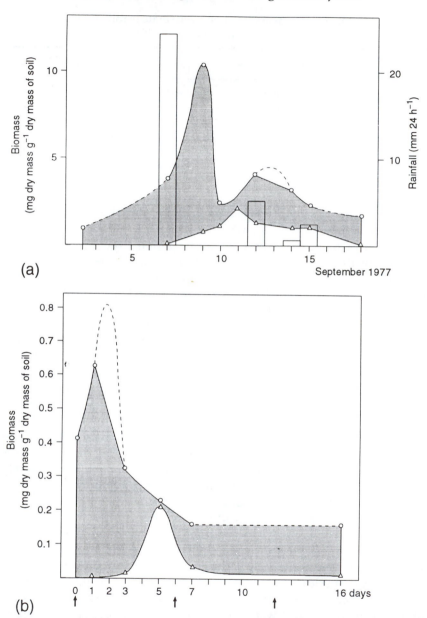

Figure 5.15 Influence of moisture on microbial (grey) and naked amoebic (white) biomass in (a) the humus layer of a podsolized pine forest soil, (b) a pot experiment with wheat plants growing in arable soil. The histograms represent rainfall, the arrows indicate times of watering. The dashed lines indicate probable development not registered because of insufficiently frequent sampling. (After Clarholm, 1981.)

processes (Allison, 1964; Gupta and Abrol, 1990). Under such conditions high salt levels could be a factor in limiting earthworm populations (Khalaf El-Duweini and Ghabbour, 1965).

5.6.3 Reclamation of wetlands

Waterlogged soils are characterized by a scarcity of soil macro-invertebrates, slow decomposition rates and an accumulation of raw organic matter leading to peat formation in extreme cases. Reclamation involving drainage, liming and seeding with pasture grasses and legumes brings about a change from essentially anaerobic to aerobic biological processes with greatly increased ecosystem productivity and major changes in invertebrate activity. The responses of earthworm populations to wetland reclamation are considered below, with particular reference to polders reclaimed from the sea and peatlands.

(a) Earthworms in reclaimed polders

Earthworms can recover from periodic flooding by sea water within a few years (Piearce, 1982), but populations are completely eliminated by prolonged flooding. Natural recolonization of reclaimed Dutch polder soils may begin fairly rapidly; Meijer (1972) and van Rhee (1969a) reported densities exceeding 200 m^{-2} in orchard soil 26 years after drainage and reclamation of a former lake site. The most successful early colonizers of reclaimed polder soils were peregrine species such as *Aporrectodea caliginosa* and, in the case of orchard soils with surface organic accumulation, *Lumbricus rubellus*. However, the rate of earthworm spread was low (4–9 m yr^{-1}) but this was greatly accelerated by artificial introduction in some sites (van Rhee, 1963, 1969a,b, 1977; Hoogerkamp, Rogäar and Eijsackers, 1983). Established population densities ranged from 140 to 250 m^{-2} (80% *A. caliginosa*, 20% *L. terrestris*) in orchard sites where populations may have been inhibited by pesticides, to as high as 750 m^{-2} in fertile grassland. These earthworm populations had a marked influence on soil structure and fertility (Chapter 6).

(b) Earthworms in reclaimed peat

Peat forms where organic deposits accumulate in situations of impeded drainage and/or high rainfall where decomposition is inhibited by saturated, anaerobic conditions and by low pH. Raised bogs were once common in lowland areas of Europe, although few now survive. In these situations, peat starts forming in natural basins where water accumulates; successive layers of peat raise the ground surface above the water table and a dome-shaped bog develops. The lower strata can comprise a broad

Table 5.10 Influence of reclamation on raised bog (after Hammond, 1979)

Unmodified profile

Location	Clonawing Td., Co. Westmeath, Grid ref. N: 49.52
Classification	Histosol
Series	Allen series
Parent material	Ombrotrophic peat
Vegetation	*Calluna* and *Sphagnum*

Horizon	Oi1 0–27 cm. Dark reddish brown; *Calluna–Sphagnum* peat fibric; poorly humified; dominantly *Calluna* remains
	Oi2 27–58 cm. Dark reddish brown; *Sphagnum* peat; fibric; poorly humified
	Oi3 58–87 cm. Dark reddish brown; *Sphagnum–Calluna* peat, fibric; poorly humified
	Oi4 87–118 cm. *Sphagnum–Calluna* peat; fibric; poorly humified

Analytical data

Horizon	Depth (m)	Field moisture (%)	Saturated moisture (%)	Ash (%)	Bulk density (g cm^{-3})	pH (H_2O)
Oi1	0–27	68.2	1548	3.0	0.061	3.42
Oi2	27–58	77.4	nd	1.0	nd	3.40
Oi3	58–87	87.5	1685	0.6	0.055	3.35
Oi4	87–118	90.0	nd	0.6	nd	3.50

nd, Not determined.

spectrum of plant remains including trees, shrubs, sedges and grasses, while the upper strata consist mainly of *Sphagnum* moss. Fens occur where continuous flushing by base-rich ground water prevents the development of oxyphilous plant species. Blanket peats are common in upland areas and in some wet lowland areas in high latitudes.

Peatland occupies 1.34 million ha or about 16% of the land area of Ireland. About 50% of this is high-level blanket bog, 26% is low-level (Atlantic) blanket bog along the western seaboard, and 24% is raised bog that occurs mainly in the central plain (Hammond, 1979), but much of this has been modified by people to a greater or lesser degree. In some cases, bogs have been mined for fuel before reclamation; in other cases, the original bog surface has been reclaimed for agriculture. The degree of reclamation is variable but generally involves drainage, the addition of calcareous mineral material to the surface ('marling') and the encouragement of palatable herbage species by burning off the native vegetation or by cultivation and reseeding. The main effects of reclamation on peat characteristics can be seen in Table 5.10. These include reduction in

Profile modified by human action

Castletown Moor Td., Co. Meath, Grid ref. N:80.79
Histosol
Gortnamona series
Ombrotrophic peat
Dactylis glomerata, Ulmaria filipindula, Urtica dioica

Oap 0–33 cm. Black; sapric; no plant remains visible; fine, strong crumb
structure; well humified; much marling carried out
Oe1 35–59 cm. Black; hemic; well-humified, fine, strong subangular structure
Oe2 59–80 cm. Strong, brown, turning rapidly on exposure to black; greasy
cyperaceous hemic peat

Analytical data

Horizon	Depth (%)	Field moisture (%)	Saturated moisture (%)	Ash (%)	Bulk density (g cm^{-3})	pH (H$_2$O)
Oap	0–33	58.2	260	53.8	0.468	7.3
Oe1	33–59	85.9	835	12.0	0.120	5.8
Oe2	59–80	85.6	1072	6.0	0.092	5.3

moisture content and increases in ash content, bulk density and pH in the surface horizons. Enhanced biological activity is reflected in the greater degree of humification and structural development seen in the modified profile. The most severe form of disruption is that associated with mechanical extraction of peat for fuel, which leaves behind a shallow cutover about 0.5–1.5 m deep comprising fen peat *in situ* and some redistributed vegetation and superficial sphagnum layers. After appropriate reclamation work involving drainage, extraction of fossil timber, deep ploughing to mix residual peat with underlying mineral soil, disc cultivation and liming these cutovers are very suitable for grass production and other agricultural uses.

Earthworms in virgin peat are scarce, with mean population densities ranging from fewer than 0.1 m^{-2} in deep peat in the Pennines, England (Svendsen, 1957a) to 12 m^{-2} in acid peaty soils in Scotland (Guild, 1948). The population mainly comprises surface-dwelling, raw-humus species such as *Lumbricus rubellus, Dendrodrilus rubidus, Dendrobaena octaedra* and *Lumbricus eiseni*; the burrowing species characteristic of mull soils tend to

Table 5.11 Occurrence of earthworm species in grassland on reclaimed peat soils (after Curry and Cotton, 1983)

	Young sites											
Site	1	2	3	4	5	6	7	8	9	10	11	12
Years since restoration	1–2	5–6	7–8	10	12	14–16	16–17	17–20	17–20	24	24	24
Dendrobaena octaedra	+					+						
Dendrodrilus rubidus			+								+	+
Eiseniella tetraedra					+			+	+	+	+	
Lumbricus castaneus						+			+			
Lumbricus eiseni	+										+	
Lumbricus rubellus					+			+		+	+	+
Lumbricus festivus			+	+		+	+		+	+	+	
Octolasion tyrtaeum	+							+	+			+
Allolobophora chlorotica		+		+								+
Aporrectodea caliginosa			+	+		+	+		+	+		+
Aporrectodea tuberculata								+				
Aporrectodea rosea									+			
Aporrectodea longa												
Lumbricus terrestris												
Octolasion cyaneum												
Satchellius mammalis												

Mature sites

Site	13	14	15	16	17	18	19	20	21	22	23	24
Years since restoration							<25					
Dendrobaena octaedra												
Dendrodrillus rubidus											+	
Eiseniella tetraedra												
Lumbricus castaneus												
Lumbricus eiseni												
Lumbricus rubellus	+											+
Lumbricus festivus						+			+		+	+
Octolasion tyrtaeum								+	+			
Allolobophora chlorotica	+		+	+	+	+	+	+	+	+	+	+
Aporrectodea caliginosa	+	+	+	+	+	+	+	+	+	+	+	+
Aporrectodea tuberculata					+		+			+		
Aporrectodea rosea	+	+	+	+	+	+	+	+	+	+	+	+
Aporrectodea longa			+	+								
Lumbricus terrestris	+	+				+	+	+	+	+	+	+
Octolasion cyaneum											+	
Satchellius mammalis		+								+		+

be absent (Guild, 1948; Boyd, 1956, 1957; Svendsen, 1957a,b). Factors that inhibit earthworm activity in peat include waterlogging, low pH and poor litter quality – high C:nutrient ratios, low N content and high levels of phenolic compounds (Swift, Heal and Anderson, 1979). By contrast, reclaimed peats can support sizeable populations. Guild (1948) reported population densities of 50–100 worms m^{-2} from improved peaty hill pastures in Scotland, including typical pasture species such as *Aporrectodea caliginosa*, *Allolobophora chlorotica*, and, in smaller numbers, *Lumbricus terrestris*, *Aporrectodea longa* and *A. rosea*, which were scarce or absent in unimproved areas. The wet-soil species *Eiseniella tetraedra*, *Dendrobaena octaedra*, and *Octolasion lacteum* were eliminated by drainage and cultivation of peaty soil in Lithuania, while *A. rosea*, *L. rubellus*, *A. chlorotica* and *L. terrestris* became the dominant species (Atlavinyté, 1976).

A survey of 24 sites in Ireland with varying reclamation histories indicated that the sites could be divided into two main groups on the basis of their earthworm faunas (Table 5.11). Site age had a major influence. Younger sites reclaimed for less than 25 years had few species and low population densities ($< 25\ m^{-2}$), although higher population densities were encountered under particularly favourable conditions. Maturer sites reclaimed for 25 years or longer had more species and population densities in the range 100–200 m^{-2}. The 15 species recorded during the survey fall into three groups. The first, characteristic of young sites, comprised species such as *Dendrobaena octaedra*, *Dendrodilus rubidus*, *Lumbricus eiseni* and *Eiseniella tetraedra*, which are typical of acid peat and raw humus habitats. *Lumbricus castaneus* and *L. rubellus* are highly mobile species that are widespread in moorland and deciduous woodland habitats. A second group comprised typical mineral soil-dwelling species such as *Aporrectodea rosea*, *A. longa* and *Octolasion cyaneum*, which are intolerant of acid conditions, and *L. terrestris*, which is more pH tolerant but is slow to colonize new habitats. This group occurred in the mature sites where soil characteristics had been radically changed. Although fairly widely distributed in maturer sites, the population densities of *L. terrestris* were low compared with those that are typical of fertile mineral soil. This species requires a considerable depth of aerobic soil, and this may be a factor limiting its success in reclaimed peat. A third group comprised species found commonly in both young and old sites – notably *Aporrectodea caliginosa* and *Allolobophora chlorotica*, which are often the dominant species in mineral grassland soils. *Aporrectodea caliginosa* is a particularly successful early colonizer and has been reported as a pioneer species in such diverse habitats as irrigated desert (Khalaf El-Duweini and Ghabbour, 1965), restored coal-mining wastes (Dunger, 1969a,b), reclaimed Dutch polders (van Rhee, 1977) and improved hill pasture in New Zealand (Stockdill, 1959, 1966). Under the most favourable conditions colonization of reclaimed peat soils can occur more rapidly and population densities can be comparable with those found in mineral grassland soils within 5 years (Curry and Boyle, 1987).

5.7 PESTICIDES AND CHEMICAL POLLUTANTS

5.7.1 Pesticides

Pesticides are not as widely used on grassland as in intensive arable farming, but grassland comprises a very significant habitat type in most countries and the potential for ecological damage resulting from the misuse of pesticides is substantial. Suppression of populations of natural enemies in grassland could have adverse consequences for pest control both within grassland and in neighbouring arable crops, while adverse effects of heavy applications of pesticides on earthworms and other soil invertebrates could have more serious consequences for soil fertility in grassland than in arable land where litter incorporation is mainly achieved by mechanical cultivation.

Hundreds of chemicals used for plant protection enter the environment every year; these produce vast numbers of breakdown products and metabolites that can be more or less toxic to non-target organisms (Menzie, 1972; Crosby, 1973; Edwards, 1973). Above-ground invertebrates will be most directly affected by pesticides applied to the crop canopy, but because a large proportion runs off into the soil (Metcalf, 1975) hemiedaphic invertebrates are also at risk. Surface-feeding invertebrates such as *Lumbricus terrestris* can ingest considerable amounts of pesticide when feeding on contaminated plant residues, while soil invertebrates in general are most likely to be affected by high-volume applications of insecticides applied for the control of soil pests. Information on the environmental effects of pesticides is still patchy, despite the accumulation of a considerable volume of literature on this topic. Some of the major conclusions relating to grassland invertebrates are summarized below.

(a) Herbicides, fungicides and fumigants

Most herbicides are probably not directly toxic to invertebrates to any great extent at normal application rates, although some compounds, including those based on triazine, urea and phenol, can have short-term inhibitory effects (Eijsackers and van der Drift, 1976). Laboratory tests have shown that paraquat and atrazine, two widely used desiccants, can have some toxic effects on collembolans (Subagja and Snider, 1981), while Pizl (1988) reported moderate toxicity in earthworms exposed to five herbicides (bentazon, bromphenoxin, bromoxynil, bromoxynil octaonate/ioxynil and atrazine) in the laboratory. However, the main effects of herbicides on soil invertebrates are probably brought about indirectly by changes in plant cover, microclimate and food supply.

Figure 5.16 shows the effects on soil-arthropod populations of two herbicides, paraquat and dalapon, applied to an old *Agrostis-Festuca* grassland sward at Lyons Estate, Celbridge, Co. Kildare, Ireland. The figure shows the average responses of the major arthropod groups and dominant species sampled on three occasions over a 10-month period. The most usual response was a reduction in numbers in herbicide-treated plots compared with untreated controls. Interestingly, oribatid mites, which are generally found to be adversely affected by habitat disturbance, were unaffected by treatment. The effect of paraquat on populations generally tended to be more severe that that of dalapon, although few of the apparent differences were statistically significant. The effects of the herbicides on the fauna appeared to be less marked in an earlier experiment where plots were surface cultivated and reseeded after herbicidal treatment (Curry, 1970); in that instance, adverse effects may have been more quickly offest by rapid establishment of the new sward. Differential responses of the fauna could reflect differential effects of the herbicides on the sward. Paraquat, for example, shows its maximum effect on the sward within 10 days whereas dalapon takes 4 to 6 weeks to achieve its maximum effect. Conrady (1986) studied the effects of atrazine and pentachlorophenol (PCP) on soil and surface arthropods in a Lolio–Cynosuretum meadow in Göttingen, Germany, and recorded depression of populations of hemiedaphic collembolans in the order of 80%. Atrazine increased the mortality rate of larval and adult staphylinid beetles, while PCP depressed the density of erigonid spiders by about 70%. Both herbicides increased the mortality of the epigeic earthworm *Lumbricus castaneus*, while individuals of the endogeic species *Allolobophora chlorotica* and *Aporrectodea rosea* aggregated in patches of low chemical concentration.

While population and community-level responses of soil invertebrates to pesticides can be documented in broad field studies of the kind described, more specific studies are required to elucidate the causal mechanisms involved in these responses and their implications for soil fertility. An example of such a study was that by Hendrix and Parmelee (1985), who were concerned with how herbicidal treatment influences the microarthropod community and the process of decomposition of grass litter in a fallow field in Georgia, USA. Litter bags containing dried Johnson grass (*Sorghum halepense*) leaves were dipped in herbicide solution at the recommended, and at 10 times the recommended, field application rates and placed in the field. At the highest treatment levels of paraquat and glyphysate the leaves showed slower weight loss, faster losses of P, Ca and Mg, and higher densities of microfloral-grazing microarthropods than untreated controls. The authors hypothesized that

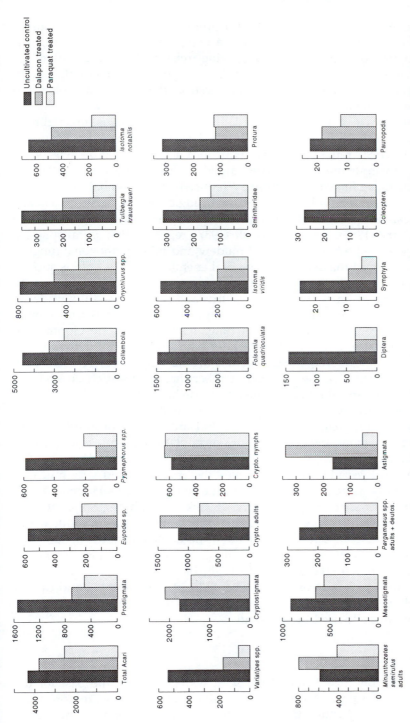

Figure 5.16 Population densities of the more-abundant arthropod groups and species in untreated grassland plots and in plots treated with herbicides. Total numbers in 24 samples each 25 cm² per treatment. Samples taken 1, 6 and 10 months post-treatment. (From Curry, 1970.)

herbicides altered the system by (1) promoting microbial utilization of additives in the formulations as carbon sources; (2) increasing the importance of microarthropod grazing relative to comminution; (3) eliminating or reducing the importance of predatory microarthropods; and (4) increasing the rate of nutrient loss from the litter via microbial and micoarthropod activity.

Fungicides can influence populations of mycophagus invertebrates indirectly by reducing their food supply. For example, Mueller, Beare and Crossley (1990) reported marked reductions in the population densities of the fungivorous mite families Oribatulidae, Oppiidae, Haplozetidae and Tarsonemidae linked with strongly reduced fungal hyphal densities, in litter bags treated with captan. Several fungicides also have insecticidal and acaricidal properties (Morgan, Anderson and Swales, 1958; Fungo and Curry, 1983). Sotherton *et al.* (1987) reported that pyrazophos, a fungicide used for the control of mildew and other pathogenic fungi in barley, reduced the densities of natural enemies of cereal aphids such as carabid and staphylinid bettles in autumn-sown cereals. Routine use of mercury-based fungicides on sports turfs and copper-based fungicides in orchards has sometimes caused the suppression of earthworm populations and a reduction in cellulolytic fungi, resulting in the development of a surface mat of undecayed plant litter (van Rhee, 1963; Pugh and Williams, 1971); however, these materials are no longer widely used. Currently, the extensive use of substituted benzimidazoles such as benomyl, carbendazim and thiophanate-methyl in orchards and arable crops can adversely affect earthworm populations in treated habitats. These materials are highly toxic to earthworms and are capable of affecting their feeding behaviour and, at high rates of application, depressing populations in the field (Keogh and Whitehead, 1975; Stringer and Wright, 1976; Keogh, 1979; Wright, 1977). However, the effects depend very much on such factors as rate of application and soil type (Lofs-Holmin, 1981). *Lumbricus terrestris* is particularly at risk through ingestion of contaminated plant residues. Application of benomyl caused a reduction in gamasine mites in Danish grassland while other microarthropods were unaffected or were indirectly affected via changes in the microflora or in predation pressure (Petersen and Krogh, 1987).

Soil fumigants such as DD, chloropicrin, methyl bromide and carbon disulphide are highly toxic to most soil invertebrates, but these materials are not normally used in grassland.

(b) Insecticides

Organochlorine insecticides have been widely used in the past for the control of soil pests. More than 5.5 million kg of DDT were applied to New Zealand pasture from 1965 to 1967 to control the grass grub

Costelytra zealandica (Kain, 1979). However, the use of organochlorines is now restricted in most countries because of their long persistence in the environment and their ability to accumulate in food chains. Estimated 95% disappearance times for DDT and dieldrin, the organochlorines that persist for longest in the soil, range from 4 to 30 years, while estimates for less-persistent compounds such as aldrin and heptachlor range from 1 to 6 years (Edwards, 1973).

Moderate to severe depressions in numbers of many invertebrate groups have been attributed to organochlorines. Numbers have been reduced by more than 50% and the effects have lasted for several years in some instances where invertebrates have been exposed to high concentrations; the effects were generally more moderate at normal agricultural rates of application (Thompson and Gore, 1972; Edwards, 1965, 1977; Edwards and Thompson, 1973). Collembolans are not affected by DDT and often increase in numbers in treated soils after depletion of their acarine predators, which are susceptible to it (Sheals, 1956; Edwards, Dennis and Empson, 1967). Collembolans are susceptible to other organochlorines, including HCH, heptachlor, aldrin, dieldrin, chlordane and endrin (Sheals, 1956; Fox, 1967; Brown, 1977; Edwards, Dennis and Empson, 1967, etc.). Most organochlorines reduce numbers of saprophagous mites, but predatory species do not seem to be much affected by aldrin and dieldrin (Edwards and Thompson, 1973). Surface-active predatory carabid beetles are adversely affected by several organochlorines (Dempster, 1967; Davis, 1968; Edwards and Thompson, 1973).

Among the Myriapoda, pauropods are most sensitive to organochlorines; symphylids are fairly tolerant to most except HCH; and diplopods are little affected by DDT but are susceptible to some others including HCH (Edwards, 1974b; Edwards and Thompson, 1973). Earthworms do not suffer any significant direct mortality as a result of exposure to organochlorine insecticides applied at agricultural rates, with the exceptions of chlordane and, to some extent, endrin (Edwards, 1980). However, residues readily accumulate in their tissues, creating hazards for earthworm-feeding vertebrates. Edwards (1973) cites concentrations of DDT in earthworm tissues up to 19 times greater than those in the soil. Also, organochlorines can have sublethal effects on earthworms that may have long-term consequences for the population. For instance, Reinecke and Venter (1985) reported an inverse relationship between the concentration of dieldrin and cocoon production, numbers of hatchlings per cocoon, hatching success and incubation period in laboratory cultures of *Eisenia fetida*, while Venter and Reinecke (1985) reported retardation of growth and development at the highest concentration of dieldrin (100 mg kg^{-1}) in the culture medium (washed cow manure).

Organophosphorous insecticides are less persistent and generally less injurious to non-target organisms than are the organochlorines. Most, with the exceptions of chlorfenvinphos and fonofos, persist for only a few weeks in soil (Edwards, 1973). Generally, only excessive doses depress arthropod numbers to the same extent as normal doses of organo-chlorines. Reductions in numbers of collembolans have been attributed to several compounds, including demeton, diazinon, disulfoton, phorate, malathion, menazon and fenitrothion (Edwards and Thompson, 1973; Martin, 1975; Brown, 1977) while mite numbers have been reduced by phorate, chlorfenvinphos and chlorpyrifos (Edwards *et al.*, 1968; Brown, 1977; Hoy, 1980). Acarine and collembolan numbers were strongly depressed for a period of at least 6 months by isofenphos applied at normal rates to permanent pasture in Denmark, with hemiedaphic species such as *Isotoma viridis* and *Brachystomella parvula* being particularly affected (Petersen and Krogh, 1987). Fonofos, parathion, phorate and disulfoton are quite toxic to predatory carabid beetles, but chlorfenvinphos is not (Edwards and Thompson, 1975). Pauropods are quite sensitive to a range of organophosphates, including chlorfenvinphos, diazinon, disulfoton, parathion and phorate (Edwards, 1974b; Edwards, Dennis and Empson, 1967; Edwards, Thompson and Beynon, 1968): symphylids and myriapods are less sensitive to organophosphates except for parathion. Most organophosphates have little effect on earthworms, except phorate and, to a lesser extent, parathion and terbufos (Edwards, 1980; Clements, Bentley and Jackson, 1986).

Several carbamates are widely used as soil insecticides in agriculture, and these are more toxic to most soil animals than organochlorines or organophosphates. Aldicarb, carbofuran and, to some extent, carbaryl are nematicidal as well as being insecticidal; these compounds have been reported to depress mite and collembolan populations in the field (Martin, 1975; Brown, 1977; Hoy, 1980; Stanton, Allen and Campion, 1981; Ellis, Clements and Bale, 1990), although, at least in the case of carbofuran, populations may recover fairly rapidly. Methiocarb is widely used as a molluscicide for the control of slugs; it is also toxic to a range of surface-dwelling arthropods including carabid beetles (Purvis and Bannon, 1992). Most carbamates, including carbaryl, carbofuran and methiocarb, are relatively toxic to earthworms (Edwards, 1980; Clements *et al.*, 1991; Parmelee *et al.*, 1990). Clements, Bentley and Jackson (1986) recorded moderate to severe depression in numbers of earthworms in grass plots treated with carbofuran at seeding time in autumn as well as in plots treated with the organophosphates terbufos and phorate, but by the following spring numbers had recovered. Aldicarb, being very soluble in water, can be taken up by earthworms in large quantities in wet soils, causing irritation that brings them to the surface where they may be eaten by birds (Edwards, 1983). However, Ellis, Clements and Bale (1990) reported only transitory effects on earthworms of 10% granules applied at 5 kg ha^{-1} at the time of pasture reseeding.

The more recently introduced synthetic pyrethroids have not as yet been thoroughly evaluated from the point of view of their influence on non-target invertebrates. Although non-persistent compared with the organochlorines, as a group they are very toxic at low concentrations and can be expected to have at least transient effects on non-target invertebrates. Pyrethroids such as deltamethrin and cypermethrin are widely used for the control of aphids and aphid-borne viral diseases in cereal crops, and several studies have indicated reductions in numbers of polyphagous predators after their use. For example, Purvis, Carter and Powell (1988) studied the effects of a pyrethroid spray applied in the autumn to winter cereals at nine sites throughout England and Wales on populations of carabid and staphylinid beetles and spiders. The numbers of Carabidae caught in pitfall traps were reduced by about 70% compared with untreated controls for a period of 2 months and, in one case, a reduction in summer populations of some autumn-breeding carabids and staphylinids was noted. Activity of linyphiid spiders on treated plots was initially reduced by about 75% and this reduction persisted until at least early summer and, for most species, well into summer (Figure 5.17). Shires (1985) investigated the effects of cypermethrin, among other insecticides (parathion methyl and DDT), on predatory beetles, earthworms and litter decomposition in spring wheat. All insecticides induced a short-term reduction in numbers of carabids and staphylinids but neither earthworm numbers nor litter decomposition rates were affected.

Routine use of anthelminthic drugs for the control of intestinal nematode parasites in cattle give rise to concern about possible adverse effects of residues in dung on the dung-decomposer fauna. Wall and Strong (1987) reported severe effects of ivermectin administered as a ruminal bolus on the dung fauna and on dung-decomposition rates, while Madsen *et al.* (1990) attributed similar retardation of dung decomposition after ivermectin administration by injection, to adverse effects on flies.

The broader implications of the effects of pesticides on non-target invertebrates for the functioning of the biological community and for plant growth and fertility in grassland are more difficult to evaluate. It is possible that the widespread use of organochlorines for the control of soil pests in New Zealand pasture could have contributed to the relatively low level of effectiveness of natural enemies in controlling populations of species such as the grass grub *Costelytra zealandica* (Kain, 1979); however, these pesticides are no longer widely used in grassland in New Zealand and the adverse effects of currently used materials on natural enemies are likely to be transitory.

Several studies have demonstrated increased sward productivity after insecticidal treatment of grassland even in the absence of overt pest damage (Chapter 6). Such increases have been attributed to suppression of herbivore activity, but it has been suggested that they may be due at

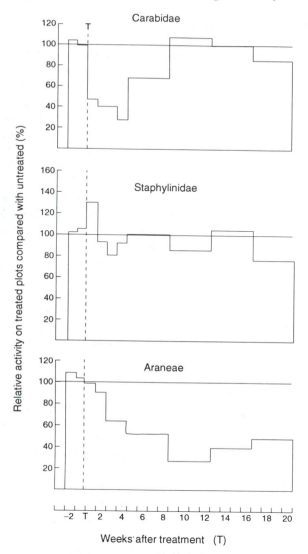

Figure 5.17 Activity of carabid and staphylinid beetles and spiders as measured by pitfall trap catches in winter cereal crops sprayed with synthetic pyrethroids compared with activity in untreated plots. (From Purvis et al., 1988.)

least in part to indirect effects on non-target soil organisms. There have been reports of stimulation of N mineralization by insecticides, possibly reflecting release of mineral N from the tissues of dead soil animals (Goring and Laskowski, 1982); some insecticides may also stimulate nitrification and N fixation (Davidson, Shackley *et al.*, 1979).

In extreme cases, direct links appear to exist between pesticide use, faunal depression and reduced rates of decomposition. Severe disruption of decomposition processes resulting in the accumulation of a surface layer of undecomposed organic matter has been noted in some instances involving the use of fungicides containing Cu and Hg and also substituted benzimidazoles (section 6.4.1 (c)). These effects have been linked with the suppression of earthworms, although reduction in cellulolytic fungi may also be a contributory factor (Pugh and Williams, 1971). Such disruption of decomposition processes can have a demonstrable effect on grassland fertility and productivity. Keogh (1979) reported a surface accumulation of dung and litter and a reduction of 23% in spring growth in sheep-grazed ryegrass pasture in New Zealand that had been treated with benomyl over two consecutive summers, reducing earthworm population densities from 620 m^{-2} to 140 m^{-2}. A surface organic layer developed and soil structure changed markedly in grass plots that had been treated frequently with large doses of insecticides (mainly phorate) to suppress pests over a period of 20 years in England (Clements *et al.*, 1991) (see Table 6.12). Earthworms had been eliminated from treated plots. The response of the grassland ecosystem to moderate pesticide treatment can be much more subtle. Malone (1969) recorded a drastic initial reduction in microarthropod numbers in an old field that had been sprayed with diazinon in the USA, and the rate of disappearance of organic matter was greater in treated than in untreated areas. The latter was attributed to stimulation of microfloral activity by the insecticide but there was no evidence to indicate whether this was a direct response to the chemical or whether it was an indirect effect mediated through reduction of microarthropod grazing pressure. In the first post-treatment growing season, species diversity, density and production of herb species were greater in the treated than in the untreated area. This might have been a response to decreased herbivore pressure on seeds and seedlings and enhanced nutrient availability resulting from microbial stimulation. Such changes in the flora would in turn influence the soil microclimate and the nature and quantity of litter returned to the soil-decomposer system.

Biocides have been used experimentally to manipulate soil-decomposer communities and this has been a useful approach in investigating the structure and functioning of such communities and in evaluating the role of various groups in ecosystem processes. This approach will be discussed further in Chapter 6.

5.7.2 Heavy metals and other industrial pollutants

Industrial extraction and smelting of metals result in the emission of small particles that contain heavy metals in various chemical forms. The particles may be deposited over a wide area, the deposition rate being greatest near the pollution source. Marked accumulation of heavy metals in litter and soil organic layers may result (Hughes, Lepp and Phipps, 1980). Heavy metal contamination of grassland may also arise from the spreading of pig slurry containing significant levels of Cu and Zn added to the food as growth promoters. Landspreading of sewage sludge can also result in high levels of metals such as Cu, Zn, Ni and Cd in soil and litter (CAST, 1976).

The impact of heavy metals on the soil ecosystem is complex and at present not well understood (Hughes, Lepp and Phipps 1980). Their biological activity is strongly influenced by factors such as chemical form, soil organic matter, the presence of chelating agents, the nature of the clay minerals and soil pH. The effects of gross metal contamination in the immediate vicinity of metal smelters can be very pronounced, especially in acid soils. Strojan (1978a) noted a marked accumulation of litter in a forest site 1 km from a zinc smelter in Finland where very high metal concentrations occurred in the litter (26 000 mg kg^{-1} Zn, 10 000 mg kg^{-1} Fe, 2300 mg kg^{-1} Pb, 900 mg kg^{-1} Cd, 34 000 mg kg^{-1} Cu). He found that CO_2 evolution and mass loss from litter samples decreased as proximity to the smelter increased. Likewise, Tyler (1975a,b) reported depressed soil enzymatic activity, reduction in organic matter decomposition and mineralization rates and litter accumulation in a spruce forest soil near an old brassworks. These processes were negatively related to Cu and Zn concentrations, nitrogen mineralization being depressed at Cu levels as low as 50 mg kg^{-1} in this acid soil. Severe depression in populations of litter microarthropods, especially oribatid mites, was reported by Strojan (1978b) in heavily contaminated forest soil, and by Watson *et al.* (1976) in the region of a lead-mining smelter. On the other hand, Williamson and Evans (1973) reported high densities of mites and collembolans in overgrown spoil heaps of disused lead mines in England.

The metals in sewage sludge appear to be relatively stable and they are not absorbed by plants to the same extent as are inorganic salts (Chang *et al.*, 1983). Sludge amendment enhances soil respiration (Mitchell *et al.*, 1978; MacGregor and Naylor, 1982) and appears to benefit earthworms (Hartenstein, 1986), at least in the short-term, although it is possible that long-term exposure may lead to adverse effects (see below).

Earthworms are particularly at risk from heavy-metal toxicity since they may ingest large quantities of contaminated soil and litter. However, Ireland (1983) concluded that these animals can tolerate relatively high levels of most heavy metals without adverse effects and there is also some evidence for biological adaptation through avoidance of heavily contaminated soil (Eijsackers, 1987). Earthworms can accumulate high concentrations of a range of metals within their bodies, and these may present a hazard to earthworm-feeding invertebrates. Metals in ionic form pose a more immediate threat to earthworms than those that are organically bound. Acute and chronic toxicity of a range of metals to *Eisenia fetida* has been demonstrated in the laboratory (Malecki, Neuhauser and Loehr, 1982; Neuhauser, Malecki and Loehr, 1984; Neuhauser et al., 1985). Significant depression in growth and reproduction occurred in *Eisenia fetida* populations exposed to some salts of Cd in concentrations as low as 50 mg kg^{-1}. Salts of other metals (Ni, Cu, Zn and Pb) were considerably less toxic. By contrast, no toxic effects of activated sludge containing up to 1500 mg kg^{-1} Cu on *E. fetida* were detected during a 4-month culturing period (Hartenstein, Neuhauser and Collier, 1980). Soil Cu and Zn levels reached 117 mg kg^{-1} and 103 mg kg^{-1}, respectively, in plots receiving heavy applications of pig slurry over a 4-year period, without apparent adverse effects on earthworms (Unwin and Lewis, 1986); however, van Rhee (1975) and Curry and Cotton (1980) concluded that earthworm populations in soils heavily contaminated with Cu from pig slurry over an extended period of time can be adversely affected in the longer term. This conclusion is supported by the work of Ma (1988), who reported reductions in the reproductive potential of earthworms related to Cu accumulation in soil and also reported that earthworms were deterred from recolonizing Cu-polluted soils. From comparisons of soils contaminated with single metals and multiple metals he concluded that Cu accounts for the greater part of the population decline in soils treated with sewage sludge or composted urban wastes.

The levels of soil Cu at which adverse effects on earthworms have been reported vary depending on soil conditions. In the unlimed acidic sandy soils studied by Ma, detrimental effects could be expected to start at about 50 mg kg^{-1}, with a drastic decline in numbers at about 100 mg kg^{-1}. Under those conditions, liming to raise the pH could alleviate the problem of Cu toxicity. By contrast, *Lumbricus terrestris* was abundant in soil containing 235–255 mg kg^{-1} Cu in the vicinity of a copper refinery, although earthworms were scarce close to the refinery where soil Cu levels were 2000–3000 mg kg^{-1} (Hunter and Johnson, 1982). Three- to four-fold accumulations of Cu and 10–20-fold accumulations of Cd occurred in the bodies of detritivorous macrofauna over levels in the contaminated soils (Hunter, Johnson and Thompson, 1987), but marked variation in metal concentration occurred throughout the year related to seasonal changes in abundance, species composition and age structure of the population.

5.8 SUMMARY

Management practices such as grazing, cutting, liming and fertilization, burning, and soil-water management greatly influence sward structure, composition and productivity and these effects are reflected in corresponding changes in the invertebrate fauna. As intensity of utilization increases, the invertebrate community becomes simplified and species able to tolerate disturbance and to exploit the greater productivity become dominant. Among the herbivores, aphids, thrips and leafhoppers are well adapted to exploit the higher leaf N content and prolonged vegetative growth associated with mown grass leys. Heavy mortality of the herbage-layer fauna occurs when the grass is cut, but recolonization is rapid and the population builds up quickly during the phase of grass regrowth. Swards that are kept short by continuous grazing support a low density of invertebrates. The higher productivity and more rapid turnover of organic matter and nutrients in managed temperate pastures is favourable for earthworms, which are the dominant component of the fauna under such conditions and which play an important part in this accelerated decomposition and mineralization cycle. A consequence of the enhanced rate of decomposition is the disappearance of the surface-litter layer and an impoverishment of the highly diversified litter community that lives in undisturbed grassland.

Amendment with organic materials has generally beneficial effects on detritivorous soil invertebrates, particularly earthworms, although heavy applications of raw organic wastes can be toxic initially to earthworms and other invertebrates. Moderate applications of mineral fertilizers generally have beneficial effects through increasing plant production, the nutrient content of plants and the return of litter to the soil, but high application rates of some materials including ammonium sulphate can be detrimental.

Fire depresses population densities of epigeic and hemiedaphic invertebrates, partly through direct mortality induced by high temperatures and partly through habitat alteration associated with removal of dead plant material and litter. However, the regrowth is of superior nutritional quality to unburned vegetation and supports a denser phytophagous fauna.

Grassland reclamation and improvement by removing barriers to biological activity such as low pH and excessive or inadequate moisture content, generally results in enhanced invertebrate activity but short-term depressions in numbers of susceptible groups may occur following disturbance associated with cultivation.

Many pesticides and heavy-metal pollutants at high levels can drastically affect susceptible groups including earthworms, but occasional doses of pesticides and moderate metal loadings associated with organic wastes do not appear to have any significant consequences; however, the long-term effects of metaliferous wastes over many years need further study.

6

Effects of grassland invertebrates on soil fertility and plant growth

6.1 INTRODUCTION

The main invertebrate activities that directly or indirectly influence soil fertility and plant growth are summarized in Table 6.1. As detritivorous forms normally comprise the greater proportion of the invertebrates biomass, their major ecosystem function in most situations is related to the decomposition and mineralization of organic matter. Invertebrates may influence those processes directly by their feeding and metabolism, and indirectly by physically altering the nature of their organic substrate and through interacting with the decomposer microflora. Larger soil invertebrates, notably earthworms, can also exert a pronounced mechanical influence on the soil environment. They can markedly accelerate decomposition processes by incorporating organic matter into the soil and can alter soil physical and chemical properties and water balance by burrowing, soil ingestion and mixing.

Herbivores can influence grassland productivity directly by consuming plant tissue and indirectly by altering rates of plant growth and senescence. The effects of herbivory on productivity are not always negative, but there are many situations in which serious pasture losses may be caused by pest species. Problems associated with evaluating such losses in economic terms and prospects for the development of effective and environmentally acceptable pest-management approaches are considered towards the end of this chapter. Invertebrates may also have an important role in the transport of other organisms (e.g. disease-causing viruses, plasmodia and bacteria, fungal spores and pollen) and these may greatly influence plant production and decomposition processes.

6.2 CONTRIBUTION TO ENERGY METABOLISM

All organisms require a constant source of food energy and the quantity of energy they utilize thus provides one basis for assessing the relative

Table 6.1 The main invertebrate groups in grassland and their ecosystem roles

Groups	Organic matter decomposition/ mineralization	Soil mixing/ affecting soil properties	Predation/ parasitism	Herbivory	Disease transmission	Pollination
Microfauna						
Protozoa	+		+			
Nematoda	+		+	+	+	
Mesofauna						
Enchytraeidae	+	+				
Acari	+		+	+	+	
Collembola	+			+		
Protura	+					
Diplura	+		+			
Pauropoda	+					
Symphyla	+			+		
Macrofauna						
Lumbricidae*	+	+		+		
Mollusca	+			+		
Isopoda	+					
Diplopoda	+			+		
Chilopoda			+			
Araneae			+			
Coleoptera	+		+	+		
Lepidoptera				+		
Diptera	+		+	+		
Thysanoptera				+		
Hymenoptera	+	+	+	+		+
Hemiptera			+	+	+	
Dermaptera	+		+			
Orthoptera	+			+		
Neuroptera			+			
Isoptera	+	+		+		

* and other earthworms

'importance' of different groups within the community and their overall contribution to ecosystem processes. A considerable volume of information is now available on the energetics of grassland invertebrate populations, much of it coming from studies carried out under the International Biological programme (IBP).

The data for lowland grassland sites are summarized in Coupland (1979) and Breymeyer and Van Dyne (1980); data for moorland and montane sites are presented in Heal and Perkins (1976) and Coulson and Whittaker (1978), while the role of invertebrates in tropical soils was reviewed by Lamotte (1975, 1979) and Lamotte *et al.* (1979). A comprehensive study of invertebrate contribution to energy flow in a Swedish fen grassland was carried out by Persson and Lohm (1977), while Hutchinson and King (1980) compared energy utilization by invertebrates and sheep in Australian pastures. An extensive review of the soil fauna of a range of habitats, their energetics and their contribution to decomposition processes is provided by Petersen and Luxton (1982), while Wiegert and Petersen (1983) review energy-transfer rates generally in insects and their allies.

In this section, the general approach to constructing energy budgets is briefly considered with reference to selected examples of grassland species, before going on to consider how energy-utilization coefficients derived from studies of individual species populations may be used to construct energy budgets for entire communities.

6.2.1 Estimating energy utilization

Holistic studies of community energetics are based on the 'trophic dynamic' concepts of Lindemann (1942), who combined the energetics approach with information on trophic structure to describe the functioning of a lake community in terms of the standing stock of energy at each trophic level and the efficiency of transfer to the next trophic level.

Since the grassland invertebrate community can comprise hundreds of species with very many trophic interactions, a complete picture of energy utilization is clearly impractical and any attempt to produce an overview must inevitably be reductionist. The usual approach is one of coupling field population and biomass data for the different trophic groups with energy-utilization parameters derived from detailed studies of representative species. Energy relationships can be described by the expressions $C = A + F$ and $A = P + R$, where C is the energy consumed, A is the energy assimilated, P is the energy fixed in tissue production, R is the energy dissipated through respiration and F is the unutilized energy (Petrusewicz, 1967).

Energy budgets for entire communities usually rely heavily on utilization coefficients derived from published studies on species representing different taxonomic, physiological and trophic groups (McNeill and Lawton, 1970; Heal and MacLean; 1975; Humphreys, 1979; Andrzejewska and Gyllenberg, 1980; Wiegert and Petersen, 1983). Respiratory metabolism may be estimated from biomass using the expression $Y = aX^b$, where Y is the oxygen consumption, X is individual live mass and a and b are constants that have been empirically determined for a range of species (Reichle, 1971; Ryszkowski, 1975; Persson and Lohm, 1977; Hutchinson and King, 1980); the other parameters of the energy equation may be calculated from R using appropriate coefficients. Some examples of studies that have provided the basic data for estimating the coefficients necessary to calculate community budgets are now considered.

(a) Dung beetle (Aphodius rufipes) larvae

In an attempt to quantify the role of *Aphodius* larvae in the decomposition of dung, Holter (1975) estimated the energy budget of a natural population living in cowpats in Denmark. Laboratory data on food utilization involving measurements of respiration, production and consumption (Holter, 1974) were combined with field data on temperature and population size to construct the energy budget.

Under laboratory conditions, the relationship between larval body weight and oxygen consumption could be described by the equation

$$Y = 0.060 \ X^{0.70},$$

where Y is the oxygen consumed (in millilitres) per larva per day and X is the live body mass (in milligrams). The relationship between temperature and respiration was described by the expression

$$\log Y = 1.069 + 0.073X - 0.011 \ X^2,$$

where Y is the percentage of respiration rate at 17 °C and X is the temperature (°C).

Manure consumption was measured by feeding larvae on dung containing chromic oxide as an indigestible marker and estimating dung consumption per unit time by determining the amount of chromic oxide ingested (Holter, 1973). Ingestion in the laboratory at 17 °C could be related to body weight by the expression

$$Y = 7.82 \ X^{0.61},$$

where Y is the dry dung (in milligrams) ingested per day and X is the dry larval body mass (in milligrams). By estimating larval consumption in the laboratory at different temperatures it was possible to derive a relationship between larval ingestion and temperatures within the range 5–24 °C; however, several sources of error may arise in the application of laboratory-derived temperature–respiration and temperature–ingestion

Table 6.2 Energy budget for a field population of dung beetle (*Aphodius rufipes*) larvae in cow pats (after Holter, 1975)

	C	P	R	A
Kilojoules per cow pat	913.4	37.2	39.3	76.5
Kilojoules per m²	19 465.2	793.7	838	1 631.8

relationships to field population data and these are discussed by Holter (1975).

Samples of dung were collected from the field site on 14 dates over a 12-month period and eggs, larvae and imagines were separated from the sample by wet sieving and $MgSO_4$ flotation. Abundance and individual mass of all instars on each sampling date were thus available for calculation of population respiration and consumption, and tissue production was calculated by the Allen graphical method (Petrusewicz and Macfadyen, 1970), which was appropriate in this instance since the oviposition period was short and the larval population could be considered as one cohort. The main features of the energy budget are summarized in Table 6.2. Holter draws attention to the remarkably high rate of energy flow through *Aphodius* compared with other terrestrial populations but points out that a cow pat represents a very concentrated source of energy with relatively high N content. Only 1.8–3.8% of the energy available was actually assimilated by *Aphodius*, but since the efficiency of assimilation (A/C) was low (8.4%) it is likely that a large proportion of the dung was ingested and egested by the larvae. A minimal estimate was that over 36% of the available dung passed through the larval populations at least once.

(b) Meadow spittlebug (Philaenus spumarius)

This is a widely distributed sucking insect in temperate grasslands. It is particularly associated with legumes and can be a serious pest of legume crops in some localities. It produces a single generation annually and overwinters in the egg stage. Hatching occurs in May in southern Michigan, USA, where its energetics were studied by Wiegert (1964). It feeds on a wide variety of plant species, preferring herbs to grasses. There are five nymphal instars and the final moult to the adult stage takes place in June. Mating begins soon after adult emergence and adult oviposition occurs in September. By late October the adults are absent from the study area. The sites studied comprised an old field with a diverse grass and herb flora and two alfalfa fields of different ages. Most spittlebug reproduction occurred on legume cropland, and high densities of nymphs occurred on alfalfa during spring and early summer. When the alfalfa crop is cut for hay the adults disperse into surrounding habitats, the population being restored by immigrants when alfalfa growth is renewed. The cycle is repeated during the second period of hay cutting. The resident population density in the old field was low but was augmented by large numbers of migrants during periods of alfalfa hay cutting.

Population densities of adults and nymphs were estimated at 3–7-day intervals in the old field during the summers (May–September) of 1959 and 1960, while population densities in the alfalfa field were monitored in 1960. Peak densities of nymphs declined from 8 to 12 m^{-2} in May to less than 1 m^{-2} in June in the old field. Adults had bimodal peaks of 10–25 m^{-2} in July and August, most being immigrants from nearby mowed alfalfa fields. The alfalfa field contained peak densities of 1280 nymphs and 466 adults m^{-2}. To determine energy flow (assimilation or intake of

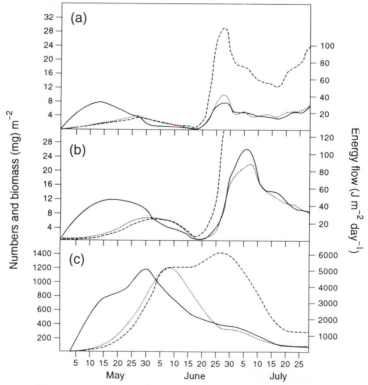

Figure 6.1 Fluctuations in numbers (——), biomass (- - -) and energy flow (. . .) through the populations of meadow spittlebug (*Philaenus spumarius*) in an old field in 1959 (a) and 1960 (b), and a small alfalfa field in 1960 (c). (After Wiegert, 1964.)

Table 6.3 Main features of population energy budget (kJ m^{-2} yr^{-1}) for meadow spittlebug (*Philaenus spumarius*) in an alfalfa field in Michigan (after Wiegert, 1964)

	Nymphs	Adults	Total
Production (growth + moulted exuviae)	67		
Respiration	51	43.5	94.6
Assimilation	118	43.5	161.5
Excretion	236	21.8	419.3
Consumption	354	65.3	419.3

metabolizable energy) through the spittlebug population the rates of xylem sap ingestion and egestion, respiration, tissue production, mortality and reproduction were determined. Figure 6.1 shows the changes in population density, biomass and energy flow in the study fields throughout the growing season. The annual energy flow was 161.5 kJ m^{-2} in the alfalfa field in 1960 (Table 6.3), compared with 2.6 kJ m^{-2} in the old field in 1959 and 5 kJ m^{-2} in the old field in 1960. Peak energy flow in both habitats was correlated with the highest growth rate of the vegetation and with the greatest concentration of amino acids in the xylem sap. The bulk of the energy uptake comprised amino acids. Nymphs assimilated 33% and adults 67% on average of the energy ingested, the difference in assimilation efficiency being associated with the excretion of organic materials required to depress surface tension and to facilitate the formation of foamy spittle in the case of the nymphs. The ecological growth efficiency in nymphs (P/C) was about 19% and about 16% for the population as a whole.

When considering how spittlebug consumption might influence vegetation production, Weigert reasoned that the effect of xylem sap consumption would be to remove a portion of the N supply essential for plant growth. On the assumption (supported by evidence provided by Weaver and Hibbs, 1952) that nitrogen removed by spittlebug feeding does not produce a change in the protein content per unit weight of vegetation, and assuming the N content of alfalfa to be 20% of dry mass, then 1 g of amino acid removed by feeding would reduce overall plant production by 5 g. Total ingestion by the nymphal population was equivalent to about 20 g amino acids, with adults accounting for a further 4 g; this resulted in an estimated reduction of plant production in the order to 120 g m^{-2}. This was within the range of losses reported by Weaver and Hibbs (1952) in a field-plot experiment comparing infested with treated plots.

(c) Isopod (Philoscia muscorum)

Philoscia muscorum was one of three isopod species inhabiting dune grassland that were studied over a period of 7 years at Spurn Head, Yorkshire, England (Hassall and Sutton, 1978; Hassall, 1983). These animals feed on dead plant material and associated microflora and the energetics of *P. muscorum* were studied over a 5-year period as part of an evaluation of their role in decomposition processes. A detailed account of population metabolism is given by Hassall (1983); the main results are summarized in the form of an average annual energy budget in Figure 6.2.

The annual respiratory losses for the two other species present,

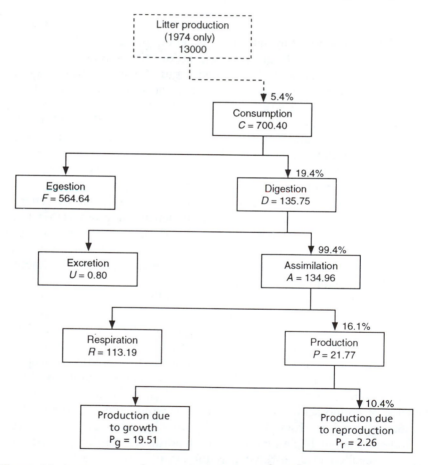

Figure 6.2 Average annual energy budget (kJ m^{-2} yr^{-1}) for a population of the isopod *Philoscia muscorum* in dune grassland. (From Hassall and Sutton, 1978.) Percentages indicate efficiencies of energy transfer.

Armadillidium vulgare and *Porcellio scaber*, were approximated and the total amount of energy dissipated by the three species was estimated to be 162.4 kJ m^{-2} yr^{-1}. The energy dissipated by the metabolic activity of *P. muscorum* represented about 0.9% of the total annual energy input in the form of plant litter, while the total energy losses arising from respiration by all three species comprised less than 1.3%. Consumption by *P. muscorum* represented about 5.4% while total energy consumption by all three species was estimated to be approximately 10% of the annual litter input. Ecological growth efficiency (P/C) was low (3.1%) compared with that of *Philaenus spumarius*, reflecting low food quality; production efficiency (P/A) was also relatively low (16.1%) compared with that of *Philaenus* nymphs (57%), indicating considerably higher energy costs associated with body maintenance.

6.2.2 Variability in food ingestion and utilization

Data on energy utilization by a range of grassland fauna are summarized in Table 6.4. The data reveal considerable variability, and it is often difficult to know the extent to which such variability reflects inherent differences between taxa or experimental error associated with estimating the different utilization parameters and with extrapolating from laboratory studies to field populations. Ingestion rates, for example, can be measured satisfactorily for larger chewing species by gravimetric means, but this approach is likely to be unreliable or impractical for small, liquid-feeding or microbial-feeding species. In such cases, radiotracer techniques may be valuable (e.g. Reichle, 1971; Luxton, 1972). Methodological difficulties apart, food ingestion rates may be influenced by factors such as developmental stage, food quality, temperature and season of the year. Some phytophagous beetles and caterpillars may ingest several times their body weight daily, while soil animals typically consume 1–10% of their body weight each day (Reichle, 1968; Andrzejewska, 1979b; Petersen and Luxton, 1982; Rushton and Hassall, 1983b).

Assimilation efficiency is also very variable, ranging from 1% to 65% among soil fauna (Petersen and Luxton, 1982), while Andrzejewska and Gyllenberg (1980) cite values ranging from 20% to 68% for grassland herbivores. Differences in assimilation efficiency do not appear to relate to trophic pattern in any obvious way, but are very strongly related to food quality. Saprophagous groups such as oligochaete worms feeding on low-quality organic matter tend to have low assimilation efficiencies and need to ingest large quantities of substrate to satisfy their food requirements.

The proportional allocation of assimilated energy to production and respiration likewise shows great variation between species and within species, depending on factors such as developmental stage, environ-

Table 6.4 Ingestion and energy utilization (kJ m^{-2} yr^{-1}) by field populations of selected grassland invertebrates (C = consumption, A = assimilation, P = production, R = respiration)

Group/Species	Habitat	C	A	P	R
Lumbricidae	Limestone grassland, England (Coulson and Whittaker, 1978)		1372	242	1130
Oligochaeta (large)	Sheep pasture, Australia (Hutchinson and King, 1980)	2500	375	103	272
Enchytraeidae	Limestone grassland, England (Coulson and Whittaker, 1978)		569	113	456
	Sheep pasture, Australia (Hutchinson and King, 1980)	1680	252	109	143
Acari (total)	Limestone grassland, England (Coulson and Whittaker, 1978)		9.6	3.3	6.3
	Sheep pasture, Australia (Hutchinson and King, 1980)	553	258	46	212
Collembola	Limestone grassland, England (Coulson and Whittaker, 1978)		14.6	6.7	7.9
Tipulidae	Sheep pasture, Australia (Hutchinson and King, 1980)	2590	1037	479	558
Scarabaeidae	Limestone grassland, England (Coulson and Whittaker, 1978)		720	385	335
Diplopoda	Sheep pasture, Australia (Hutchinson and King, 1980)	3000	925	400	525
Philoscia muscorum	Dune grassland, England (Hassall, 1978)	700.4	135	21.8	113.2
Total millipedes	Sheep pasture, Australia (Hutchinson and King, 1980)	1050	367	159	208
Nematoda	Limestone grassland, England (Coulson and Whittaker, 1978)		39.7	14.6	25.1
	Sheep pasture, Australia (Hutchinson and King, 1980)	273	164	62	102
Acrididae					
Melanoplus spp.	Alfalfa field, USA (Wiegert, 1965)	151	55.5	19.1	36.4
Melanoplus sanguinipes	Grassland, Tennessee, USA (Van Hook, 1971)	378	216	109	107
Hemiptera					
Leptopterna dolabrata	Limestone grassland, England (McNeill, 1971)	2.5	0.8	0.4	0.4
Cicadella viridis	Grassland, Poland (Andrzejewska, 1967)	271		23	
Philaenus spumarius	Alfalfa field, USA (Wiegert, 1964)	419	161	66	95
Neophilaenus lineatus	Grassy heath, England (Hinton, 1971)	9.5	4.0	1.5	2.5
Collembola					
Isotoma trispinata	Grassland, Japan (Tanaka, 1970)		202.2	91.7	110.5
Onychiurus procampatus	Limestone grassland, England (Hale, 1980)		5.5	2.1	3.4
O. tricampatus	Limestone grassland, England (Hale, 1980)		3.5	1.2	2.3
Isoptera					
Macrotermes subhyalinus	Savannah, Senegambia (Lepage, 1972)		44.4	13.5	30.9
Trinervitermes trinervoicus	Savannah, Senegambia (Lepage, 1972)		9.2	3.9	5.3

mental conditions, life-history strategy, method of feeding and food availability. Production efficiencies (P/A) cited for grassland invertebrates range from 8% to 57% (Andrzejewska and Gyllenberg, 1980; Petersen and Luxton, 1982). Production efficiency is higher for invertebrates as a group than for vertebrates, and is higher in non-social than in social insects (McNeill and Lawton, 1970; Humphreys, 1979); there do not appear to be clear-cut differences between invertebrate trophic groups (Wiegert and Petersen, 1983). Petersen and Luxton (1982) suggested that two groupings of detritivorous invertebrates may be distinguished on the basis of production efficiencies. One group with relatively high $P: R$ ratios comprises organisms that are compelled to store energy in body tissues to enable them to survive adverse or non-feeding periods. This group includes some oribatid mites, collembolans, tipulid larvae and millipedes. A second group with relatively lower $P: R$ ratios includes animals with a high metabolic rate during a relatively short period of activity, such as ants. These authors also note that $P: R$ ratios generally tend to be low for soil invertebrates, but tend to be greatest (> 0.5) at climatic extremes where respiration is likely to be restricted by severe environmental conditions and where more energy must be stored to withstand adverse conditions. Ecological growth efficiency (P/C) is generally less than 20%.

6.2.3 Energetics of entire communities

The starting-point for computing energy budgets for entire communities is usually a list of species or higher taxonomic groups with estimates of their population densities and biomass. The taxa are assigned to trophic groups on the basis of their feeding habits. Where these span more than one trophic level proportional assignments are made as appropriate. Community respiration (R) can be computed with some degree of confidence using appropriate biomass:respiration relationships (section 6.2.1) and adjustments can be made for temperature variations in the field using appropriate Q_{10} values. Calculations of respiration are usually based on respirometry studies of inactive animals under laboratory conditions, and corrections should be made for heat loss associated with growth and activity in the field (Hutchinson and King, 1980). Independent estimation of the other parameters of community energy budgets is rarely attempted; these are usually derived from R using conversion coefficients deemed to be appropriate for the various taxonomic, physiological or trophic groups. Examples of some of the relationships frequently used are given in Table 6.5. Considerable discrepancies are apparent between the mean values proposed for the various trophic groups by different authors, reflecting differences between the data sets used. Community budgets calculated with the aid of such relationships are inevitably crude; however, they can yield valuable insights into pat-

Table 6.5 Mean assimilation and production efficiencies (in %) of different invertebrate categories

Author	Invertebrate categories	A/C	P/A	R/A
Heal and MacLean (1975)	Invertebrate herbivores	40	40	60
	Invertebrate carnivores	80	30	70
	Invertebrate saprovores	20	40	60
	Invertebrate microbivores	30	40	60
Humphreys (1979)	Non-insect invertebrates		25	75
	Non-insect herbivores		21	79
	Non-insect carnivores		28	72
	Non-insect saprovores		36	64
	Non-social insects		41	59
	Non-social herbivores		39	61
	Non-social carnivores		56	44
	Non-social saprovores		47	53
Wiegert and Peterson (1983)	Insects and other arthropods			
	Herbivores		44	56
	Carnivores		38	62
	Saprovores		27	73

terns of energy utilization within and between different communities, as will be seen from the examples considered below.

(a) Fen grassland, Sweden

The soil fauna and above-ground arthropods of an abandoned grass field on fen peat soil at Spikbole in central Sweden were sampled at intervals over a year (Persson and Lohm, 1977). Mean abundance, biomass and respiration were estimated for about 250 species, but only the pooled data for major taxonomic and feeding categories are considered here.

The mean annual biomass of invertebrates (annelids plus arthropods) was 10.8 g dry mass m^{-2}, with above-ground arthropods comprising less than 2% of this (Figure 6.3). Earthworms and beetles were the dominant groups in terms of biomass. Below-ground animals respired 837 kJ m^{-2} yr^{-1}, while above-ground arthropod respiration amounted to a further 27 kJ m^{-2}. The lumbricid dominance of respiration was proportionately less than their dominance in terms of biomass, reflecting their low respiration rate per unit mass, while Diptera and enchytraeid worms contributed relatively more to invertebrate respiration than to biomass. Saprophagous

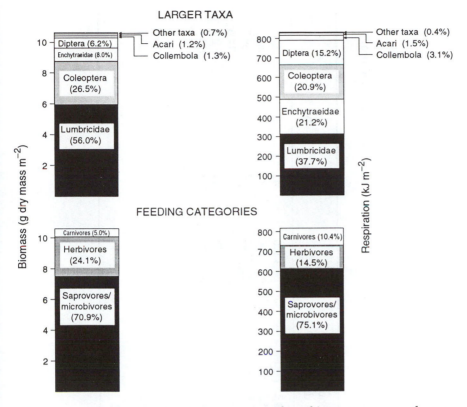

Figure 6.3 Contribution of major taxonomic and trophic groups to annual mean biomass and respiration of below-ground annelids and arthropods in fen grassland, Sweden. (After Persson and Lohm, 1977.)

forms such as microbivores dominated biomass and respiration, accounting for 628 kJ m^{-2} yr^{-1} (75% of the total animal respiration) while herbivores and predators accounted for 15% and 10%, respectively.

Above-ground herbivore respiration was equivalent to 0.2% of above-ground net primary production (NPP), while herbivore consumption as estimated according to the $R{:}C$ ratio suggested by Heal and MacLean (1975) was equivalent to a removal of 0.8% of NPP. Nematodes had not been included in the study, but when an estimate was included for plant-feeding forms based on literature data the total respiration of below-ground herbivores was estimated at 150 kJ m^{-2} yr^{-1}, corresponding with consumption of 628 kJ m^{-2} yr^{-1}, which was equivalent to 5.3% of NPP. Thus, total estimated herbivore consumption amounted to 6.1% of NPP. Total annelid and arthropod respiration in the soil was estimated to comprise 6–7% of total heterotrophic soil respiration at Spikbole, while

Table 6.6 Population densities, biomass and energetics of invertebrates in limestone grassland at Moor House, England (from Coulson and Whittaker, 1978)

	Numbers (m^{-2})	Biomass (g dry mass m^{-2})	R	P (kJ m^{-2} yr^{-1})	A
Lumbricidae	390	23.2	1130	242	137.2
Enchytraeidae	80 000	4.1	456	113	569
Nematoda	3.3 × 10^6	0.14	25.1	14.6	39.7
Collembola	46 000	0.15	7.9	6.7	14.6
Acari	33 000	0.35	19.7	11.2	0.9
Tipulidae	120	4.1	335	385	720
Carabidae	3	0.007	0.4	0.4	0.8
Araneida	40	0.003	0.8	0.4	1.2
Lepidoptera	<1	–	–	–	–
Hemiptera	100	0.010	2.1	1.3	3.4
Total		32	1977	775	2752

an adjustment for groups not measured – protozoans, rotifers, tardigrades and molluscs – might increase this figure to about 8%.

(b) Moorland grassland, England

Studies at Moor House, Yorkshire, were carried out on a variety of habitats, ranging from shallow mineral soils to peat (Coulson and Whittaker, 1978). Moor House is an area of high rainfall, with relatively low soil temperatures. Blanket bog covers 80% of the study area, but there are also extensive areas of limestone grassland, which are grazed by sheep, and these have a well-diversified soil fauna similar in composition to that of lowland grasslands (Table 6.6). Sheep consumed 80% of aboveground primary production (c. 40% of NPP), but they accounted for only about a third of total energy assimilation by animals. Sheep dung was a major source of food for decomposers, comprising almost 60% of the organic matter transferred to them annually.

Climatic and soil conditions at Moor House are favourable for soil fauna, with saprophagic lumbricid earthworms, enchytraeids and tipulid larvae being dominant in terms of biomass and energy flow. Invertebrate biomass was about three times greater at Moor House than at Spikbole, and their contribution to ecosystem energetics was correspondingly greater with invertebrate respiration representing 21.3% of NPP.

(c) Sheep-grazed pasture, Australia

One of the few studies to consider relationships between invertebrates and domestic animals was that reported by Hutchinson and King (1980), who compiled energy budgets for sheep and invertebrate groups in pastures grazed at three stocking densities at Armidale, New South Wales, Australia. The stocking rates were 10, 20 and 30 sheep ha^{-1} and the responses of 16 invertebrate groups to the different intensities of grazing were monitored over a 3-year period. Energy losses attributable to maintenance metabolism (R_1) were calculated from fresh body mass (W) based on the relationship aW^b using appropriate values for the constants a and b and Q_{10} values for temperature correction selected from the literature. Estimates for heat losses associated with feeding activity and with converting assimilated energy into tissue growth were added to R_1 values to calculate total respiratory losses of actively feeding and growing animals (R_2).

The 16 invertebrate groups present comprised members with differing methods of feeding, so they were apportioned to different trophic categories for the purpose of compiling energy budgets on the basis of fractional allocations that reflected the known feeding habits of their constituent families and dominant species. Energy budgets for sheep and

Table 6.7 Annual energy losses (R_2) (MJ ha^{-1} of invertebrates and sheep in Australian pasture (after Hutchinson and King, 1980)

| | Numbers of sheep (ha^{-1}) | | |
	10	20	30
Mesofauna	6 810	4 360	1 170
Macrofauna	11 490	11 840	5 010
Total invertebrates	18 300	16 200	6 180
Sheep	44 200	83 900	86 500

Table 6.8 Percentage energy losses (R_2) by the various invertebrate groups in Australian pasture (after Hutchinson and King, 1980)

| | Numbers of sheep (ha^{-1}) | | |
	10	20	30
Herbivores	28.5	28.5	28
Decomposers	63.5	66	63
Predators	8	5.5	9

invertebrates at the three stocking densities are presented in Figure 6.4. Some general aspects of the ways in which grazing intensity altered patterns of herbage utilization and litter return have already been considered in Chapter 5. Insofar as respiratory metabolism can be viewed as an indicator of success in deriving energy from the grassland system, the relative magnitudes of R_2 values associated with different invertebrate groups and with sheep under the three stocking regimes are of interest (Table 6.7).

There was a notable decline in invertebrate respiration at the highest stocking density, with the largely hemiedaphic mesofauna (Collembola, Acari, Enchytraeidae, Coleoptera larvae, etc.) being particularly affected. The contribution of the mesofauna to total invertebrate respiration declined from 37% at the lowest stocking level to 19% at the highest. The proportional contribution of the three trophic groups to invertebrate respiration remained relatively constant (Table 6.8), indicating that they were all similarly influenced by grazing pressure.

Notable features of the energy budgets (Figure 6.4) are the ways in which stocking density altered the relative magnitude of the different pathways of energy flow. There was a marked drop in ingestion by invertebrate herbivores from 21 GJ ha^{-1} yr^{-2} at 10 sheep ha^{-1} to 7.4 at

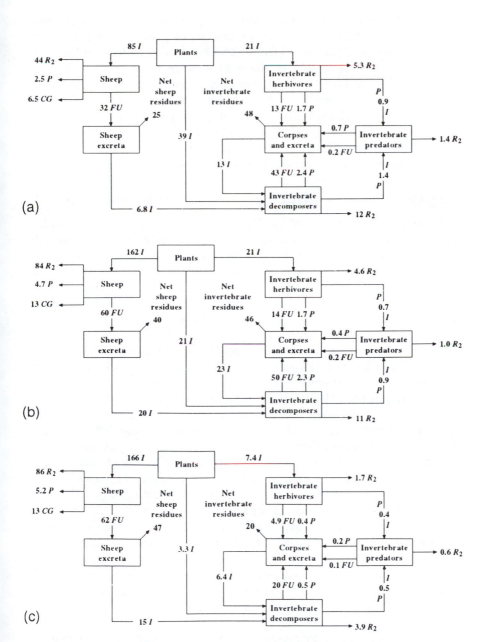

Figure 6.4 Energy budgets (GJ ha^{-1} yr^{-1}) for sheep and invertebrates at three stocking densities, Armidale, New South Wales: (a) 10 sheep ha^{-1}; (b) 20 sheep ha^{-1}; (c) 30 sheep ha^{-1}. I = intake, FU = excreta, CG = combustible gases, R_2 = respiratory losses, P = production. (From Hutchinson and King, 1980.)

30 sheep ha^{-1}, while litter ingestion by saprophagic invertebrates dropped from 39 to 3.3 GJ ha^{-1}. The importance of plant-litter return to the soil via sheep faeces increased from 32 GJ ha^{-1} to 62 GJ ha^{-1} at the same time. An interesting feature of the data was the importance of invertebrate corpses and excreta, which, together, were a more significant route of litter return to the soil than sheep faeces at all but the highest stocking density. This material would form a more evenly distributed resource for decomposers than sheep dung since the camping behaviour of Merino sheep leads to an uneven distribution of their excreta and significant dislocation of nutrients under conditions of intensive stocking.

6.2.4 Contribution of invertebrates to total grassland energetics

Few studies have attempted to estimate energy utilization by the entire grassland invertebrate community but available information suggests that its direct contribution to grassland energetics is relatively small. Invertebrate herbivore consumption as a proportion of net primary production in terrestrial ecosystems may be as low as 1% and rarely exceeds 10% (Wiegert and Evans, 1967; Sinclair, 1975; Gibbs, 1976). Estimates from grassland include 6.1% for Swedish fen grassland (Persson and Lohm, 1977), 7% for a meadow in Finland (Törmälä, 1982), 5–9% for Polish meadows (Breymeyer, 1971) and 10% for a *Festuca/Andropogon* old field in Kentucky (Van Hook, 1971). However, as will be discussed later, herbivore–plant interactions are complex and there is no simple correlation between tissue loss due to herbivory and plant growth.

Estimates for total invertebrate respiration as a proportion of heterotrophic metabolism range from 6–7% in Swedish grassland (Persson and Lohm, 1977) to about 20% in grassland with a large faunal biomass (Macfadyen, 1963; Coulson and Whittaker, 1978), a relatively modest contribution to the total energy economy of grassland. However, a very high proportion of plant litter may be ingested by saprophages. This may be at least 20–30% of the total organic matter input in most temperate soils (Peterson and Luxton, 1982) and undoubtedly much more in earthworm-rich soils. Although only a small proportion of litter may be assimilated, passage through the alimentary tract can considerably alter the characteristics and subsequent decomposition of this material. This is one of several ways in which soil-invertebrate decomposers may exert an indirect influence on soil processes and soil fertility.

6.3 INFLUENCE ON DECOMPOSITION AND MINERALIZATION

6.3.1 Litter and dung consumption

Earthworms are the major litter consumers in most temperate grasslands. Consumption rates as high as 80 mg dry mass g^{-1} fresh mass d^{-1} were recorded by Needham (1957) for *Lumbricus terrestris* in laboratory culture, the average rate being about 27 mg. Van Rhee (1963) reported mean daily grass consumption rates of 12–17 mg g^{-1} by six earthworm species under seminatural conditions. Assuming an average consumption rate of 15 mg dry mass g^{-1} fresh mass d^{-1}, an earthworm population with a fresh biomass of 100 g m^{-2} active for 200 days could consume 300 g dry mass of grass litter a year, a considerable proportion of the total litter return to the soil in most temperate grasslands. Isopods and millipedes may also consume significant quantities of litter in rough grassland (Macfadyen, 1963; Hassall and Sutton, 1978). In tropical soils, termites consume large quantities of litter that they concentrate in their mounds, thus depleting large areas of organic matter (Josens, 1972; Wood, 1978; Wood and Sands, 1978; Lee, 1979). Termites possess a gut microflora with the capacity to digest cellulose while some (Macrotermitinae) cultivate fungus gardens to increase the digestibility of organic matter (Wood, 1978). Thus, unlike earthworms, they have a high assimilation efficiency and their faeces are a poor substrate for other decomposers.

Enzymes capable of digesting plant structural polysaccharides have been recorded from many groups of invertebrates including isopods, millipedes, molluscs, dipterous larvae, beetle larvae and oribatid mites (Nielsen, 1962; Luxton, 1972; Hartenstein, 1982); however, the direct role of soil invertebrates with the exception of termites in primary litter decomposition is probably small. Their main contribution is probably that of rendering litter more amenable to microbial decomposition by ingesting and fragmenting it and by incorporating it into the soil. The role of earthworms in the latter process is particularly important, as can be seen when their activities are disrupted. Suppression of earthworms by pesticides results in the development of a surface layer of undecayed organic matter (van Rhee, 1963; Keogh and Whitehead, 1975), while, conversely, accumulations of surface dung and dead grass rapidly disappear when earthworms become established in pastures where they were not previously present (Barley and Kleinig, 1964; Stockdill and Cossens, 1966; Hoogerkamp, Rogäar and Eijsackers, 1983).

Coprophagous invertebrates may make only a minor contribution to community energetics in dung pats (Section 6.2.1), but the role of dipterous larvae, dung beetles and earthworms in the decomposition of dung and in its incorporation into the soil is crucial. The absence of indigenous fauna adapted to cope with the large quantities of ungulate dung produced in intensively grazed sown pastures in Australia is a major factor contributing to the problem of dung accumulation. This results in pasture fowling, sward deterioration, immobilization of plant nutrients in faeces, and significant loss of N through volatilization. Extensive efforts have been made to introduce coprophagous beetles in attempts to alleviate these problems (Gillard, 1967; Ferrar, 1973). Dipterous larvae consumed about 16% of the total quantity of dung produced by grazing sheep in a Polish mountain pasture (Olechowicz, 1976b). Holter (1979) investigated the contribution of *Aphodius* to the decomposition of cattle dung in Denmark. *Aphodius rufipes* and other night-flying scarabaeid beetles were selectively excluded by covering some of the pats with gauze during the first 5–7 nights of exposure and the rates of disappearance of these pats were compared with those that had not been covered. The results suggested that larval *Aphodius* were responsible for about 14–20% of mass loss from the pats but the main contribution of the beetles to decomposition was in mixing dung into the soil and in stimulating microbial activity. However, in this study the main agents in dung decomposition were earthworms (notably *Aporrectodea longa*), which accounted for about 50% of dung disappearance from the pasture surface. Dung is readily consumed by most earthworm species. Guild (1955) calculated average consumption rates of cow dung by *Aporrectodea longa, A. caliginosa* and *Lumbricus rubellus* in culture of 16–40 g dry mass per worm per year while Barley (1959b) reported that *A. caliginosa* consumed 80 mg dry mass g^{-1} body mass d^{-1} when dung was the only food offered. When other organic materials (dead grass, clover leaves, dead roots) were also available this species consumed 1% of its own mass of dry dung per day and Barley calculated that a field population of 80 g biomass m^{-2} active for 150 days under Australian conditions would consume 120 g m^{-2} of dung per year, assuming an equivalent rate of dung selection in the field.

6.3.2 Effects of fauna on rates of litter decomposition

Decomposition is a complex process and decomposition rates are influenced by many factors that include the characteristics of the organic matter itself, particularly its quality as a resource for decomposer organisms (Swift, Heal and Anderson, 1979). The contribution of the fauna to this process has received considerable attention but is still inadequately defined in many respects. An experimental approach that has yielded some insight into the role of the fauna is the use of chemical and physical techniques selectively to exclude various groups of biota from experimental litter. Different-size classes of invertebrates may be excluded by using mesh of different aperture size, while various biocides have been used in attempts to eliminate or suppress different organisms. In practice, however, biocides are not sufficiently selective to allow more than very general conclusions to be drawn (Ingham, 1985; Ingham *et al.*, 1986), while mechanical methods at best allow a crude partitioning of macrofaunal and/or mesofaunal effects from those of other biotic and abiotic agencies.

Generally, studies with deciduous woodland litter in temperate soils have indicated moderate to marked retardation of decomposition rates when invertebrates are excluded or suppressed (Edwards and Heath, 1963; Crossley and Witkamp, 1964; Curry, Kelly and Bolger, 1985). Seastedt (1984) reviewed the literature with reference to mesofauna and reported an average retardation in decomposition rate of 23% when they were excluded, but the data showed a wide range of variation, ranging from 0 to 70%. Curry (1969b) reported only a negligible effect of excluding meso- and macrofauna on mass loss from *Festuca/Agrostis* litter confined in litterbags, while Zlotin (1971) concluded that the fauna played only a minor role in grass-litter decomposition under the severe climatic conditions of steppe grassland, where 67% of grass-litter decomposition was directly attributed to abiotic factors. Vossbrinck, Coleman and Woolley (1979) used a combination of biocides and physical (mesh size) exclusion techniques in an attempt to partition abiotic and biotic factors involved in grass-litter decomposition under semiarid prairie grassland conditions. Blue grama grass litter was confined in litter bags with 53-μm mesh to exclude mesofauna or in 1-mm mesh bags to allow access to mesofauna, while in one (abiotic) treatment a solution of $HgCl_2$ plus $CuSO_4$ was applied to prevent microbial activity. After 9 months, 7.2% of the litter had disappeared from the abiotic treatment, 15.2% from the mesofauna-excluded treatment and 29.4% from the treatment to which mesofauna had access. Earthworm exclusion had a marked effect on grass-litter decomposition rates in England (Dickinson, 1983): 65% of the litter had disappeared within 85 days from litter bags to which earthworms had access compared with 30–40% when earthworms were excluded.

Very little of the mass loss from litter bags can be directly attributed to faunal metabolism; enhanced mass loss has been attributed mainly to the comminution of litter by faunal feeding and to its removal from the litter bags either by the animals themselves or by leaching, and to enhanced microbial activity. Indeed, it has been suggested that the soil fauna can indirectly influence pathways of energy and mass transfer in soils to a degree that may be an order of magnitude higher than direct faunal contribution to energy and nutrient fluxes (Anderson, 1988).

The view has long been held that soil animals can exert a major influence on decomposition processes through their interactions with the microflora. Macfadyen (1964) viewed the relationship between soil mesofauna and microorganisms as a symbiotic one, with microorganisms providing the enzymes required to break down plant structural compounds, while the animals enhance and promote microbial activity in a variety of ways. Subsequent work has tended to support this view, although it is now clear that high levels of feeding by microbivorous animals can suppress microbial activity under certain circumstances.

Ingestion and incorporation of litter into the soil by macroinvertebrates brings it into an environment that is more favourable for microbial activity, and this may be the major contribution to decomposition by litter feeders such as anécique Lumbricidae. Litter fragmentation during maceration and passage through the gut could be expected to facilitate leaching of soluble nutrients and to increase the surface area available for microbial colonization. This surface area effect was investigated by Hassall and Sutton (1978), who studied the effects of passage of litter through the digestive tract of isopods on the surface area of decaying leaves of *Hippophae rhamnoides* (sea buckthorn) and *Poa pratensis* (smooth meadow-grass). They calculated that total surface area was increased by factors ranging from $\times 3.5$ to $\times 4.4$ over a range of four species occurring in dune grassland at Spurn Head, Yorkshire, England. Other ways in which invertebrates are considered to stimulate microbial activity include browsing on senescent colonies, promotion of the germination of fungal spores on passage through the gut, and dissemination of microbial propagules throughout the soil and litter (van der Drift and Witkamp, 1960; Macfadyen, 1964; Harding and Stottard, 1974), although it has proved difficult to quantify these effects. The faeces of detritivorous invertebrates often appear to be a more suitable substrate for micro-organisms than the litter on which they feed, and enhanced microbial activity has been recorded from the faeces of the millipede *Glomeris marginata* and other soil animals (Anderson and Bignell, 1980; Hanlon, 1981). However, it does not always appear to be the case that faeces decompose faster than non-ingested litter. Hassal, Turner and Rands (1987) concluded from laboratory studies on the influence of isopods on the decomposition of *Betula pendula* (silver birch) leaf litter that frag-mentation and digestion of litter *per se* does not necessarily accelerate subsequent decomposition, while Griffiths, Wood and Cheshire (1989) reported that the faeces of *Porcellio scaber* fed on ^{14}C-labelled *Limna gibba* (pondweed) actually decomposed at a slower rate than the plant tissue from which it originated.

Earthworm casts tend to support greater microbial activity than the surrounding soil, and this probably reflects selective feeding on organic residues and their associated microflora (Barley and Jennings, 1959; Went, 1963; Loquet, 1978; Satchell, 1983). In a detailed study of microbial activity associated with earthworms in grassland in France, Loquet *et al.* (1977) found differences in bacterial populations between casts and sur-face soil, and also between the 'drilosphere' – the burrow walls lined by earthworm casts – and the bulk soil. Compared with soil from the top 6-cm horizon, casts contained more cellulolytic, hemicellulolytic, amylolytic and nitrifying bacteria and fewer denitrifying bacteria. Burrow walls to a mean depth of 30 cm contained more *Azotobacter*, more N-fixing bacteria, and more denitrifying, proteolytic and ammonifying bacteria than soils not influenced by earthworms. Enrichment of earthworm casts with organic matter and microbial populations was reflected in a correlated

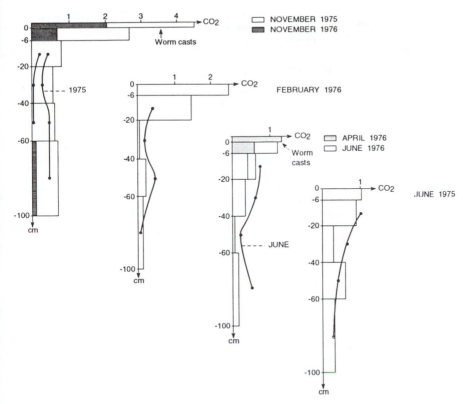

Figure 6.5 Effects of earthworms on evolution of CO_2 (mg g^{-1} dry soil wk^{-1}) from soil, worm casts and the drilosphere (burrow linings) at different depths and in different seasons. Histograms represent soil and casts; curved lines represent the drilosphere. (From Loquet, 1978.)

increase in respiratory and enzymatic activity but results for the drilosphere were less clear-cut (Figures 6.5 and 6.6).

Microcosm experiments carried out under controlled conditions offer evidence that invertebrates can significantly influence the decomposition of organic matter and mineralization rates by feeding on microorganisms. Examples of such studies, and the possible implications of microbial grazing for nutrient cycling, are considered in the next section.

6.3.3 Soil fauna and nutrient cycling

As in the case of energy utilization, invertebrates can contribute both directly and indirectly to nutrient fluxes. The direct effects are exerted through trophic transfer within the food web and are concerned with feeding, excretion and tissue production and turnover, while the indirect

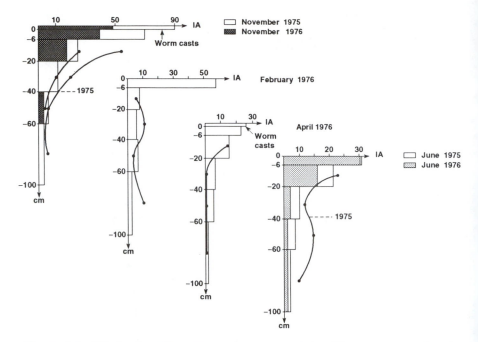

Figure 6.6 Effects of earthworms on invertase activity (IA; mg sugar reduced g^{-1} dry soil d^{-1}) in soil, worm casts and the drilosphere. Histograms represent soil and casts; curved lines represent the drilosphere. (From Loquet, 1978.)

effects relate to the influence of the fauna on the activities of micro-organisms and on the physicochemical properties of the soil-litter habitat generally. Nitrogen and phosphorus have received most attention because these elements can frequently be limiting in grassland ecosystems. The effects of the fauna on energy metabolism and nutrient cycling are related, but not necessarily in a linear fashion. Different mineral nutrients may be utilized with different efficiency, but, generally speaking, detritivorous invertebrates appear to be more successful in extracting nutrients from food substrate rather than energy. Table 6.9 gives estimates of faunal contribution to N mineralization in a range of sites for which N budgets have been compiled.

The data indicate a generally greater role for the fauna in N mineralization than their metabolic activity would suggest. This has been attributed to the relatively low production efficiency of invertebrates generally and hence their relatively low requirement for N compared with respiratory C (Anderson *et al.*, 1981a). Excess N and other nutrients are returned to the soil through defecation and excretion, much of the N being excreted as NH$_3$, urea or amino acids. Much of this nutrient supply may be rapidly immobilized in microbial biomass, which in turn may be

Table 6.9 Contribution of soil fauna to N mineralization in a range of soils

	Net mineralization $(g\ N\ m^{-2}\ yr^{-1})$	Percent mineralized by the fauna
Meadow fescue ley, Sweden (Paustian *et al.*, 1990)	21.4	16 ⎫
Unfertilized barley, Sweden (Paustian *et al.*, 1990)	8.0	27 ⎬ (mainly microfauna)
Lucerne, Sweden (Paustian *et al.*, 1990)	14.7	25 ⎭
Manured arable land, Denmark (Andersen, 1983)	30	10 (earthworms) 5–12 (nematodes)
Grassland, New Zealand (Keogh, 1979)	60	20 (earthworms)
Shortgrass prairie, USA (Hunt *et al.*, 1987)	7.8	37 (mainly microfauna)

fed on by microbivorous fauna; this process results in the rapid turnover of a small pool of nutrients that may be of considerable importance for plant growth (Coleman, Reid and Cole, 1983). Microbivorous protozoans and nematodes appear to be particularly significant in this context, and the relative importance of microbivory in N mineralization appears to increase as the system becomes poorer in N (Rosswall and Paustian, 1984). The contribution of the microfauna, and of microbial grazing mesofauna such as collembolans, to mineralization processes will be considered further in section 6.3.3(b).

(a) Earthworms and nutrient cycling

Earthworms ingest dead plant material that usually has a C: N ratio of 20:1 or more and convert it into body tissue of low C: N ratio (about 5:1). Thus earthworm tissues represent a rich source of N that is rapidly mineralized after death. The N utilized by earthworms is returned to the soil as protein in dead tissue, as waste products of metabolism excreted in urine, as mucoproteins in intestinal mucus added to ingested soil and litter and in mucus excreted onto the body surface for lubrication and maintenance of surface moisture for respiration. Assuming annual tissue production to be two to five times the mean biomass and tissue N content to be 10–12.5% of dry mass, over a range of earthworm dry mass from 5 $g\ m^{-2}$ to 45 $g\ m^{-2}$, the inputs of N from dead tissue could range from less than 10 up to 225 kg ha^{-1} (Lee, 1983a). Most of this N is rapidly released by bacterial decomposition (Satchell, 1967), while most of the N excreted in urine is in the form of ammonia or urea (Laverack, 1963). The relative proportions of ammonia and urea vary between species and within species in response to changes in factors such as nutritional level, temperature, soil texture and, particularly, water availability. Estimates of N output in ammonia and urea for fasting earthworms under experimental conditions range from 97–159 µg g^{-1} day^{-1} for *Aporrectodea caliginosa* and *Lumbricus terrestris* to 308–391 µg g^{-1} for *Eisenia fetida* and *Metaphire californica* (Needham, 1957; Khalaf El-Duweini and Ghabbour, 1971). Assuming the biomass of lumbricid earthworms to range from 50 to 150 g wet mass m^{-2} in temperate soils, an average output of N in ammonia and urea of 200 µg g^{-1} day^{-1} and an activity period of 6 months per year, Lee calculated that N excretion by earthworms would commonly be in the range 18–50 kg ha^{-1} yr^{-1}, with possibly a further 18–35 kg ha^{-1} contribution from non-lumbricid earthworms in some regions. Losses of N in mucoproteins secreted by epidermal glands appear to be of the same order of magnitude as losses in urine (Needham, 1957). Few workers have attempted to quantify N losses in the form of intestinal mucus, but Lavelle (1988) calculated that intestinal mucus could add 6–16% of readily assimilable organic matter to ingested tropical soil and litter. Lee's calculations suggest that earthworms could contribute up to 300 kg N ha^{-1} yr^{-1} in mineral or readily mineralizable form to the soil-nutrient pool in earthworm-rich soils, although Syers and Springett (1984) suggested a more conservative range (18–92 kg ha^{-1}) for grassland. However, Keogh (1979) estimated that earthworms contributed 109–147 kg mineral N ha^{-1} to productive New Zealand pasture, roughly 20% of the total estimated annual turnover of 600 kg from organic to mineral N.

Almost all estimates of N turnover by earthworm populations have been obtained indirectly, combining laboratory estimations of urine and mucoprotein excretion with field biomass data and estimates for tissue turnover rates of unknown reliability. One of the few attempts to quantify N turnover rates directly in the field was that reported by Ferrière and Bouché (1985), who used a double-labelling technique (a colour dye and ^{15}N) to calculate rates of N flow through an adult population of *Nicodrilus longus* (= *Aporrectodea longa*) in undisturbed permanent pasture in France. Labelled worms were reintroduced into the pasture and the levels of initial N remaining in the bodies of individuals recaptured at intervals were determined. The rate of N flow in the field was calculated to be 105 mg g^{-1} body N d^{-1}, indicating much higher levels of N excretion by active worms in the field than those based on the laboratory studies referred to earlier. Extrapolating these results to the total earthworm faunal biomass of 153 g m^{-2}, Bouché, Ferrière and Soto (1987) calculated that the annual N output of the earthworm population could be as high as 800 kg ha^{-1} yr^{-1}. However, this is probably an overestimate since it is unlikely that such a high rate of turnover could be ascribed to anything other than a very labile fraction of the total earthworm N pool. Preliminary results from ^{15}N studies on *Lumbricus terrestris* currently under way in this laboratory do not indicate such major discrepancies between direct and indirect estimates for N turnover.

Thus, available estimates suggest that the direct role of earthworms in N mineralization can range from being very modest to being quite substantial. What is not clear is whether this mineralization would occur microbially in the absence of earthworms anyhow, although there is considerable evidence that when earthworms are absent conditions for microbial activity are often unfavourable. Much of the earthworm-derived N is probably assimilated by microorganisms, which are in turn consumed by earthworms and other soil invertebrates, resulting in rapid recycling of N (and other nutrients) within the soil biotic community. However, there is evidence that a high proportion of earthworm N is taken up by plants during periods of high N demand: Bouché and Ferrière (1986) reported that ^{15}N-labelled N from earthworms was taken up rapidly and almost entirely by grassland plants in spring. Two phases of assimilation could be distinguished, one comprising excretion products (ammonia, urea, etc.), which were almost immediately available to plants, and the other comprising mucus that requires microbial degradation before it can be assimilated.

The indirect contribution of earthworms to N mineralization is difficult to quantity, but in view of their capacity to process plant residues and to alter soil conditions it is likely to be considerable. Since probably less than 10% on average of ingested organic matter is actually assimilated by earthworms, organic matter is voided in casts having undergone considerable physical change but little or no chemical change. Because of organic-matter enrichment, added intestinal mucus, and enhanced microbial activity, casts contain higher concentrations of available N and other plant nutrients than soil contains (Syers and Springett, 1983). Earthworm casts collected from Irish grassland typically contain about 0.02% more mineral N than unworked soil; thus, casts from a moderate grassland population turning over 20 kg soil m^{-2} yr^{-1} (section 6.4.1(a)) could add about 40 kg mineral N ha^{-1} yr^{-1} to pasture.

Earthworm body tissues contain about 0.7% P and 0.9% S, and Syers and Springett (1984) estimated that tissue turnover could contribute 2–9 kg ha^{-1} to the annual cycling of these elements in New Zealand pasture, while their casts comprise a rich source of available nutrients for plant growth. Hutchinson and King (1982) estimated that invertebrate corpses would yield 1.4 kg P ha^{-1} yr^{-1} in sown Australian pasture, while their excreta would contribute a further 9.6 kg ha^{-1} yr^{-1}. The total amount of P cycled from residues was estimated to be 40 kg ha^{-1} yr^{-1}, of which invertebrate corpses and excreta contribute 11 kg (23%) compared with 14

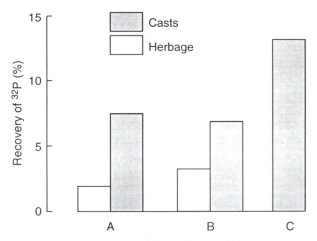

Figure 6.7 Uptake by ryegrass of ^{32}P added in labelled herbage or in labelled earthworm casts to the surface of soil of low-P status (A), soil of high-P status (B) and sand (C). (After Mansell, Syers and Gregg, 1981.)

kg from sheep excreta. Sharpley and Syers (1976, 1977) reported that earthworm casts on the surface in New Zealand pasture had roughly four times more loosely bound inorganic P and twice as much loosely bound organic P as underlying soil. Using ^{32}P-labelled ryegrass, Mansell, Syers and Gregg (1981) demonstrated that passage of ingested material through earthworms increased the short-term availability of P by a factor of two to three. Recovery of ^{32}P from casts by ryegrass growing in sand and soil with very low available P was greater than in the case of ryegrass growing in soil with higher available P (Figure 6.7), indicating that effects of earthworms on nutrient uptake are likely to be most important in soils of low nutrient status. Earthworms also increase the availability to plants of mineral nutrients applied in forms which have low water solubility, such as P in mineral rock phosphate and Ca in ground limestone. This results from mixing and redistribution of such materials into the soil, bringing them into closer contact with roots (Springett, 1983), although in the case of phosphate rock the main effect appears to relate to the increased contact between rock particles and soil surfaces during their passage through the earthworm, and also to the coating of phosphate-rich particles in the soil with earthworm castings (Mackay *et al.*, 1982; Syers and Springett, 1983). Also, by creating channels that improve root distribution earthworms can facilitate greater exploitation of the soil and in this way increase the size of the nutrient pool available to plants.

The role of earthworms in nutrient cycling has been studied in mountain pasture ecosystems in Poland (Czerwínski *et al.*, 1974; Nowak,

1975; Petal *et al.*, 1977), where comparisons were made between sheep-grazed pasture of low fertility and areas of high fertility that had received high inputs of sheep dung during periods when they had been used as sheepfolds. As in the New Zealand studies, there was a large increase (4–10-fold) in assimilable P on passage of litter through the alimentary tract; this was considered to be particularly significant in mountain pastures deficient in P. Earthworm casts also had higher contents of Ca, Mg and assimilable K (Table 6.10).

The quantities of elements made available by earthworms in the heavily manured areas were estimated to increase the level of available P in the soil by 2.48 g m^{-2} and the level of K by 10.8 g m^{-2} for a growing season. Corresponding estimates by Graff (1971) for a mown meadow in Germany were 3.42 g m^{-2} P and 11.15 g m^{-2} K. In the Polish mountain-pasture studies, the role of earthworms in mineralization was proportionately greater in the poorer soil. This was indicated by proportionately greater stimulation of microbial activity and proportionately greater increases in the levels of available nutrients in casts over control soil in the pasture than in the heavily manured areas (Table 6.10).

In addition to influencing mineralization, earthworm activity may also contribute to the process of humification. In the Polish mountain-pasture, the growth of humus in casts in the heavily manured site was 145 g m^{-2} per season while in the pasture it was 28 g m^{-2} per season (Czerwinski *et al.*, 1974). Nowak (1975) reviewed the role of earthworms in a series of Polish meadows and pastures of varying fertility and concluded that in nutrient-poor meadows a predominance of small earthworms and enchytraeid worms stimulate mineralization, whereas in fertile meadows earthworms tend to retard the degradation of organic matter by storing relatively more in their own tissues and by promoting humification.

Table 6.10 Some chemical properties of control soil and worm casts in Polish mountain sheep pasture (adapted from Czerwinski *et al.*, 1974)

	Exchangeable cations (mEq 100 g^{-1})			Content (mg 100 g^{-1})	
	Ca	Mg	K	K$_2$O	P$_2$O$_5$
Sheep-grazed pasture					
Control	9.4	1.97	0.23	9.4	0.8
Casts	15.2	3.48	1.18	64.0	8.7
Heavily manured pasture					
Control	11.0	2.05	0.97	46.4	1.6
Casts	13.5	2.75	1.52	68.3	6.6

(b) Microfloral–faunal interactions

Because invertebrates may ingest relatively high proportions of microbial production in most ecosystems – 30–60% in pine forest (Persson *et al.*, 1980), 30–90% in arable land in Sweden (Paustian *et al.*, 1990) – the effects of microbial grazing on decomposition and mineralization processes are potentially important. Two major pathways of decomposition, 'fast' and 'slow', have been recognized when considering microfloral–faunal interactions (Anderson, Coleman and Cole, 1981; Coleman, Reid and Cole, 1983). Fast-cycle decomposition involves the utilization of relatively simple organic compounds such as amino acids and various mono- and oligosaccharides arising from root exudation and sloughed-off root cells, while the slow cycle involves the utilization of less-labile organic matter present in plant cell walls, in arthropod exoskeletons and in the walls of fungal hyphae. Microflora are the primary agents in both types of decomposition processes, but they generally immobilize nutrients in their own tissues. Invertebrates, on the other hand, have relatively low production efficiencies and utilize only a small proportion of ingested nutrients, excreting significant quantities in mineralized form. The interactions between bacteria and bacterial-feeding microfauna have been deemed to be of greatest significance in fast-cycle decomposition, while in the case of slow-cycle decomposition the emphasis has been mainly on fungi and fungivorous mesofauna. In the field, the two types of cycles are interlinked by microflora that can utilize a range of substrates and by fauna that ingest both bacteria and fungi.

Microcosm experiments carried out under controlled conditions have given useful insights into the potential significance of microbial grazing. In laboratory soil-microcosm experiments with glucose added as a fast carbon source, Coleman *et al.* (1977) found that both an amoeboid protozoan and a bacterial-feeding nematode increased CO_2 production and N and P mineralization as compared with 'control' microcosms, in which microbial grazers were absent (Figure 6.8). Under more complex conditions involving interaction between omnivorous nematodes and amoebae, respiration and N mineralization were found to be further enhanced despite decreased bacterial numbers, indicating a stimulatory effect of microbial grazing on substrate utilization and N mineralization (Coleman *et al.*, 1978; Anderson *et al.*, 1981).

Figure 6.8 Influence of protozoan and nematode grazing on respiration and mineralization in laboratory soil microcosms. (a) Cumulative CO_2–C evolved; (b) bicarbonate-extractable inorganic P (P_i); (c) extractable NH_4^+–N. Bacteria alone; (———); bacteria plus *Acanthamoeba* (- - - -); bacteria plus *Mesodiplogaster*, a bacteriophagic nematode (· · · ·).

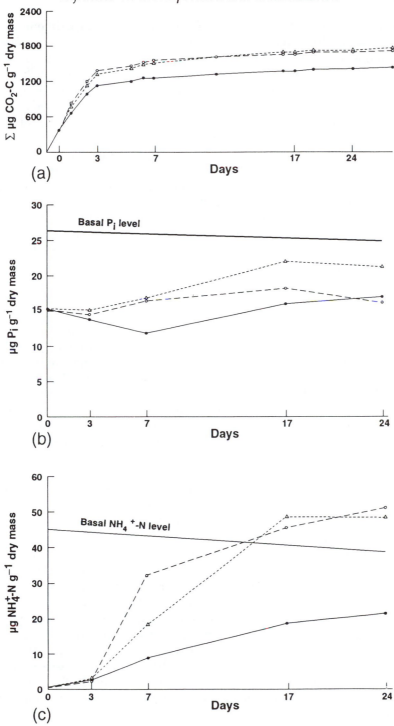

In a series of laboratory and field microcosm experiments with decomposing deciduous litter, Anderson and his coworkers demonstrated positive effects of a range of litter-dwelling invertebrates (isopods, millipedes, collembolans, lumbricid earthworms (*Lumbricus rubellus*) and enchytraeid worms) on rates of nutrient leaching (Anderson and Ineson, 1983; Anderson, Ineson and Huish, 1983a,b; Anderson *et al.*, 1985). Macrofauna, particularly, *L. rubellus*, had a greater effect on nutrient release than mesofauna and much of this effect appeared to be brought about indirectly. In the case of *Lumbricus rubellus*, the rate of increase in ammonium release was up to 80 times that in control microcosms (Anderson, Ineson and Huish, 1983b); smaller increases (up to four times) occurred in Ca, K and Na release. Experiments with ^{15}N-labelled millipedes showed that only about 7% of N released was attributable to their excretion (Anderson, Ineson and Huish, 1983a); the authors suggest that the enhanced rate of N mineralization may have come about in part through the activities of ammonifying bacteria in the animal guts and possibly through stimulation of autotrophic nitrification.

Microbial grazing does not necessarily lead to enhanced decomposition rates and mineral release in all circumstances; the effects of grazers will vary depending on such factors as their density and feeding habits. For example, Hanlon and Anderson (1979) reported that fungal grazing by *Folsomia candida* at low density (five animals per culture) stimulates microbial activity in decomposing leaf litter, but at higher density (10–20 per culture) microbial activity was depressed and there was a shift in balance between fungal and bacterial populations. Also, different categories of collembolans appear to differ in their influence on organic matter mineralization. Van Amelsvoort, van Dongen and van der Werff (1988) concluded that larger, hemiedaphic collembolans such as Isotomidae, associated with relatively fresh organic matter with a high decomposition rate and feeding selectively on nutritious fungal hyphae, stimulate mineralization processes. By contrast, smaller, euedaphic, species such as *Tullbergia krausbaueri* and *Neelus minimus* living in finer textural soil with older organic matter and low fungal activity ingest organic and mineral particles non-selectively and derive their nutrition largely from the bacteria associated with these particles; by mixing organic materials with mineral soils they promote the process of humification. Trofymow and Coleman (1982) reported inhibition of CO_2 evolution by a fungal-feeding nematode in microcosm culture in contrast to a bacterial-feeding nematode that stimulated bacterial activity. They suggested that the method of feeding of the nematodes may govern the overall response: holophagic feeding on bacteria will release nutrients while piercing of mycelia by nematode stylets is likely to immobilize nutrients by leaving large segments of fungi inactive or with empty cell walls.

Many of the earlier microcosm studies could be said to lack realism in that they did not include growing plants, but this deficiency has been remedied in a number of more recent studies. Examples include that by Clarholm (1985), who grew wheat plants in a laboratory experiment and showed that in the presence of C derived from roots bacteria can mineralize N from soil organic matter to support their own growth (Figure 6.9). In treatments that included bacterial-feeding protozoans, N taken up by the plants increased by 75%, indicating that grazing of the bacteria is necessary before N is released for plant growth. Similar trends were reported by Kuikman *et al.* (1990) although the effect of protozoans on N mineralization was less marked. These authors reported 36–53% increases in ^{14}C turnover and 9–17% increases in N uptake by wheat plants, attributable to protozoan grazing. Ingham *et al.* (1985) reported that blue grama grass (*Bouteloua gracilis*) growing in soil microcosms with bacteria and bacterial-feeding nematodes grew faster and initially took up more N than did plants in soil with bacteria alone. This was attributed to increased N mineralization by bacteria, NH_4^+–N excretion by nematodes, and greater initial exploitation of soil by plant roots (Figure 6.10). Addition of fungal-feeding nematodes to cultures with fungus did not increase plant growth or N uptake because these nematodes excreted less NH_4^+–N than did bacterial-feeding nematodes and because the N mineralized by the fungus alone was sufficient for plant growth. Ingham *et al.* concluded that additional mineralization due to microbial grazing may

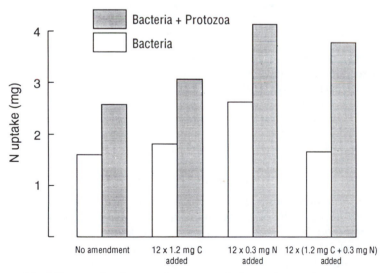

Figure 6.9 Nitrogen in shoots plus roots of three wheat plants grown for 6 weeks in soil microcosms with and without added C (glucose) and/or N (NH_4 NO_3). The initial plant N content (in seeds) is subtracted. (After Clarholm, 1985.)

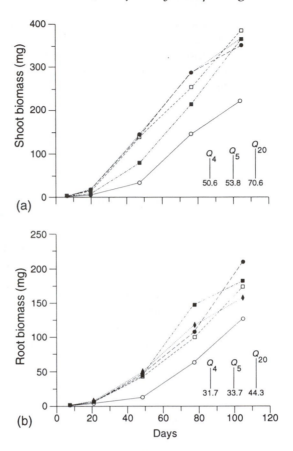

Figure 6.10 Shoot biomass (a) and root biomass (b) of *Bouteloua gracilis* grown in soil microcosms with different biological treatments. plant (○); plant + bacteria (■); plant + bacteria + bacterial feeding nematode (◆); plant + bacteria + fungus (□); plant + bacteria + fungus + fungal feeding nematode (●). Q_4, Q_5, Q_{20} represent Tukey's HSD test of significance ($P < 0.05$) for comparison between dates for the same treatment, between treatments on the same date, and between treatments on different dates, respectively. (From Ingham *et al.*, 1985.)

be significant for increasing plant growth only when mineralization by microflora alone is insufficient to meet plant requirements. However, they considered that microbial grazers may perform important regulatory functions at critical times in the growth of plants (e.g. by increasing N mineralization during short periods of ideal conditions for plant growth).

There is no simple answer to the question of how to relate the results of highly simplified microcosm studies to the more complex field situation. Anderson *et al.* (1985) developed a regression model aimed at quantifying the effects of soil and litter macrofauna on N mineralization in woodland as a function of temperature and animal biomass for different 'resource quality' litter and humus materials and sites, but the model gave poor agreement with recorded field data and underestimated animal effects. Apart from the total quantities of nutrients mobilized by faunal activity, there is the important question of the timing of their release relative to available sinks and particularly the potential for plant uptake. The work of Bouché and Ferrière (1986) indicates rapid and almost complete plant uptake of earthworm-derived N by pasture plants in spring, but most of the N input from dead worm tissue may be in late summer and autumn when the plant demand for N is likely to be reduced (Christensen, 1988). Furthermore, casts appear to be significant sources of N loss through denitrification (Svensson, Boström and Klemedtson, 1986; Elliott, Knight and Anderson, 1990). It is often suggested that invertebrates contribute to the regulation of nutrient mobilization/immobilization and nutrient conservation in undisturbed ecosystems (e.g. Ausmus, Edwards and Witkamp, 1976; Verhoef and Brussaard, 1990), and there is some evidence to support this hypothesis. Invertebrate biomass can represent a significant nutrient pool in immature ecosystems: Hassall and Sutton (1985) concluded that substantial quantities of Na, K, Ca, Mg and Cu are immobilized in isopod biomass in dune grassland, thereby prolonging their retention and increasing the probability of their being reabsorbed by plants. Field studies in oak woodland in the UK indicate that up to 80% of animal-mobilized N is taken up by plant roots and that animals apparently facilitate plant uptake of mineral N, although the mechanism involved is unknown (Anderson *et al.*, 1985). Losses of N in winter were lower from lysimeters containing animals at the same site, suggesting that animals can play a role in N conservation at this time.

Conceptual models of detrital food webs have proven useful for integrating data from laboratory and field studies to describe the role of organisms in nutrient cycling. By these means, Ingham *et al.* (1986) were able to interpret changes in a range of parameters (plant root and shoot biomass, microbial and invertebrate population densities, and changes in soil-N levels in prairie grassland) on the basis of laboratory predator–prey microcosm studies. Drawing on a wide array of data from various sources, Hunt *et al.* (1987) calculated N-flux rates through the detrital food web of short-grass prairie (Figure 3.2). Information utilized included population sizes, food preferences, nitrogen contents, lifespans, assimilation efficiencies, production:assimilation ratios and decomposition rates of decomposer organisms. Plants were estimated to take up 7.6 g N m^{-2} yr^{-1}; bacteria mineralized most N (4.5 g m^{-2} yr^{-1}), followed by the fauna (2.9 g) and fungi (0.3 g). Bacterial-feeding amoebae and nematodes accounted for over 83% of N mineralized by the fauna, the high mineralization rates (14 times their biomass N per year) of these microfauna being attributed to relatively high biomass-turnover rates and to the fact that they mineralize a high fraction of N consumed. Fauna were estimated to contribute an additional 2.3 g N m^{-2} yr^{-1} to the 'labile substrate' pool in the form of faeces, dead bodies, etc., much of this being rapidly converted by bacteria to ammonia.

Food-web simulation modelling is probably the best-available means of describing the role of soil organisms in decomposition and mineralization, although this approach has some limitations (Verhoef and Brussaard, 1990). For example, the method does not take account of the ways in which detritivorous invertebrates can indirectly influence decomposition by altering the physicochemical conditions under which the process takes place.

6.4 INFLUENCE ON SOIL PROPERTIES

Large invertebrates, notably earthworms in mesic temperate soils and termites and ants in arid regions, have the capacity to exert a considerable influence on soil structure and fertility by their feeding and mixing activities and by their movements in the soil.

6.4.1 Earthworms and soil properties

(a) Soil mixing and turnover

The contribution of earthworms to soil fertility through burrowing and soil mixing has received considerable attention since the pioneer work of Müller (1878) and Darwin (1881). The extent of soil working depends on several factors, including the abundance and composition of the earthworm fauna, soil properties such as organic matter and texture, moisture content and temperature.

Assimilation efficiencies of geophagous earthworms relying on soil organic matter for food are very low – less than 2.5% in the case of *Aporrectodea rosea* in England (Bolton and Phillipson, 1976) and about 9% in the case of *Millsonia anomala* in tropical savannah (Lavelle, 1974) – so relatively large amounts of soil must be ingested to satisfy nutritional requirements, especially if the organic matter content is low. The relationship between organic matter content and ingestion rate was demonstrated by Martin (1982), who measured the burrowing rates of three earthworm species in laboratory cultures of sandy soil with measured amounts of organic matter added in the form of powdered grass leaves. Growth rates were highest in soils with most organic matter; consumption of soil increased as the proportion of organic matter declined until a particular level was reached (1.1 g kg^{-1} for *Allolobophora* spp., 4.4 g kg^{-1} for *Lumbricus rubellus*); below this level consumption declined. *Lumbricus rubellus* was unable to increase its rate of feeding to the same extent as *Allolobophora* spp. as the organic matter level declined, consistent with the fact that this species only thrives in organic-rich habitats. The quality of the organic matter is also important: Lavelle, Schaefer and Zadi (1989) reared *Millsonia anomala* in savannah soil low in native organic matter (1 to 2%) enriched with organic matter of different quality. Soil ingestion rates declined when the quality of the added organic matter was high, while increased soil-ingestion rates were linked with low-quality organic resources.

Estimates for daily rates of soil consumption range from 200–300 mg dry mass g^{-1} fresh mass by *Aporrectodea caliginosa* in a New Zealand loam soil (Barley, 1959b) to 1–5 g dry mass g^{-1} by *A. rosea* in beech woodland soil in England (Bolton and Phillipson, 1976), while Martin (1982) reported ingestion rates of 1.6–3.6 g g^{-1} for *Lumbricus rubellus* and 1.1–1.6 g g^{-1} for *Allolobophora* spp. in his sand/organic matter cultures. Estimates for the quantities of soil ingested by field populations in grasslands under conditions suitable for earthworm activity include 3–6 kg dry mass m^{-2} annually in New Zealand pasture (Barley, 1959b), 12 kg m^{-2} yr^{-1} in a French pasture (Bouché, 1977), 7 and 11 kg m^{-2} in Swedish grass and lucerne leys respectively (Boström, 1988a) and 80–100 kg m^{-2} in humid tropical savannah (Lavelle, 1974). Soil ingestion during burrowing and subsequent egestion in the form of casts represent a gradual mixing and relocation of material within the soil profile, mostly within the top 10–20 cm of the soil layer. In soils with high populations of deep-burrowing species, such as *Lumbricus terrestris* and *Aporrectodea longa*, numerous earthworm burrows will be found to a much greater depth, but burrowing below the organic-rich top 10–20-cm layer is less extensive and consequently earthworm activity does not lead to the degree of profile inversion associated with termites in the tropics (Lee, 1985).

In a mixed-species grassland community, the combined action of anécique species incorporating surface plant litter and anécique and endogeic species working the soil, results in an intimate mixing of organic and mineral components that profoundly influences soil characteristics. Ingested soil and organic materials are mixed with intestinal secretions during passage through the gut and are voided as semiliquid casts. Some of this material is used to line the walls of the developing burrows, the remainder being deposited at or near the soil surface or internally within the soil. Surface casting leads to a slow turnover of soil, the amount depending on the species present and their activity, food supply, soil type and weather conditions. *Aporrectodea longa, Lumbricus terrestris and A. caliginosa* appear to be the main surface casters in temperate soils. Estimates of the percentage of total casts deposited on the surface in the case of *A. caliginosa* are about 10% in New Zealand pasture (Barley, 1959c) and

20–50% in grass and lucerne leys in Sweden (Boström, 1988a), but these estimates vary considerably depending on the time of the year and other factors. Surface casting in temperate grassland may range from less than 1 kg m^{-2} yr^{-1} in poorer, light-textured soils, where earthworm numbers are low and a high proportion of cast material is required for lining the burrow, up to 7 kg ha^{-1} in clay soils; about 3 kg may be average for medium-textured soils (Evans, 1948; Evans and Guild, 1947; Bouché, 1977). Barley (1959c) reported a much lower rate of surface casting (0.2–0.3 kg m^{-2} yr^{-1}) in a New Zealand pasture that lacked the endogeic, surface-casting species *A. longa* and *L. terrestris*. In tropical soils, a range of 1–28 kg m^{-2} may be encountered, but the species involved in surface casting are mainly epigées and their activities are confined to the superficial soil layers (Lee, 1983b). Barley estimated that the *A. caliginosa* population in a New Zealand pasture could completely work the topsoil to a depth of 15 cm in a period of 60 years; this process might take as long as 100 years in moisture-limited mollisols (Buol, Hole and McCracken, 1973).

(b) Incorporation of fertilizers and pesticides

Earthworms can improve soil fertility and productivity by incorporating surface-applied fertilizers and agrichemicals into the soil. This has been referred to earlier in the context of mineral rock phosphate (P), where the main effect of earthworms appears to be one of bringing phosphate particles into intimate contact with soil. Stockdill and Cossens (1966) found that in the absence of earthworms lime applied at rates of up to 5 t ha^{-1} to New Zealand pasture remained in the surface 2.5 cm, whereas in the presence of *A. caliginosa* it was mixed throughout the top 20 cm of soil within a period of 4–5 years. Further work in New Zealand confirmed the role of earthworms in lime incorporation but drew attention to differences in the effects of different species (Springett, 1983). In laboratory cultures, *A. caliginosa* and *L. rubellus* tended to mix lime laterally as might be expected on the basis of their horizontal burrowing habits, whereas *A. longa* tended to mix it vertically. When *A. longa* was added to a field site where *A. caliginosa* and *L. rubellus* were already established it increased the vertical mixing of lime in the profile (Figure 6.11).

Stockdill and Cossens (1966) showed that *A. caliginosa* increased the downward movement of DDT and its effectiveness in the control of the grass grub, *Costelytra zealandica*, in New Zealand pasture.

(c) Soil structure

Continuous processing by earthworms is recognized as having a major influence on soil development (see section 1.9.3) although there is still

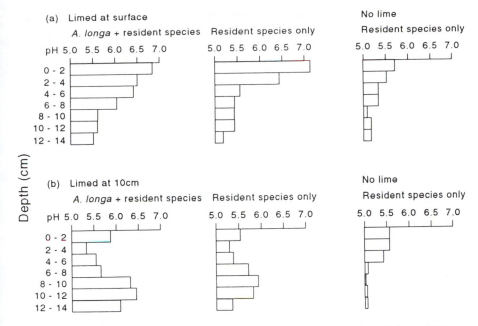

Figure 6.11 Effects of the earthworm *Aporrectodea longa* on the mixing of lime into the profile in a New Zealand pasture as indicated by changes in pH to a depth of 14 cm. (From Springett, 1983.)

some uncertainty as to the mechanisms involved. The overall effect of earthworms on soil porosity is relatively minor: earthworm burrows normally represent less than 1% of total soil porosity (Lavelle, Schaefer and Zadi, 1989). However, by altering the pore-size distribution earthworms can markedly influence a range of soil physical properties, as discussed below. An important attribute of fertile mull soils is their granular structure, which provides the conditions of moisture and aeration needed for optimum plant rooting and biological activity. Studies in reclaimed soils have given insight into the ways in which earthworms can influence structural development. Van Rhee (1977) and Rogäar and Boswinkel (1978) reported that reclaimed polder soils that had been worked by earthworms had a better structure than soil without earthworms. Stewart and Scullion (1988) provided evidence from greenhouse experiments supported by some field data to show that earthworms play a crucial role in the structural rehabilitation of loam soils after open-cast mining in Wales (Figure 6.12). Contrary to the widely held view that grass roots are the primary agent in bringing about granulation in grassland soil, their results indicate that grass by itself, even in the presence of added organic matter, can not achieve significant granulation of structurally degraded soil. Granulation was achieved by earthworms in treatments that had received a single addition of organic matter initially, but this effect was transitory; however, the combination of earthworms with growing grass produced sustained granular structure. The mechanisms by which earthworms influence soil granulation are not fully understood but appear to involve alteration of pore size distribution by burrowing; incorporation and comminution of fresh organic residues, bringing clay particles into intimate contact with the products of organic decay; mixing of mineral and organic materials with calcium from calciferous glands and intest-

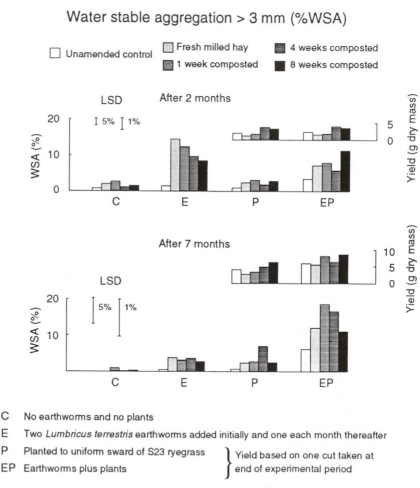

Figure 6.12 Effects of earthworms and organic matter on water stable aggregates (WSA) in silty clay loam soil of poor structure in greenhouse experiments. LSD = Least significant difference. (From Stewart and Scullion, 1988.)

inal secretions that may have a cementing effect; enrichment of casts with mucus and urine and the products of enhanced microbial activity; and the shape of the casts themselves (Syers and Springett, 1983; Stewart and Scullion, 1988).

Modification of soil structure by earthworms influences soil properties such as soil aeration, water infiltration and conductivity, water-holding capacity, and root distribution. Enhanced aeration after earthworm establishment was reported by Noble, Gordon and Kleinig (1970) in irrigated pastures in southern Australia, by Stockdill (1966) for New Zealand pasture and by van Rhee (1977), Rogäar and Boswinkel (1978) and Hoogerkamp, Rogäar and Eijsackers (1983) for reclaimed Dutch polder soils. Stockdill and Cossens (1966) reported improved water-holding capacity in New Zealand grassland following earthworm introduction. This effect is probably attributable to a change in pore-size distribution brought about by casting and redistribution of organic matter; this would increase the proportion of fine- and medium-sized pores (0.003–0.6 mm diameter) that are important for water storage (Syers and Springett, 1983). Improved water infiltration in worm-worked soils has been demonstrated by Stockdill and Cossens (1966) and Sharpley, Syers and Springett (1979) for New Zealand soils, and by Hoogerkamp, Rogäar and Eijsackers (1983) for Dutch polders; this effect may not be of much significance under normal weather conditions but can be important in the case of heavy soils during periods of heavy rainfall.

Table 6.11 demonstrates the influence of earthworms on moisture and aeration in the top 20 cm of reclaimed Dutch polder grassland after 8 years of earthworm activity. The working of the topsoil by earthworms resulted in better aeration, higher infiltration capacity and water conductivity and reduced water stagnation during wet periods. Channels created by earthworm burrowing (2–11 cm diameter) appear to be the main factors influencing infiltration and aeration. Channels open to the surface are considered to be primarily responsible for improved infiltration, but Hoogerkamp, Rogäar and Eijsackers (1983) consider that the voids left by the granular structure are also important in water transport in Dutch polder grassland, especially in the surface layer where open burrows are virtually absent. The effects of earthworms on total porosity and water infiltration are likely to be greatest in soils that have a high proportion of surface-feeding (anécique) and surface-casting species, whereas endogées and species that cast predominantly within the soil have the greatest effect on pore-size distribution within the soil, tending to reduce macroporosity and to produce more fine pores, which are important for water storage.

Earthworm activity increases root mass and the extent of soil penetration by roots. This was observed in New Zealand pasture by Stockdill

Table 6.11 Effects of earthworms on (a) water infiltration capacity, (b) conductivity for water at saturation, (c) conductivity for air and (d) oxygen diffusion at pF 2 in reclaimed polder grassland soil without earthworms and after 8–10 years of earthworm activity (after Hoogerkamp, Rogäar and Eijsackers, 1983)

(a) Infiltration capacity (m 24 h^{-1})

	Without worms	With worms
Biddinghuizen	0.039	4.6
Swifterbant	0.047	6.4

(b) Conductivity for water (m 24 h^{-1})

	Without worms		With worms	
Depth (cm)	Mean	Range	Mean	Range
0–10	0.95	0.46–1.50	22.2	1.0–88.4
10–20	0.42	0.11–0.64	6.5	0.8–28.6
20–30	14.00	1.15–19.20	4.2	0.4–25.6

(c) Conductivity for air (cm^{-2} × 10^{-8})

	Without worms		With worms	
Depth (cm)	Mean	Range	Mean	Range
0–10	8.40	4.7–12.5	54.3	2.9–114.8
10–20	7.10	2.9–13.7	36.0	20.5–57.4
20–30	92.20	49.2–143.5	19.6	8.4–68.9

(d) O$_2$ diffusion (cm^2 s^{-1} × 10^{-2})

	Without worms		With worms	
Depth (cm)	Mean	Range	Mean	Range
0–10	0.27	0.15–0.37	0.68	0.33–1.06
10–20	0.28	0.18–0.38	0.57	0.39–0.81
20–30	0.38	0.71–0.93	0.33	0.28–0.37

and Cossens (1966) and in Dutch polder grassland by Hoogerkamp, Rogäar and Eijsackers (1983). On worm plots, grass roots grew directly into the mineral soil with a concentration in the upper few centimetres, whereas in the absence of worms rooting tended to be largely restricted to the surface mat of decaying organic matter. Improved root growth probably partly reflects improved conditions of aeration, moisture and nutrient supply, but is also attributable to decreased penetration resistance and easier root penetration in earthworm-worked soil (Figure 6.13). Laboratory experiments conducted by Edwards and Lofty (1978) using undisturbed soil monoliths from zero cultivated arable land showed that the effect of earthworms on rooting depended on the species present: shallow-working species promoted root growth in the surface soil layer while deep-burrowing species promoted deeper rooting. Earthworm channels are probably of less significance for root growth in soils with established granular structure than in poorly structured, newly reclaimed or compacted soils. Rhizotron observations carried out by Carpenter (1985) indicated that less than 10% of the root growth of grasses normally occurred in earthworm burrows in mature grasslands although occasionally the amount was greater.

The effects of earthworms on soil structure are seen most dramatically

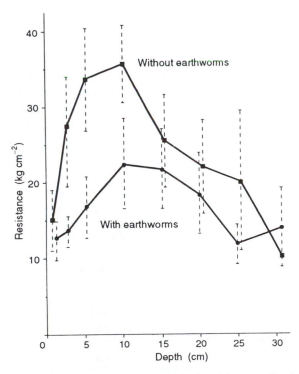

Figure 6.13 Penetration resistance of reclaimed polder grassland soil with and without earthworms (mean values and standard deviations). (From Hoogerkamp, Rogäar and Eijsackers, 1983.)

when earthworms become established in sites where they were pre-viously scarce or absent. These effects include the disappearance of the surface mat of accumulated organic material and improvement of var-ious aspects of soil structure, as observed in the case of New Zealand pasture (Stockdill and Cossens, 1966) and Dutch polders (van Rhee, 1977; Hoogerkamp, Rogäar and Eijsackers, 1983). When earthworms are eliminated from soils where they were abundant an equally dramatic deterioration in soil condition occurs, resulting in accumulation of organic residues on the surface and reversion of soil conditions from mull to mor (van Rhee, 1963; Clements, Murray and Sturdy, 1991). Table 6.12 shows the consequences of eliminating earthworms by repeated application of insecticides (mainly phorate) to ryegrass plots in England over a period of 20 years. A marked accumulation of surface litter occurred in insecticide-treated plots with a concomitant decrease in soil organic matter levels. Considerably increased compaction was noted on treated plots. This was reflected in increased penetration resistance, bulk density and shear strength and lower hydraulic conductivity, water infiltration rate, and moisture content. As a result, root mass was con-siderably lower in the treated plots. Surprisingly, the impact of these changes in soil charac-teristics on grass yields was negligible (Clements, Bentley and Jackson, 1990).

Marked effects of earthworms on the characteristics of fen peat in central Ireland reclaimed after industrial peat harvesting have been observed (Boyle, 1990). The reclamation process involved mixing residual peat (*c.* 0.5 m) with underlying mineral soil of highly degraded structure. Earthworms markedly accelerated the rate of humification of the peat and the micromorphology of worm-worked material was radically

Table 6.12 Effects of repeated application of insecticides to perennial ryegrass plots receiving two levels of nitrogen fertilization on various soil parameters (after Clements, Murray and Sturdy, 1991 and *Clements, 1982)

	Low N (untreated)	Low N (treated)	High N (untreated)	High N (treated)
Surface leaf litter (g m^{-2})	87	3219	272	3730
Bulk density (g cm^{-3})	1.40	1.57	1.39	1.65
Shear strength (kPa)	25.2	64.3	45.3	67.8
Hydraulic conductivity (m d^{-1})*	17.8	0.7	20.1	1.4
Initial infiltration rates (ml min^{-1})*	68.7	4.7	–	–
Soil organic matter (%)	5.49	4.75	5.90	4.53
Penetrometer (kPa)	38.4	83.8	63.4	111
Soil moisture (%)	21.4	15.2	20.1	14.3

Figure 6.14 Influence of earthworms on the micromorphology of deep-ploughed reclaimed peat (x 15). The section is dominated by finely divided earthworm faecal material. (From Boyle, 1990.)

altered (Figure 6.14), reflecting the intimate mixing of organic and mineral components on passage through the earthworm gut.

There have been references in the literature to some adverse effects of earthworms on pasture. In Dutch polders where earthworms had become established some damage by treading and soiling of grass during wet weather was reported, whereas in fields without earthworms where there was a well-developed mat poaching and soiling were almost unknown (Hoogerkamp, Rogäar and Eijsackers, 1983). Earthworm activity could also contribute to surface runoff resulting from dispersion of fresh surface casts by raindrop impact (Shipitalo and Protz, 1988). Sharpley, Syers and Springett (1979) compared hill grassland plots in New Zealand where earthworms were active with plots where earthworms had been eliminated with carbaryl. Higher sediment loads were recorded from untreated plots, arising from erosion of surface casts, while concentrations of soluble N and P were four to eight times lower in runoff from these plots. Thus it appears that earthworm casting can contribute to soil erosion on slopes, but rapid incorporation of litter into the soil by earthworms reduces loss of nutrients by leaching and surface runoff.

6.4.2 Earthworms and plant growth

There is evidence to indicate that improvements in soil structure and fertility brought about by earthworms result in improved plant growth. Increases in dry-mass yield of grass, clover and cereals ranging from 28 to 1000% have been attributed to earthworms in a range of studies carried out in small containers or pots (Hopp and Slater, 1948; Waters, 1951; Nielson, 1951; van Rhee, 1965; Curry and Boyle, 1987; Temirov and Valiakhmedov, 1988). However, field data definitively relating earthworm activity to plant growth are more difficult to find since few controlled experiments have been carried out. One such experiment was reported by Edwards and Lofty (1980), who inoculated DD-sterilized plots of direct drilled cereals with 'normal' populations of earthworms and reported increased plant population, root penetration and biomass, and shoot biomass as a result. There are many instances where a relationship between earthworm activity and increased grass growth in the field is strongly indicated. For example, increased dry-matter production of herbage has been reported following establishment of lumbricid earthworms in improved New Zealand pasture (Nielson, 1951; Stockdill, 1959; Stockdill and Cossens, 1966) and reclaimed Dutch polder (Hoogerkamp, Rogäar and Eijsackers, 1983). The initial increase in grass production in New Zealand pasture was up to 70%, probably mainly as a consequence of the breakdown of the accumulated surface mat with the accompanying release of plant nutrients. Subsequently, a more modest sustained increase in the order of 25–30% was reported (Lacy, 1977).

Such increases in plant growth are generally attributed to earthworm-induced changes in soil physical and chemical properties, although it is possible that additional factors may be involved, at least in some situations. Earthworms are known to produce indole compounds with plant hormonal activity (Nielson, 1965; Springett and Syers, 1979), but there is no evidence that these exert a major effect on plant growth in the field. They also consume significant amounts of plant root material (Carpenter, 1985; Baylis, Cherrett and Ford, 1986), and it may be that this root pruning has the effect of removing senescent roots and stimulating new root development, thus enhancing nutrient uptake and plant growth (Davidson, 1979).

6.4.3 Termites and soil properties

Termites use large quantities of soil in the construction of nests and associated structures – runways, galleries, etc. The quantities of soil involved in these structures may be considerable in relatively undisturbed tropical forest where larger mound-building species are abundant. Meyer (1960), for example, reported that *Macrotermes* mounds covered 30% of the soil surface in a site in central Zaire and contained more than 2800 t soil ha^{-1}. The quantities involved are much smaller in open savannah where large mounds are scarce and small mound-building species predominate. Lepage (1972) estimated that *Bellicositermes bellicosus* used 28–35 t soil ha^{-1} depending on their density. Laker *et al*. (1982) estimated the annual turnover of soil by *Trinervitermes trinervoides* for mound construction and related activities in South Africa to be 350 kg ha^{-1}.

Humivorous species that derive their nutrition from soil organic matter consume relatively large quantities of soil. Josens (1972) estimated 15 t ha^{-1} yr^{-1} consumption of surface soil by humivores in humid savannah at Lamto, Ivory Coast while Hébrant (1970) calculated that the *Cubitermes exiguus* population in a savannah in Zaire would have to consume 17 t soil ha^{-1} yr^{-1} to obtain its cellulose requirements.

Large quantities of litter are collected by foraging termites and concentrated in termite mounds. Josens (1972) estimated that four species of subterranean fungal-growing species at Lamto incorporated 170 g litter m^{-2} yr^{-1} into their terminaria while Collins (1981) calculated that termites removed 836 kg leaf litter ha^{-1} yr^{-1} (24% of the total) from Guinea savannah in Nigeria. Fungus-growing Macrotermitinae consumed 95% of this. This concentration and processing of litter in termite mounds is a major pathway for litter decomposition in the tropics, but it has the effect of depleting organic matter stocks over large areas and, because of the high assimilation efficiency made possible by the gut microflora, the resulting faeces is a poor substrate for other decomposers. In addition, the activities of humivorous termites further deplete the content of soil organic matter. Termite excreta are used in mound construction and, in the case of Macrotermitinae, as a substrate for fungi growing in fungal gardens, a process that promotes further decomposition. Thus, much termite excreta only become available to other soil decomposers when nests are abandoned, possibly after several decades.

Termites can play an important part in the removal of dung from the surface in arid grasslands. Ferrar and Watson (1970) described how scavenging species, mainly *Amitermes* spp., attack dry dung in northern Queensland and can destroy an average-sized pat within 3 months, thereby releasing substantial areas for fresh pasture growth.

Subsoil tends to be used in preference to topsoil for mound construction (Lee and Wood, 1971a; Laker *et al*., 1982), and this can lead to a gradual inversion of the soil profile as mounds disintegrate. The rate of erosion of mound material is slow; however, there are examples in northern Australia and elsewhere of soil profiles with distinct surface horizons derived from termite mounds over thousands of years. Lee and Wood (1971b) give estimates of surface-soil accumulation from mound erosion in West Africa and northern Australia ranging from 0.0125 to 0.2 mm yr^{-1}.

Smaller soil particles are selected for nest building, and mounds tend to have higher silt and clay content than the soil from which they are constructed. Pockets of topsoil derived from such mounds may be unfavourable for growth in arid areas (Laker *et al.*, 1982). Mound soil typically has higher bulk density than surrounding soil, although the associated galleries increase soil porosity and water infiltration (Lal, 1988). Mounds and other termite structures are usually found to have elevated C:N ratios and higher concentrations of exchangeable bases such as Ca, Mg and K than surrounding soil, reflecting their content of faecal material of plant origin (Lee and Wood, 1971a; Gupta, Rajvanshi and Singh, 1981; Laker *et al.*, 1982). Mounds of soil-feeding *Cubitermes* spp. in Nigeria had higher levels of available P than did surrounding soil and this could be important in reducing P fixation in tropical soils (Wood, Johnson and Anderson, 1983). However, plant nutrients in mounds are withheld from circulation in the plant–soil system while the mounds remain inhabited, and Wood and Sands (1978) conclude that the contribution of mound erosion to the soil-nutrient pool is negligible compared with normal processes of plant decomposition and leaching from foliage.

6.4.4 Ants and soil properties

Ants also modify their habitat through nest construction although their influence on soil properties is usually less apparent than that of earthworms or termites. Nevertheless, because they are more widely distributed they move more soil world-wide than do earthworms or termites (Hölldobler and Wilson, 1990). Leaf-cutting ants (Attini) are considered to have had a major influence on soil development in Costa Rica (Alvarado, Berish and Peralta, 1981). Ant mounds are constructed from mineral and plant materials, food remains and excreta bound together by mandibular gland secretions, while the walls of underground chambers and galleries are constructed from ant-worked organic materials (Petal, 1978, 1980).

Ants use coarser soil material for nest construction and pack it more loosely than do termites (Lal, 1988). Because of · this, and organic amendment, ant-worked soil typically has lower bulk density than non-worked soil (Rogers, 1972), and pH, P and K content, and exchangeable Ca, Mg, K and Na cation levels are usually elevated (Czerwinski, Jakubczyk and Petal, 1971). The degree of soil modification depends on nest size and density. Rogers (1972) estimated the amount of soil brought to the surface through mound building by a population of *Pogonomyrmex occidentalis* in a short-grass prairie site at 0.3–0.8 g m^{-2}, while the amount of soil similarly processed by a population of *Lasius niger* in an old field in Georgia was estimated to approach 86 g m^{-2} (Talbot, 1953). Weber (1966) calculated that soil ants may bring as much as 40 t ha^{-1} of subsoil to the surface in Brazil. Ant nests covered 10–11% of the grassland surface in a site in Berkshire where *Lasius flavus* was abundant (Waloff and Blackith, 1962), whereas in Polish pasture with a low density of *Myrmica* spp. and *Lasius niger* nests the area covered was only 0.1–1.0% (Petal, 1980). *Myrmica* spp. construct small, temporary nests 12–13 cm in diameter and at most 2–3 cm high and they move nests 2–3 times a year, whereas *Lasius niger* typically constructs mounds 20–30 cm in diameter and 6–20 cm high that are inhabited by the same colony for many years (Petal, 1974, 1980). Ants generally tend to be most abundant in unmanaged or lightly managed grasslands. Increased disturbance associated with intensive management is accompanied by a decline in their importance, with large nest builders such as *Lasius* spp. being particularly affected (Petal, 1974, 1980). Notable exceptions are the small leaf-cutting *Acromyrmex* spp., which appear to undergo a significant increase in numbers in South America when grassland is extended at the expense of forest and when attempts are made to improve pasture yield (Cherrett, 1986). These ants make small, inconspicuous nests, while nests of the related *Atta* spp. are large and spectacular. Their surfaces constitute bare and unproductive soil, which, at a nest density of 4–5 ha^{-1}, could cover 4% of pasture area (Jonkman, 1977). Once nests die, their surfaces become invaded by cacti and woody plants that provide foci for reversion to open woodland (Jonkman, 1978). The internal volume of *Atta* nests may reach 5 m^3 (Jonkman, 1977), and when the nests die and collapse considerable subsidence occurs, creating hazards for livestock and vehicles.

6.5 EFFECTS OF HERBIVORES ON GRASSLAND PRODUCTIVITY

6.5.1 Tissue consumption and damage: general aspects

At non-outbreak population levels, consumption by invertebrate herbivores usually falls within the range of 1–10% of net primary production in natural ecosystems when measured in terms of plant dry mass or energy content (Wiegert and Evans, 1967; Sinclair, 1975; Gibbs, 1976); however, consumption *per se* is a poor indicator of overall impact since the effect of a given level of feeding on sward productivity will be determined by a range of factors, including the nature of the tissue fed on, the age of the plant attacked, and the location of feeding on the plant. Grasses are well adapted to withstand moderate levels of grazing but are particularly susceptible to damage during the seedling stage, and pests such as slugs and frit fly (*Oscinella frit*) can cause severe damage if attack occurs prior to tillering. Damage to meristematic and actively growing tissue is potentially more serious for the plant than damage to mature tissue, and most herbivores preferentially feed on young tissues because of their higher nutritive value. The damage done by stem-boring Diptera such as frit fly larvae, which kill the central shoot, and by leatherjackets, wireworms, cutworms and slugs feeding at or close to the soil level bears little relationship to the amount of tissue actually consumed, since leaves or shoots may be severed at the point of attack, resulting in the death of a considerable amount of plant material. The relationship between tissue consumed and tissue destroyed has been considered by several workers in the case of grasshoppers (Acrididae). Grass-feeding acridids graze on leaf blades at different heights. The part of the leaf above the point of feeding often withers and falls off, resulting in the death of several times more leaf biomass than that consumed. The ratio of leaf material destroyed to leaf material consumed is affected by the shape of the leaf, the ratio tending to be greater in the case of plants with long, thin leaves such as sedges (*Carex* spp.) than in the case of short, broad-leaved grasses. Grasshoppers were estimated to destroy 6–15 times more grass than they consumed in Polish pasture (Andrzejewska and Wojcik, 1970), while much lower destruction: consumption ratios (0.5–2.5) have been reported for prairie grasshoppers in the USA (Mitchell and Pfadt, 1974).

The impact of sap-sucking insects on plant production can likewise be much greater than their actual tissue consumption would indicate. Viruses transmitted by aphids can have very serious effects on plant production (section 2.14.3(c)), while plants may also suffer various kinds of physiological disorders arising from the toxicity of saliva injected during feeding (van Emden *et al.*, 1969; Vickerman and Wratten, 1979). A severe check on plant growth as a result of wilting may occur in plants attacked by xylem-feeding species, which tend to remove large quantities of water. Andrzejewska (1967) reported that the planthopper *Cicadella viridis* consumed 250 times its own weight of xylem sap in 24 hours. This resulted in yellowing and withering of affected plants, and when grass was infested at densities typical of field populations plant biomass was lowered by 25% compared with control plants within 2 weeks. On the other hand, Ricou and Duval (1969) also reported wilting of grass infested with leafhoppers but did not record any effect on plant dry-matter production.

Herbivorous insect groups with the potential to cause serious and widespread damage to grassland include grasshoppers, termites and ants. Grasshoppers are a widely distributed and frequently damaging group of herbivores, and several studies have attempted to quantify their impact in terms of the percentage of net primary production they consume or destroy. Andrzejewska and Wojcik (1970) estimated this to be 8–14% in Polish meadows. Yield reductions of 20–70% in range grassland in the western USA have been attributed to grasshoppers (Nerney, 1958); Mitchell and Pfadt (1974) gave an estimate of 20–25% for typical populations in short-grass prairie, whereas Rodell (1977) arrived at a much smaller figure (0.2–0.4%) for a similar ecosystem. White (1974) estimated that grasshoppers removed less than 3% of plant production in alpine tussock grassland in New Zealand, and Sinclair (1975) estimated that grasshoppers took 4–8% of annual grass production in the Serengeti grassland ecosystem, Tanzania. Grass-harvesting termites may remove substantial quantities of grass in arid grasslands: estimates ranging from 1 to 30% of grass production have been given for South African veldt (Hartwig, 1955; Nel and Hewitt, 1969), and from 4 to 30% in East African savannah (Sands, 1965b; Ohiagu and Wood, 1976). Up to 20% of grass cover may be removed by termites throughout large areas of the Middle East and Asia (Sands, 1977) and in dry pasture in Texas (Bodine and Ueckert, 1975). Leaf-cutting ants (Attini) can be very damaging to improved pastures in South America where *Acromyrmex* spp. can be particularly abundant (Cherrett, 1986). It is difficult to quantify their effects on productivity, but Cherrett cites studies by Amante (1972) and others indicating significant reduction of carrying capacity for stock by *Atta* spp. Ants were estimated to consume 3.2% of shoot production in an old field in South Carolina, USA (Wiegert and Evans, 1967), while seed-harvesting ants such as *Nessor, Veromessor* and *Pogonomyrmex* spp. may remove 2–10% of seeds in desert and prairie habitats (Whitford, 1978).

6.5.2 Grass response to elimination of invertebrate herbivores

The impact of herbivores on grassland productivity may be inferred from studies where grass growth in plots without herbivores is compared with that in plots where herbivores are present. In the case of vertebrates, exclusion can be effected by fencing, but in the case of invertebrates exclusion is not feasible and the usual approach is to suppress activity by means of insecticides. This often results in a substantial increase in grass production, even in situations where there is no overt sign of pest damage. Grass-yield responses to a range of insecticides (organochlorines and thiometon applied as sprays on the vegetation, granular HCH applied to the litter) in unmanaged and managed meadow plots in Poland were assessed by Andrzejewska and Wojcik (1971). The insecticides were applied singly or in various combinations. The total grass harvested in three cuts taken between 21 July and 17 August in plots that had received all the insecticides in the managed meadow was 1365 g m^{-2} compared with 1060 g m^{-2} in the control plot (i.e. there was an overall yield increase of 29% in the insecticide-treated plots). The two dominant herbivore groups in the managed meadow were planthoppers (Auchenorrhyncha) and grasshoppers (Acrididae), with an estimated total consumption of 6.3 g m^{-2} for the period June–October. The authors concluded that the losses of plant biomass were mainly attributable to a reduction in plant growth rate and in the rate of recovery after harvesting, resulting in a loss of plant production many times greater than the actual amount of food eaten.

Comparable responses to insecticidal application have been recorded for lowland ryegrass swards in Britain (Clements, 1980; Clements, French

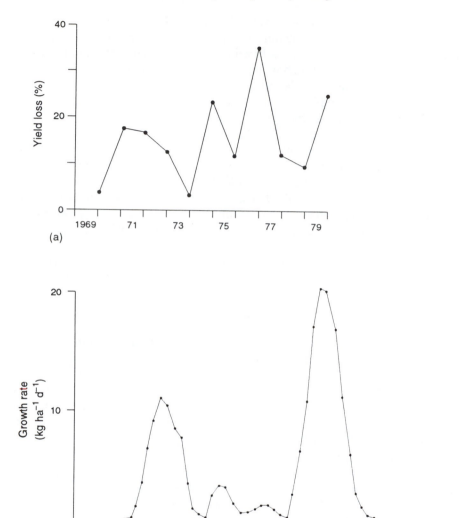

Figure 6.15 Reductions in grass yield attributed to insect pests at Hurley, England. (a) Annual losses over a period of 11 years; (b) seasonal differences in growth rates between insecticide-treated and untreated plots. (After Clements, 1980.)

and Henderson, 1981; Henderson and Clements, 1977a). Long-term experiments carried out for 11 years indicated average annual responses to insecticides of 15%, the responses being greatest in spring and autumn (Figure 6.15).

Table 6.13 Yield responses (Kg ha^{-1} dry matter) to insecticidal treatments at 14 lowland grassland sites in Britain, measured over 2 consecutive years of treatment in each case (from Henderson and Clements, 1977a) * Significant response ($P < 0.05$)

Site	First year		Second year	
	Control plots	Treated plots: difference from control	Control plots	Treated plots: difference from control
1. Yorkshire	10 414	+15	5 572	+286
2. Lancashire	9 150	+1010	6 176	−338
3. Kent	9 623	+582	10 338	+973
4. Devon	6 855	+1 642*	7 002	+1 628*
5. Lincolnshire	13 354	+1 075*	9 069	+2 927*
6. Cheshire	12 629	+198	8 895	+1 477*
7. Leicestershire	9 189	+151	7 979	+689
8. Shropshire	12 896	+1218*	8 343	+1 661*
9. Northamptonshire	10 898	+290	7 988	+1 533*
10. Herefordshire	12 118	+724*	8 990	+833*
11. Gloucestershire	14 807	+1 691	12 429	+2 138*
12. Essex	8 755	−168	6 517	+1 205
13. Wiltshire	11 634	+1 382*	10 671	+2 171*
14. Somerset	11 652	+1 443*	10 951	+1590*

* Significant increase in annual dry-matter yield ($P < 0.05$)

Treatment with insecticides resulted in increased dry-matter yields of up to 32% at 13 out of 14 lowland ryegrass sites studied throughout England and Wales (Table 6.13); the yield responses were correlated with reductions of stem-boring Diptera, aphids and other sap-sucking Hemiptera. Subsequent studies at 13 upland permanent pasture sites yielded significant responses to pesticides at only three (Clements and Henderson, 1980). The lack of response in upland sites was associated with low incidence of frit fly and a much lower proportion of ryegrass in upland swards. Further experiments carried out at 27 sites in northern England encompassing old permanent pasture, temporary swards and newly sown leys resulted in increased herbage yield or better seedling establishment in pesticide-treated plots at most sites and it was concluded that pest damage to grassland in the north of England is widespread (Clements, French and Henderson, 1981). Pest losses in lowland swards with a high proportion of ryegrass appear to be mainly due to frit fly (Clements, 1980), while yield responses in pastures with a low ryegrass content appear to be due to the control of other pests, notably *Tipula* spp. in wetter soils. Blackshaw (1984) estimated that an average increase in herbage production of more than 0.5 t ha^{-1} could be achieved by controlling leatherjacket populations in Northern Ireland grassland.

Major losses in production have been attributed to plant-feeding nematodes in prairie grasslands: Smolik and Lewis (1982) reported yield responses of 22–37% to nematicides under a range of conditions.

Some authors have questioned the attribution of increases in foliar mass in response to insecticides to the killing of herbivorous insects. Davidson, Schackley *et al.* (1979) recorded a 30% increase in perennial ryegrass yield in pasture treated with the insecticides lindane, dieldrin and azinphosethyl, but similar increases were obtained in pot experiments in the absence of pests. Soil respiration rate was increased for 1 week by the insecticides, suggesting that the killing of soil insects increased the substrate available to decomposers, resulting in a temporary surge of nutrients. Schackley and co-workers suggested that this increased availability of nutrients could account for about 30% of the enhanced grass yield normally attributed to killing injurious insects in such studies. On the other hand, Henderson and Clements (1977a) carried out glasshouse pot experiments using sterilized soil and found that the insecticides used in their field experiments (an aldrin soil drench and periodic applications of phorate granules) had no effect on ryegrass yield in the absence of invertebrates. In any case, any growth response arising from increased nutrient supply from decaying invertebrate corpses is likely to be short-lived and it is difficult to see how this could explain enhanced growth over the entire growing season or over several years as reported in some studies.

Thus, insecticidal experiments offer strong evidence that herbivores can significantly influence grass growth even at 'normal' levels of abundance. Based on data from extensive field studies, Clements *et al.* (1990b) concluded that between 11% and 15% of potential herbage production of lowland grass in the UK is lost to pests, amounting to a potential loss in excess of £500 million a year. However, it is also apparent that herbivory does not always result in a decrease in production and that moderate levels of feeding may sometimes have a stimulatory effect on plant growth. The interrelationships between invertebrates and their host plants are complex, and the outcome of these interrelationships in terms of plant growth is difficult to predict. Some of the main ways in which invertebrate feeding may influence plant growth processes and sward composition are considered in the next section.

6.5.3 Plant–herbivore interactions in grassland

The importance of grazing in maintaining grassland has long been recognized: in the absence of grazing or mechanical defoliation grass becomes moribund, shrubs can become established and succession of seminatural grassland towards woodland can proceed. The ways in which grazing may influence grassland include altering the general structure of the habitat and patterns of nutrient cycling, altering plant growth rates and death rates, influencing seed production and dispersal, and influencing competitive ability of the plants and the floral composition of the sward (Crawley, 1983). Grassland vegetation has a considerable capacity to compensate for tissue loss due to herbivory and McNaughton (1979) suggested nine mechanisms that may be involved in this process of compensation:

1. Increase in the photosynthetic rate of the remaining tissue.
2. Reallocation of substrate within the plant.
3. Increase in light penetration through the diminished canopy.
4. Removal of less-functional tissue that still requires resources.
5. Reduction of leaf senescence rates.
6. Redistribution of growth hormones.
7. Enhancement of soil water conservation.
8. Redistribution and recycling of nutrients.
9. Introduction of growth-promoting substances.

Yield increases in agricultural crops following moderate levels of herbivory have been attributed to the suppression of apical dominance in plants growing under conditions of adequate nutrient supply early in the season (Harris, 1974). In the case of recently established leys, this could be reflected in increased tillering if tillers are not killed and tillering capacity is not decreased.

(a) Nature of response to grazing

It has long been realized from empirical studies that vegetative grass growth could be prolonged and that grassland productivity could be increased by manipulating stocking density and the duration of grazing and recovery periods. The relationship between grass productivity and grazing pressure has been formalized in the herbivore optimization hypothesis. This states that net primary production (NPP) is initially increased by herbivory until a maximum point is reached, at which there is optimum grazing intensity; then NPP declines to become negative as the plant is unable to compensate for further defoliation (Figure 6.16).

Dyer *et al.* (1982) considered the relationship between plant productivity and level of grazing, in the context of how great a change in relative growth rate is required to increase primary production at a given level of grazing with relative growth rate (r), is defined as

$$r = \frac{1}{w} \cdot \frac{dw}{dt},$$

where w is the dry weight of the plant or plant part and t is time. Figure 6.17 indicates the response of a hypothetical plant to grazing; the

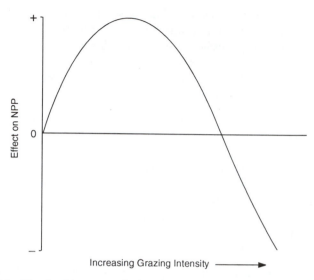

Figure 6.16 The herbivore optimization curve. (After Dyer, 1975 and Mc-Naughton, 1979.)

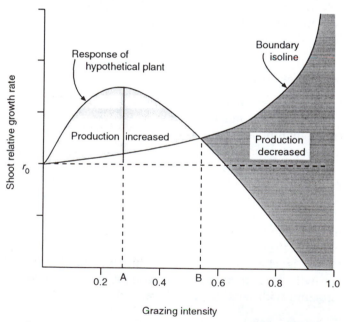

Figure 6.17 Boundary conditions (relative growth rate) at which the production of a grazed plant will equal that of an ungrazed plant at various grazing intensities. A and B represent, respectively, the grazing intensities at which primary production is maximized and the maximum sustainable grazing intensity beyond which production decreases. (From Dyer *et al.*, 1982.)

boundary line along which the productivity of a grazed and an ungrazed plant are equal. Production is optimized at the grazing intensity where the plant's relative growth rate response is at the highest point above the boundary line (A), and the highest grazing intensity at which production is not decreased occurs where the two curves intersect (B). Data from defoliation experiments (Hilbert *et al.*, cited by Dyer *et al.*, 1982) indicate that the shape of the boundary curve is a complex function of the three variables, *r* (relative growth rate), *G* (proportion of biomass removed by grazing) and *t* (time). Figure 6.17 indicates that a plant may respond to grazing with an increased relative growth rate but this may be below the boundary line and hence declines relative to an ungrazed plant over the same time interval. However, if more time is allowed for recovery the increase in relative growth rate may be sufficient to increase production. Likewise, plants growing at rates close to their maximum potential relative growth rate can (potentially) sustain less grazing than plants with growth rates further below their maximum potential.

Positive responses to grazing do not necessarily always occur; the response to defoliation can vary widely depending on a range of ecological factors. For example, Georgiadis *et al.* (1989) recorded significant overcompensation for defoliation in African savannah grassland plants grown in pots only when plant growth rates were limited by the availability of soil water at the time of defoliation.

Most of the experimental data on the growth response of grass to herbivory have been derived from mechanical defoliation or from ungulate grazing experiments, but there are also indications of positive responses to invertebrate herbivory. Andrzejewska and Wojcik (1970) concluded that grasshopper feeding can increase grass production by stimulating more intensive growth, particularly during the period of maximum grass growth early in summer. Moderate levels of root grazing by soil herbivores such as scarab larvae can remove moribund root tissue and stimulate the development of new roots, thus increasing root efficiency (Davidson, 1979).

(b) Growth-promoting effects

Although enhanced plant growth has generally been considered in terms of response to defoliation, the possibility of a direct stimulatory effect involving growth-promoting substances transmitted to the plant by the herbivore has also been raised. Such growth factors may be produced widely in vertebrates but little is known of how they affect plant growth. Growth factors could be introduced into a plant via saliva, but Dyer *et al.* (1982) were unable to detect a response to bison (*Bison bison*) saliva in laboratory defoliation experiments with blue grama grass (*Bouteloua gracilis*). However, Dyer (1980) did detect significantly increased growth of young sorghum seedlings treated with epidermal growth factor (EGF) from mouse submaxillary gland; similar effects were obtained with human EGF (Dyer *et al.*, 1982). Dyer and Bokhari (1976) reported much higher regrowth potential in *Bouteloua gracilis* on which grasshoppers (*Melanoplus sanguinipes*) had been feeding than in plants that had been clipped, suggesting a growth-promoting effect, and a substance that is probably EGF has been discovered in oral exudates and crop contents of grasshoppers, the first report of such a growth factor in insects (Dyer *et al.*, 1982). Earthworms are known to produce indole compounds that influence plant growth (Nielson, 1965), while Springett and Syers (1979) reported that casts of *Lumbricus rubellus* influenced the patterns of routine growth of ryegrass seedlings in a pot experiment in a manner that suggested a hormone-like effect. Roots tended to grow upwards into casts applied to the soil surface, while in the absence of casts the normal pattern of root growth was downwards. Casts of *Aporrectodea caliginosa* did not elicit a similar response, but shoot growth in the presence of casts of both species increased to a greater degree than would be expected on the basis of increased nutrient availability alone.

(c) Factors influencing regrowth potential

The potential of a sward to regrow and its tolerance for invertebrate damage are influenced by factors such as the size and health of the plants attacked, the growing conditions, the degree to which the plants are already under stress, and the timing of attack in relation to the phenology of the plants. Davidson (1969) reported that 50% of the roots of pasture plants that were not being defoliated could be lost due to scarabaeid larval feeding before there was any significant reduction in foliage yield. Ridsdill-Smith (1977) obtained similar results in a glasshouse pot experiment with perennial ryegrass, but foliage yield was significantly reduced when plants were regularly defoliated to simulate grazing. Larval feeding reduced root growth by an amount that was much greater than that removed in feeding: consumption represented only 5–14% of the reduction in root yield over the range of larval densities used in the experiments. There was evidence of water stress in the plants brought about by larval feeding, and damage resulting from scarabaeid and other root-feeding pests in pasture is observed to be greater when soil moisture content is low.

Overall herbage consumption by slugs in upland grassland in the UK was relatively low – less than 7% of estimated sheep intake (Lutman, 1977) – but slugs feed most intensively during the mid-October to April period when primary production is at a standstill and the vegetation has died back. By further depleting photosynthetic tissue at a critical stage in sward growth, slug feeding could delay the onset of growth in the spring.

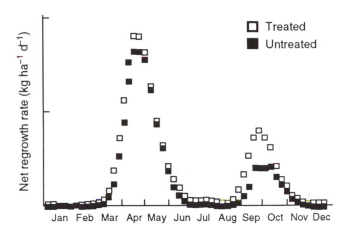

Figure 6.18 Net regrowth rate of perennial ryegrass in plots treated monthly with the systemic insecticide phorate compared with that in untreated plots during 1976. (From Clements, Chapman and Henderson, 1983.)

Rodell (1977) studied the impact of grasshoppers on the short-grass prairie ecosystem with the aid of a simulation model (section 6.6.2) and concluded that while their consumption and damage normally amounted to less than 1% of NPP, if grasshoppers were introduced 1 week earlier and at slightly higher population densities NPP would be reduced by as much as 14%, highlighting the potentially important effects of herbivory at times when plant production is low. Likewise Clements, Chapman and Henderson (1983) reported greatest losses in perennial ryegrass yields associated with frit fly larval feeding in April and October. Herbage net regrowth rates on plots treated monthly with the systemic insecticide phorate were compared with those of untreated plots (Figure 6.18). Differences between the treated and untreated plots were greatest in the period late March to late April and from late August to late October. Although larval populations were relatively low during the spring period (Figure 6.19) even a small population could have a large effect because tiller numbers are relatively low at this time of year. The autumn period of damage coincided with a period of high larval abundance at a time when the rate of tillering is reduced, and the number of new tillers formed probably cannot compensate for those damaged by stem-boring larvae.

The way in which insect damage can be relatively more important in a stressed habitat is also demonstrated in the case of termite feeding in tropical grasslands. When grass cover is drastically reduced by heavy grazing, termites may then eat the remainder, including root stocks, resulting in denudation of soil and soil erosion (Sands, 1977).

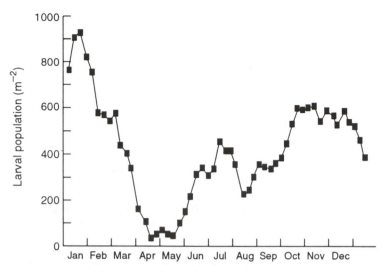

Figure 6.19   Populations of stem-boring dipterous larvae in untreated perennial ryegrass swards during 1976. (From Clements, Chapman and Henderson, 1983.)

6.5.4 Effects of invertebrate herbivory on sward composition

Many herbivorous invertebrate species are oligophagous and show strong preferences for particular plant species or for plants in particular physiological stages (Chapter 4). Examples include grass-feeding Auchenorrhyncha where food quality and plant N status in particular appear to play an important part in host-plant suitability (Waloff, 1980; Prestidge, 1982a, etc.), and grass-feeding termites (*Hodotermes* and *Trinervitermes* spp.), which show distinct preferences for different grass species related to mechanical and physiological characteristics (Sands, 1961; Nel and Hewitt, 1969; Ohiagu and Wood, 1976). Many grasshopper species show clear preference for different species of grass (Gyllenberg, 1969; Bernays and Chapman, 1970; Mulkern, 1967, 1972; Chapman, 1974), preferences being related to physical characteristics such as silica content and the presence of appropriate phagostimulant or deterrent chemicals.

Selective feeding on preferred plant species can reduce their competitive ability, alter botanical composition, and reduce sward productivity. Total herbage consumption by grasshoppers in New Zealand alpine tussock grassland is low, but much of the feeding is concentrated on important ground-cover species of low biomass (White, 1974, 1979a). Up to 59% of these plants can be removed by grasshopper feeding, resulting in loss of up to 11% in total ground cover with consequent deterioration of the habitat. Ryegrass swards are particularly susceptible to attack by frit fly, which can facilitate the invasion and establishment of unsown grasses and broad-leaved weeds (Clements and Henderson, 1979). Italian and hybrid varieties of ryegrass are valued for their high-yielding characteristics, but they lack persistence and their contribution to the sward drops off rapidly after about 2 years under European conditions. Henderson and Clements (1979) reported marked improvement in the persistence of these varieties in insecticide-treated plots compared with non-treated plots (Table 6.14). This improved persistence was attributed mainly to the control of shoot-mining frit fly, and the authors concluded that frit fly attack can be a major cause of lack of persistence in these ryegrass cultivars. Further experiments confirmed that the establishment, persistence and yield of Italian ryegrass can be greatly increased by the application of pesticides (Henderson and Clements, 1980; Clements, Murray and Sturdy, 1991).

On the positive side, weed-feeding species may limit the distribution of pasture weeds. A well-documented and dramatic example is that of the prickly pear cactus (*Opuntia* spp.), which was estimated to infest some 24 million ha of grassland in Australia in 1925 (Harris, 1973). The imported moth *Cactoblastis cactorum* was released in 1926, and within 4 years it had destroyed the two main species of *Opuntia* over large areas; since then, the prickly pear cactus has been restricted to isolated plants and scattered clumps in Australia. A second example is St John's wort (*Hypericum perforatum*), which infested considerable areas of range land in Australia and western USA – almost 1 million ha, in California alone – before being controlled over much of its range by leaf-feeding beetles (*Chrysolinia* spp.) (Harris, 1973; Chew, 1974). Satisfactory control of this weed by introduced insects was also reported from South Africa (Gordon and Kluge, 1991). Control was mainly attributable to *Chrysolina quadrigemina*, which was effective in destroying dense stands of the weed, and to a gall midge *Zeuxidiplosis giardi*, which damaged seedlings in moist habitats.

6.5.5 Invertebrates as vectors of plant diseases and pollen

The role of invertebrates as vectors of plant disease has received little attention in grassland compared with arable crops. Aphids can transmit cocksfoot streak virus (CfSV) and barley yellow dwarf virus (BYDV), and the latter can cause considerable loss in ryegrass crops (Plumb, 1978). The planthopper *Javesella pellucida* can also be of some importance in transmitting grass virus disease where it is abundant (Ossiannilsson, 1966). The eriophyid mite *Abacarus hystrix* transmits ryegrass mosaic virus in Europe (Mulligan, 1960).

Insect pollination is of little significance in perennial grass swards, but lack of effective pollination can be a major factor contributing to poor legume seed crops (Parker, Batra and Tepedino, 1987). Most managed

Table 6.14 Changes in sward composition associated with insecticidal treatments; percentage of sown species (dry-matter basis) harvested on untreated (0) and treated (+) plots (after Henderson and Clements, 1979)

Ryegrass species	Variety	Insecticide	Aug. '74	June '76	Oct. '76	May '77
Lolium perenne	S.24	0	–	96	84	77
		+	–	99	99	89
L. perenne	Reveille	0	77	99	95	93
		+	52	99	100	100
L. perenne	Taptoe	0	61	99	98	95
		+	62	100	98	99
L. perenne	Barlenna	0	87	100	96	96
		+	45	100	98	100
L. perenne	S.23	0	–	99	93	95
		+	–	100	99	99
L. perenne	Barpastra	0	56	99	97	98
		+	56	100	99	99
L. perenne × *L. multiflorum*	Grasslands Manawa	0	84	93	37	52
		+	65	100	100	99
L. multiflorum	RvP	0	81	94	43	52
		+	69	100	100	100
L. multiflorum	Baroldi	0	82	97	21	44
		+	81	100	100	100
Dactylis glomerata	S.37	0	65	100	100	99
		+	46	100	100	100
Phleum pratense	S.352	0	–	92	87	89
		+	–	82	94	85

crop pollination is done with the honey bee (*Apis mellifera*) but some species of solitary bees can also be effective pollinators. *Melitta leporina* and *Andrena ovatula* are the most common and widespread species associated with legumes throughout Europe, while managed alkali bees (*Nomia melanderi*) and leaf-cutter bees (*Megachile rotundata*) are currently being used commercially in extensive areas of seed crop production in the USA (Parker, Batra and Tepedino, 1987).

6.5.6 Herbivores and organic matter turnover

Herbivorous invertebrates can also contribute to the turnover of organic matter and nutrient cycling in grassland. Their excreta and corpses are readily mineralized and may comprise small but relatively concentrated and labile pools of nutrients such as P (Andrzejewska, 1979b; Hutchinson and King, 1982). By their destructive feeding, grasshoppers can influence rates of organic matter turnover (Andrzejewska and Wojick, 1970; Mitchell and Pfadt, 1974; Rodell, 1977), although this effect is normally quantitatively small. Rodell (1977) concluded from simulation-modelling studies (section 6.6.2) that grasshoppers would increase the annual rate of C turnover in the surface litter of short-grass prairie by at most 2.24%; however, over a period of years this might have a significant effect on nutrient cycling.

6.5.7 Invertebrate herbivores and long-term community dynamics

In systems such as deciduous forests, where insects frequently comprise the dominant herbivores, it has been suggested that their activities can influence patterns of nutrient cycling and plant succession and that they can exert a regulatory function in ecosystem processes (Chew, 1974; Mattson and Addy, 1975; Schowalter, 1981). The role of ungulate herbivores in the maintenance of much of the world's grasslands and in the regulation of grassland ecosystem processes is widely acknowledged (McNaughton, 1976, 1979; Owen and Wiegert, 1976; Dyer et al., 1982). By comparison, the role of invertebrate herbivores is probably relatively small. However, as already seen, invertebrate herbivores can have marked effects on botanical composition and on ecosystem processes at a more local level.

6.6 PEST MANAGEMENT AND CONTROL IN GRASSLAND

Although invertebrate herbivores may often depress grass growth, the level of utilization of grassland is generally not sufficient to justify costly pest-control measures except perhaps to deal with sporadic pest outbreaks in established pasture or to protect vulnerable new leys. The pest-management approach of utilizing a range of factors to maintain the pest populations below levels causing economic injury seems to have potential for the control of at least some recurrent pests. The high initial cost of developing such programmes inhibits the more rapid acceptance of this approach (Hussey, 1984); however, once a programme has been developed it offers the prospect of long-term effective control at low cost and at minimum risk to the environment.

The pest-management approach has been used with considerable success for the control of aphids in irrigated alfalfa, a valuable forage crop in central California (Flint and van den Bosch, 1981). Very severe damage was caused to this crop by the spotted alfalfa aphid (*Therioaphis trifolii*), which became established there in the mid 1950s. The problem was aggravated by the use of insecticides that depressed populations of natural enemies, allowing the resurgence of spotted alfalfa aphid and. outbreaks of secondary pest species. An integrated management programme was successfully developed for this pest; it included introduction and release of three hymenopterous parasites, strip harvesting of the crop to provide refuges for natural enemies, irrigation at appropriate times to enhance the activity of a virulent fungal disease, use of a selective aphicide where necessary, and the introduction of resistant alfalfa varieties.

Pest management requires a detailed knowledge of the biology and ecology of the pest species with detailed information on the factors influencing their abundance under various conditions, the ways in which economic loss may be caused under different circumstances, and their responses to the various control strategies. There are very few, if any, examples of grassland pests for which adequate information is currently available, but there are sufficient data in a number of cases to provide an empirical basis for at least partial pest management. Clear Hill *et al.* (1990), for example, show how frit fly damage to newly sown grass in the UK can be controlled using a combination of low doses of pesticide, resistant cultivars and naturally occurring hymenopterous parasites.

A good deal of progress has been made with the development of pest-management strategies for the control of some pasture pests in New Zealand. Improved New Zealand pastures sown with exotic grass and clover cultivars are subject to recurring attack by a range of pests, the most damaging being the indigenous grass grub (*Costelytra zealandica*) and porina (*Wiseana* spp.), and the exotic black beetle (*Heteronychus arator*), Australian soldier fly (*Inopus rubriceps*) and Argentine stem weevil (*Listronotus bonariensis*). Organochlorine insecticides, notably DDT, provided a cheap, persistent and effective means of control and were widely used in the 1950s, but following their withdrawal no satisfactory chemical substitutes have been found and considerable attention has been given to developing alternative approaches.

Before control options can be adequately evaluated, reliable information on losses due to insect pests is required. Some recent developments in methods of assessing pasture loss are considered below.

6.6.1 Pest density, pasture damage, economic loss

It is difficult to determine valid economic threshold levels for pasture pests. There is, first, the problem of estimating likely loss in sward production associated with a particular level of pest population density, and there is the further problem of quantifying the potential economic loss. In contrast to the arable cropping situation, where the consequences of a given level of reduction in yield quality or quantity can be quantified within the prevailing marker price structure, the economic consequences of a given level of yield reduction in grass will vary depending on the circumstances, including the management system and the efficiency of grassland utilization. Thus, the impact of invertebrate herbivores in grassland needs to be considered in terms of how they affect animal production under a range of grazing pressures (Kain and Atkinson, 1975; Pottinger, 1976; Kain, 1979; Roberts, 1979a; Roberts and Ridsdill-Smith, 1979). Invertebrates affect animal production only when they cause herbage production to fall below animal requirements during periods when food is limiting (Kain, 1979; Waterhouse, 1979) or when losses are sufficient to force a farmer to reduce an optimum stocking rate (Wightman and Whitford, 1979). Few studies have adequately taken account of the animal-production dimension but an example of an attempt in this direction is the study by Roberts and Ridsdill-Smith (1979), which examined the interrelationship between the abundance of the scarab *Sericesthis geminata*, the area of pasture damaged, and sheep production in Australian pasture.

Most studies have been confined to the relationships between pest density and pasture damage, and estimates of economic loss have been based on assumed herbage values. Blackshaw (1984) derived a linear relationship between herbage yield and leatherjacket numbers in Northern Ireland pasture on the basis of growth response to leatherjacket control by chlorpyrifos. He used this information in association with the probabilities of occurrence of different population levels derived from field-population census data over a period of 19 years to compute the expected economic returns from six management options (Blackshaw, 1985):

1. No action.
2. March (routine): routine spraying in early March of all grassland.
3. March (forecast): all grassland sprayed in high-risk years only in March.
4. March (monitoring): only populations above the economic threshold in March treated.

5. September (routine): all grassland sprayed in late September.
6. September (monitoring): only populations above the economic threshold in September treated. This option was hypothetical since at present there is no practical means of estimating population levels of *Tipula paludosa* in the field in September.

Figure 6.20 shows the financial margins over the 'no action' option expected to be achieved at different herbage values (£ kg^{-1}) by the different alternatives. However, there are some assumptions involved in the calculations that may not always be valid. For example, subsequent work (Blackshaw and Newbold, 1987) utilizing data from eight sites, four in Northern Ireland and four in Scotland, indicated a curvilinear relationship between leatherjacket loss and population size rather than a linear one (Figure 6.21). A practical alternative to the use of insecticides in some situations may be to increase fertilizer use to encourage compensatory grass growth. Blackshaw and Newbold (1987) found in their sites that yield increases in response to chlorpyrifos applied for

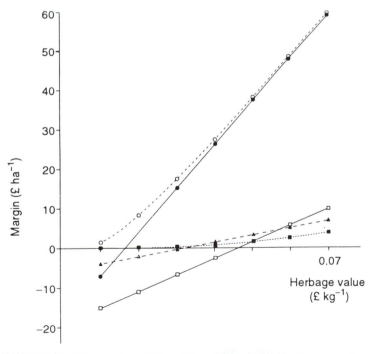

Figure 6.20 Financial margins achieved by different management options over 'no action'. ○, September (monitoring); ●, September (routine); □, March (routine); ▲, March (forecast); ■, March (monitoring) (see text for description of options). (From Blackshaw, 1985.)

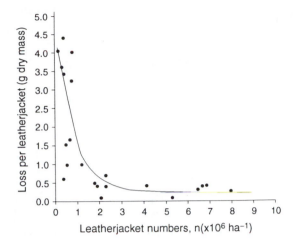

Figure 6.21 Relationship between pasture loss and leatherjacket population density in grassland. Loss = $0.123 + 4.03 \, (0.325)^n$; the percentage variance explained = 63.7. (From Blackshaw and Newbold, 1987.)

leatherjacket control in November were equivalent to the response obtained from 75 kg N applied in the third week of March.

Relationships between pest density and pasture damage have been established for grass grub (Fenemore, 1966; Kain, 1975) and porina (McLaren and Crump, 1969) in New Zealand. When considering such relationships it is important to know how yield loss varies with growth rate of herbage mass. For surface-feeding pasture pests such as porina (*Wiseana* spp.) two broad types of behaviour or 'functional responses' to herbage mass are encountered (Barlow, 1985a). A pest that grazes herbage would be expected to remove a constant amount irrespective of the herbage mass or the growth rate, in which case the effects of a given density could be equated to a certain number of stock units regardless of stocking density. By contrast, a pest that denudes vegetation, severing leaves and tillers and removing growing points, will cause losses proportional to the standing herbage mass, to the growth rate of the pasture and to the stocking rate. In the case of porina, data from insecticide trials (McLaren and Crump, 1969) give equally good correlations between pest density and either absolute pasture yield or percentage yield loss, but Barlow (1985a) concluded from a pot experiment that porina is a 'denuding' pest rather than a 'grazing' one. Herbage removal increased with increasing herbage mass up to 2000 kg ha^{-1} dry matter (Figure 6.22) and the percentage removal stayed constant and similar to the percentage removal of plant cover within this range of herbage mass. On sheep

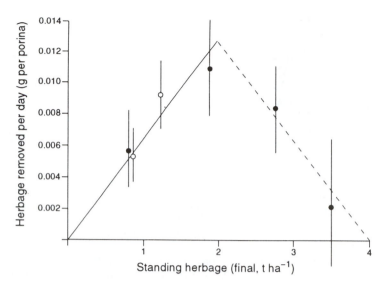

Figure 6.22 Relationship between net rate of herbage removal by porina cater-pillars and standing herbage mass. ●, Ryegrass/clover pots; ○, browntop pots; ± standard errors. (From Barlow, 1985a.)

pastures, therefore, porina could be expected to cause greater revenue losses at high stocking rates than at lower ones.

6.6.2 Models for pest population/economic loss forecasting

Computer modelling is a potentially useful tool for decision making in relation to pasture pest control, offering prospects for better timing and improved effectiveness of control procedures and for minimizing pesti-cide use and associated environmental consequences. This approach has not as yet been widely adopted, but there are several instances where considerable progress has been made in the building of models to describe population processes and, in some cases, the economic impact of major pests under given conditions. At present many of these models are primarily research tools, but some can be useful aids in designing stra-tegies for pest management and pest control even at their current state of development. Relatively simple key-factor type models constructed on the basis of life-table studies of grass grub and black beetle in New Zealand have been found to describe accurately the general patterns of population change (East, King and Watson, 1981) (Chapter 4). These models can be used in association with population monitoring to deter-mine the need for control measures and the optimum timing of appli-cation of transient insecticides. Also, by identifying the key factors responsible for population change this approach may lead to methods of manipulating these mortalities to control the pest, such as the use of summer grazing management for the control of the grass grub *Costelytra zealandica* (section 6.6.3 (c)). A more complex, systems approach, was adopted by Davidson, Wiseman and Wolfe (1970) for pasture scarabs, (e.g. *Sericesthis nigrolineata*) involving a series of individual models covering the main stages and processes in the life cycle. Wightman (1979) and Wightman and Whitford (1979) integrated energetics and population dynamics approaches to model the impact of the grass grub on pasture.

Rodell (1977) used simulation modelling to study the impact of grasshoppers on short-grass prairie in the USA. The systems approach was adopted, emphasizing the main components or compartments of the

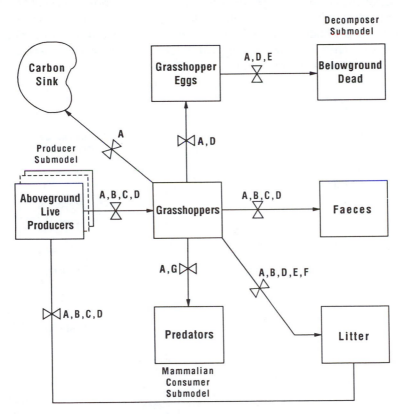

Figure 6.23 Flow diagram of grasshopper submodel and its interaction with other submodels of the short-grass prairie ecosystem. The arrows represent carbon flows. A-G represent factors that influence the flows: A = grasshopper carbon (g m^{-2}); B = producer carbon (g m^{-2}); C = producer phenology; D = temperature; E = precipitation; F = relative humidity; G = predator carbon (g m^{-2}). (From Rodell, 1977.)

ecological system and the interactions between them. The grasshopper model was designed so that it could be used as a submodel to interact with other components of a grassland-ecosystem model developed for the US/IBP Grassland Biome project. The major objective of the model was to simulate the biomass dynamics of the grasshopper population; additional objectives were to consider the effect of grasshoppers on the functioning of the total system and to estimate energy flow through the grasshopper population. The model was a state variable, process-orientated one that considered the interactions between grasshoppers and producer, decomposer and other consumer components in terms of carbon flow.

Figure 6.23 illustrates the interactions between the grasshopper model and other relevant models by means of a compartment flow diagram. Grasshoppers influence producers via food selection, herbivory and destructive feeding damage; they interact with decomposers via death and faeces production and with other consumers through competition or predation. The primary driving variables influencing the grasshopper model were air temperature and precipitation.

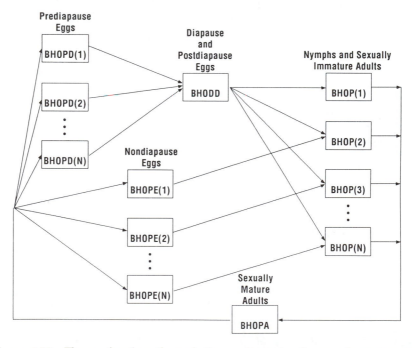

Figure 6.24 Flows of carbon through the components of a grasshopper population. BHOPD (1–N), BHOPE (1–N) and BHOP (1–N) represent different age groups of pre-diapause eggs, non-diapause eggs, and nymphs and sexually immature adults, respectively. BHODD represents diapause or post-diapause eggs; BHOPA represents sexually mature adults. (After Rodell, 1977.)

The flows of carbon through the grasshopper population are represented in Figure 6.24. Five different categories within the life cycle were recognized: pre-diapause eggs, non-diapause eggs, diapause and post-diapause eggs, nymphs and sexually immature adults, and sexually mature adults, and the flows of carbon within and between these groups associated with the processes of hatching, development and sexual maturation, egg laying and diapause were considered.

Among the conclusions that were drawn from the simulations were that the population was not food-limited in short-grass prairie, but was governed primarily by weather. Grasshopper consumption and feeding damage accounted for the removal of only 0.4% and 0.2% of net primary production in 1970 and 1971, respectively; however, where grasshoppers were introduced 1 week earlier and at a higher density in a 1970 simulation the result was a reduction of 14% in net primary productivity, indicating a potentially greater impact under conditions which favour earlier grasshopper activity at a time when plant production is low.

Although this model was not designed to quantify economic impact nor to provide a basis for pest management, it could be used to further these objectives by, for instance, simulating the effects of changing environmental parameters known to influence grasshopper populations (Table 6.15). In this example, simulation in which cold-induced egg mortality was varied showed marked deviation of grasshopper biomass from the 'standard' simulation, indicating high population sensitivity to egg mortality (Table 6.15); on the other hand, the simulation suggests that embryonic development is relatively insensitive to changes in soil moisture while oviposition would appear to be relatively insensitive to air temperature.

An obviously important feature of any model with potential application in the field of pest management is its capacity for prediction (i.e. its ability to mimic the dynamics of a field population). The results of a validation exercise in which model results are compared with independently estimated field population data are shown in Figure 6.25. The simulated population dynamics fall within the 95% confidence intervals of the field observation 73% of the time, but from September onwards the simulated and observed values deviate: the simulated population does not die as fast in the autumn as the field population, suggesting that adult mortality is underestimated or that some other factor such as migration should be included. However, the model did correctly predict year-to-year changes in population size.

White (1979b) used a comparable modelling approach to study the impact of grasshoppers on alpine tussock grassland in New Zealand. Although the systems approach is undoubtedly a useful research tool, to date the contribution of complex simulation models to pest management has been disappointing. There is some doubt as to whether they can ever

Table 6.15 Simulation results of grasshopper peak biomass arising from altering weather and soil conditions (after Rodell, 1977)

	Condition	Life-cycle stage and process involved	Max. grasshopper biomass $(g\ cm^{-2})$	Deviation from 'standard' simulation (%)
I	Inadequate soil moisture 25% of time	Eggs: embryonic development	0.0033	−26.7
II	Cold soil temperature	Eggs: survival		
	25% of the time		0.0008	−82.2
	Never		0.0367	+716.0
III	Precipitation	Nymphs: survival		
	50% of the time		0.0014	−68.9
	25% of the time		0.0027	−40.0
	Never		0.0057	+26.7
IV	Air temperature			
	Daily max −5 °C	Nymphs: development	0.0029	−35.6
	−3 °C		0.0036	−20.0
	+3 °C	Immature adults:	0.0063	+40.0
	+5 °C	sexual maturation	0.0099	+120.0
V	Air temperature	Adults: oviposition		
	Cool 25% of time		0.0041	−8.9
	Maximum		0.0054	+20.0
Standard simulation			0.0045	

predict population trends with sufficient reliability to be of practical value, and it may be that for many applications simpler, empirically based approaches are more useful.

An interesting conceptual approach to the problem of relating levels of pest infestation to economic damage under varying conditions of sward growth and utilization has been developed by Barlow (1987). He extended the predator–prey approach adopted by Noy-Meir (1975, 1976) to develop grazing models with two herbivores that could be used to assess the effects of pasture pests on the stability and productivity of pasture growing at different rates and grazed at different stocking levels. The form of the pest-damage function was described for two types of pests, grazing herbivores and denuders of pasture area. Two extreme situations were considered – one in which pest consumption is unaffected by vegetation availability, the second in which pest consumption is affected in the same way by vegetation availability as is consumption by stock.

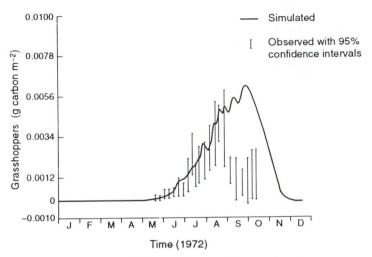

Figure 6.25 Simulated model of grasshopper biomass dynamics compared with field observations for 1972. (After Rodell, 1977.)

Figure 6.26 shows the effects of grazing pests such as rabbits and denuding pests such as porina (*Wiseana* spp.) on the stability and productivity of the vegetation in a New Zealand sheep pasture. The $\overset{*}{V}$, $\overset{*}{H}$ isoclines represent vegetation levels and stocking rates at which the vegetation stabilizes (i.e. at which growth equals consumption) while the $\overset{*}{P}$, $\overset{*}{H}$ isoclines represent equilibrium productivity/stocking rate conditions. In the example considered, the grazing pest whose consumption is affected by vegetation availability in the same way as is that of sheep has no effect on the stability of the system, because it is equivalent in its effect to additional stock units, and the shape of the vegetation isocline in Figure 6.26a retains the same 'continuously stable' form. The effect of this type of pest on productivity is to displace the curve downwards and to the left. Important consequences of this are that absolute reduction in productivity by the pest is greater at a high initial stocking rate (e.g. the optimum of 15 ha^{-1} in Figure 6.26b) than at a lower one and that in those circumstances productivity is maximized by reducing the stocking rate. The effect of the pest on productivity is non-linear and the steeper the right-hand side of the productivity/stocking rate curve the greater the non-linearity and the effect of the pest.

Where the efficiency of the grazing pest is unaffected by vegetation availability, the effect is destabilizing (Figure 6.26a). The system is 'discontinuously stable', with a maximum sustainable stocking rate in the presence of the pest (10.6 in this example). There is a lower critical level of $\overset{*}{V}$ (about 200 kg ha^{-1}) below which if the vegetation mass falls no reduction in stocking rate will permit recovery. Productivity is likewise

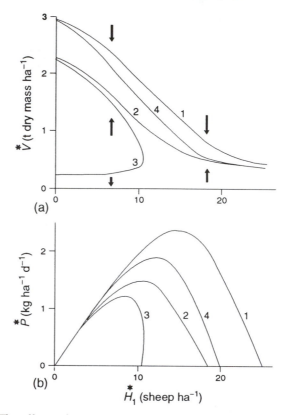

Figure 6.26 The effects of grazing pests such as rabbits and denuding pests such as porina (a) on the equilibrium vegetation ($\overset{*}{V}$)/stocking rate ($\overset{*}{H_1}$) relationship and (b) on the equilibrium productivity ($\overset{*}{P}$)/stocking rate ($\overset{*}{H_1}$) relationship. Curve 1, no pests; 2, grazing pests at 60 ha^{-1}, consumption affected by vegetation availability in same way as that of sheep; 3, grazing pests at 60 ha^{-1}, consumption unaffected by vegetation availability; 4, denuding pests removing 20% of the pasture area. The arrows indicate direction of vegetation change. (After Barlow, 1987.)

dramatically affected, with the pest's impact being greatly increased and greater non-linearity in the pest density/productivity relationship than in the former case.

In the case of denuding pests that remove a certain proportion of the total area of palatable vegetation, the effects on pasture and animal productivity are assumed to be proportional: there is no effect on the system's stability (Figure 6.26a) and the productivity/stocking rate curve is displaced downwards and to the left, with no significant change in shape (Figure 6.26b). Losses are considerably greater at a high stocking rate than at a low one, and the effect on productivity is increasingly non-linear with increasing pest density.

The shapes of the relationships between productivity and pest density are shown in Figure 6.27a for different stocking conditions and for grazing pests that respond to changes in vegetation availability in the same way as do sheep. At fixed stocking rates, losses per hectare are due to a reduction in animal productivity and tend to be non-linearly related to pest density (curves 1 and 2), whereas varying the stocking rate so that it is optimal at all pest densities gives a linear relationship (curve 3). In this situation, losses arising from stocking-rate reductions are directly

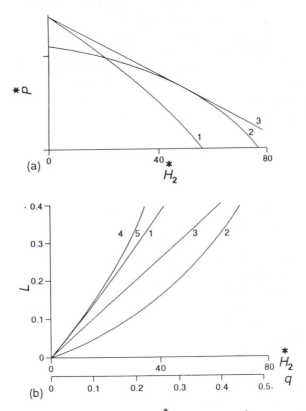

Figure 6.27 (a) Effect of grazing pests ($\overset{*}{H}_2$, rabbits ha^{-1}) on equilibrium ($\overset{*}{P}$, kg weight gain ha^{-1} d^{-1}) at different stocking rates. Curve 1, fixed, optimum 15 sheep ha^{-1}; 2, fixed, 75% optimum (11.25 ha^{-1}); 3, variable, optimum at all pest densities. (b) The damage functions for grazing and denuding pests: proportional losses in productivity (L) in relation to grazing pest density (H^*_2, ha^{-1}) or proportional loss in pasture cover (q). Curves 1, 2, 3 as above (grazing pests, consumption affected in the same way as sheep by availability of vegetation); 4, grazing pests, consumption unaffected by availability of vegetation, stocking rate optimum; 5, denuding pest, stocking rate optimum. Curve 3 also represents the effect of a denuding pest on stocking rate if per capita productivity is maintained, irrespective of the initial stocking rate. (After Barlow, 1987.)

related to pest density or proportional loss in pasture cover in the case of denuder pests. The damage functions indicate that losses are considerably greater at the high stocking rate than at the lower and varied rates, and increase non-linearly with pest density.

These damage functions (Figure 6.27b) can be used to make approximate estimates of economic loss due to pest attack: the proportional reductions in productivity can also be interpreted as proportional reductions in gross margin per hectare. A common short cut in assessing the impact of pasture pests has been to cost their effect as an equivalent reduction in stocking rate, and the grazing model presented in Figure 6.27b suggests that this can give a useful average estimate even though the true effect may be a reduction in productivity per sheep at a fixed stocking rate. Barlow (1985b) developed a simulation model to translate pasture losses from porina into revenue losses for a sheep breeding system that enables control thresholds to be established as functions of stocking rate, stocking intensity and control cost.

6.6.3 Strategies for pest management

(a) Biological control

There are not many examples of classical biological control of grassland invertebrate pests, although there are undoubtedly situations in which natural control could be improved by judicious introduction and release of biocontrol agents to augment the local complex, particularly in the case of invasive pest species that may have escaped from their natural enemies. Examples include the lucerne flea, *Sminthurus viridis*, and the white snail *Theba pisana*. The bdellid mite *Thoribdella lapidaria* exerts effective biological control of *S. viridis* over much of its range in South Australia (Wallace, 1967); it has also been introduced into South Africa in attempts to control this pest in Cape Province (Wallace and Walters, 1974). *T. lapidaria* does not occur in drier parts of the range of *S. viridis* in South Australia, and another bdellid species, *Nemologus capillatus*, has been introduced in attempts to control it in such areas.

The snail *Theba pisana*, a native of the Mediterranean region and western Europe, has become an important agricultural pest in southeastern and southwestern Australia (Baker, 1986; Baker and Vogelzang, 1988), where it may commonly be found in grassland. Conventional chemical control with molluscicides is expensive as well as being environmentally damaging. Several natural enemies such as parasitic sarcophagid and sciomyzid flies are known to occur in Europe and there may be potential for introducing some of these to Australia.

The biological-control approach has been applied with some success for the control of nuisance dung-breeding flies such as the Australian bush fly, *Musca vetustissima*. When sufficiently abundant, scarabaeine dung beetles can substantially reduce survival of such flies through competition but native dung beetle species are ineffective and several species better adapted to bovine dung have been introduced into Australia at various times. Nine out of 13 African and European species introduced into southwestern Australia, a region with similar Mediterranean-type climate, have become established, most being widespread and seasonally abundant (Ridsdill-Smith and Matthiessen, 1988). Ridsdill-Smith and Matthiessen investigated the impact of two introduced, summer-active, species (*Oniticellus pallipes* and *Onthophagus binodis*) on bush fly populations in an area of southwestern Australia where previously only a single, winter-active species (*Onthophagus ferox*) occurred. Peak populations of adult flies occur in late December–early January. The abundance of *O. ferox* declines in early December and this species is thus of limited effectiveness for *M. vetustissima* control, while the major summer emergence of *O. binodis* and *O. pallipes* is in late December–early January and their main period of abundance is from January to May. Comparison of fly populations in the 2 years before with those in the 3 years after the establishment of the introduced beetles showed little difference in November population densities, but *M. vetustissima* abundance fell to 12% of its previous level in January, halving the duration of the major fly problem in the area. There appears to be further scope for improving early-season control by introducing additional dung beetles that are abundant in spring.

Enzootic nuclear polyhedrosis virus (NPV) is widely distributed in porina (*Wiseana* spp.) in older New Zealand pastures and can maintain larval densities below the economic threshold in situations where a stable host–virus equilibrium has been established (Kalmakoff, 1979). Farm management practices that allow a build-up of the virus at the soil surface and thus enhance effectiveness, include intensive rotational stocking of young pastures and oversowing rather than cultivation for sward establishment or improvement.

Biotoxins produced by *Bacillus thuringiensis* (Bt) appear to have considerable potential for the control of pasture pests such as grass grub *Costelytra zealandica* (Wigley and Chilcott, 1990). Currently, the use of Bt products against soil pests is limited by such factors as host-range specificity, problems of delivery and short residual activity, but there are good prospects that these difficulties can be overcome by recombinant DNA technology (Gelernter, 1990).

There is also increasing interest in entomopathogenic nematodes such as *Steinernema* and *Heterorhabditis* spp. as potential biocontrol agents for grassland soil pests. They are well adapted to life in the soil, and they are able to locate and parasitize a wide range of insect hosts that are killed very quickly by their associated bacteria (Poinar, 1979). While encouraging results have been reported from some field studies on their use against grass grub in New Zealand (Jackson and Trought, 1982) and other chafer larvae in the USA (e.g. Forschler and Gardner, 1991), results from field trials generally have been variable and unpredictable (Georgis and Gaugler, 1991). Improved mass rearing and genetic selection technology can be expected to improve their performance, but the main impediment to their use in pest-management programmes at present may be lack of basic information on their ecology in the soil (Gaugler, 1988).

(b) Plant resistance

In the case of oligophagous pests there may be some scope for altering the botanical composition of the sward in favour of species that are unsuitable as food plants, or it may be possible in some instances to use varieties that are resistant to or can tolerate pest damage. Clovers are very favourable host plants for grass grub; lucerne, *Lotus pedunculatus* and *Phalaris tuberosa* are resistant; and cocksfoot and tall fescue are tolerant (Kain, East and Douglas, 1979; Russell *et al.*, 1979). Paspalum and perennial ryegrass favour black beetle (*Heteronychus arator*) while *Phalaris tuberosa*, lucerne and *Lotus pedunculatus* are unfavourable (King *et al.*, 1975; King and East, 1979; King, Mercer and Meekings, 1981a). Legumes favour white-fringed weevil (*Graphognathus leucoloma*), while grasses are unfavourable (East, 1976, 1977; King and East, 1979). Soldier fly larvae (*Inopus rubriceps*) grow better on a range of grasses than on legumes (Gerard, 1979). Certain grass species including *Lolium*, *Poa* and *Anthoxanthum* spp. appear to support greater densities of shoot fly larvae (mainly frit fly, *Oscinella frit*) that others (Southwood and Jepson, 1962), while Italian and hybrid ryegrass cultivars are particularly susceptible to this pest (Henderson and Clements, 1979). Resistant cultivars of lucerne play an important role in the management of spotted alfalfa aphid (*Therioaphis trifolii*) in California (Flint and van den Bosch, 1981), while varietal resistance is seen to have a key role in the management of aphid, stem nematode and other pests of lucerne in Australia and New Zealand (Turner and Franzmann, 1979; Dunbier *et al.*, 1979).

The scope for utilizing plant resistance for pest management in intensive ley farming may be limited in many cases since characteristics that confer natural resistance to pests (particularly digestibility-reducing physical defences) are eliminated or at least diluted during the process of selection of agronomically valuable cultivars. However, in some cases it may be possible to introduce genes that confer resistance against selected pests. Another way in which resistance to insects might be improved is by exploiting fungal endophytes of grasses (Siegel, Latch and Johnson, 1987). Endophytes are widespread in some economically important temperate grasses such as tall fescue and ryegrass. Infected grasses produce an array of chemicals with a wide range of biological activity, which can include resistance to some insect herbivores (but not frit fly larvae, according to Lewis and Clements, 1986). Work carried out in New Zealand over the past decade points to the key role that endophytic fungi can play in the interactions between some grassland invertebrate pest species and their host plants. Argentine stem weevil (*Listronotus bonariensis*), currently regarded as the most serious pest of improved pastures in New Zealand, has received most attention (Prestidge, Barker and Pottinger, 1991). The endophytic fungus, *Acremonium lolii*, confers resistance to attack, and the availability of *A. lolii*-free ryegrass tillers as feeding and oviposition sites has a major influence on the size and dynamics of the *L. bonariensis* population. Toxins produced by the fungus negatively influence larval survival and behaviour and adult feeding and oviposition. In reseeded pastures with low levels of endophyte infection, high larval numbers and considerable damage to ryegrass can occur; as endophyte-free ryegrass is eliminated by selective feeding the population declines to a stable, low level. Endophytes can also adversely affect the health of grazing animals, causing conditions such as 'ryegrass staggers' but, despite this risk, there has been a big increase in the practice of sowing endophyte-infected ryegrass seed in New Zealand. Considerable progress has also been made with the selection of *A. lolii* strains which confer resistance to the pest but which do not produce the toxins which are responsible for animal ill-health (Fletcher, Popay and Tapper, 1991).

(c) Altering grazing or cutting regime

There appears to be considerable potential for controlling some soil pests by adjusting grazing or cutting management to increase environmental stress during susceptible stages in the life cycle. As seen in Chapter 5, increases in stocking rate or in defoliation intensity generally reduce numbers and biomass of phytophagous grassland invertebrates, although the effects of increasing grazing intensity are more severe on species feeding on the above-ground parts of pasture plants than on those living in the more buffered soil environment. Nevertheless, several important soil-dwelling pests such as soldier fly, grass grub and some other chafers, and probably porina, do appear to be influenced by changes in stocking rates (East and Pottinger, 1983). The effects of stocking rate appear to relate mainly to defoliation and trampling that may kill invertebrates directly or may modify their living space, microclimate or food supply.

Summer mortality of larvae has been shown to be a key factor determining population changes in grass grub populations in New Zealand, and East and Willoughby (1980) reported the results of a series of experiments undertaken to determine whether this factor could be manipulated by the control of summer pasture cover through grazing

Invertebrates, soil fertility and plant growth

Table 6.16 Effects of grazing and mowing treatments on soil moisture and temperature and on populations of grass grub (*Costelytra zealandica*) at Rukuhia, near Hamilton, New Zealand (after East, 1979b and East and Willoughby, 1980)

	Grazing						Mowing			
	Hard			Light			Short		Long	
Soil moisture (%)*	19			24			22		26	
Soil temperature (°C)†	18–31			21–25			24–33		23–27	
Grass grub population‡	N	M	W	N	M	W	N	W	N	W
Eggs (November '77)	428			465						
		90			55					
Larvae (March '78)	44		51	209			30	84	170	111
		43			20					
Larvae (July '78)	25		74	168			110		19	
		68			61					
Pupae + adults (October '78)	8			66						
Mortality of total generation		98			87					
Eggs (November '78)	47			356						

*Average of 0–15 cm values, January–March 1978.
†Range of daily maxima at depth of 7.5 cm, January–March 1978.
‡N = grass grub numbers (m^{-2}); M = per cent mortality between successive stages; W = average larval mass (mg).

management. Table 6.16 gives the results of life-table studies of grass grub populations at Rukuhia near Hamilton, New Zealand, in plots set-stocked with sheep under a hard grazing regime, which maintained the pasture height at 2–4 cm, and under a lax grazing regime, where the average pasture height was 15–20 cm (East, 1979b; East and Willoughby, 1980). The treatments began in October 1977 and were continued until November 1978. The table also contains data from a parallel mowing experiment at the same site. Mortality from November to March was significantly higher ($P < 0.01$) in the hard-grazed treatment than in the lightly grazed one while the growth rate of larvae was significantly lower. Similar trends were apparent in the mowing treatment. During this period the soil was drier and hotter when the grass was grazed or cut short compared with the soil under the longer grass. Mortality in subsequent periods did not differ significantly between the short and longer grass, but the marked difference in summer mortality between the grazing treatments resulted in a major reduction in the numbers of

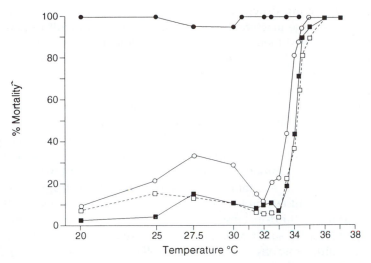

Figure 6.28 Effects of temperature on mortality of grass grubs held at different soil moisture levels: 5% (●); 10% (○); 15% (■); 20% (□). (From East and Willoughby, 1980.)

autumn larvae – the damaging stage of the life cycle – and in the numbers of eggs produced in the next generation. In both grazing and mowing treatments, the maximum daily soil temperature in January at 7.5 cm (the average depth of grass grub larvae) reached the lethal range (Kain, 1975) under the short grass but not under the longer grass.

Temperature-induced mortality was considered to be the primary factor responsible for the trends observed, with less importance being attributed to differences in soil moisture. However, the potential for soil moisture to modify the effects of soil temperature was demonstrated in a laboratory experiment in which field-collected second-instar and third-instar grass grub were exposed to various moisture and temperature levels representative of the range encountered in the field during summer and autumn (Figure 6.28). Mortalities at 30% and 40% soil moisture were similar to those at 15% and 20% and are not shown in the figure. At 5% moisture content, mortality was virtually 100% at all temperatures; below 32 °C, mortality at 10% moisture was much lower than at 5% but was significantly greater ($P < 0.05$) than at higher soil-moisture levels. The lethal temperature threshold was 33–34 °C at 15–20% moisture and was 0.5–1 °C lower when soil moisture was reduced to 10%. Under field conditions, soil moisture did not appear to be as important as temperature as a probable cause of high summer mortality of larvae, but it was of major importance because of its inverse relationship with temperature and because it can influence the vertical distribution of larvae and hence their risk of being exposed to lethal temperatures. Larvae respond to drought by moving down to depths of 7.5–10 cm or more in the soil where they become inactive, but dry soils are more difficult to penetrate than moist soils and there appears to be a decline in larval activity as soil moisture increases; hence it seems unlikely that larvae are able to escape completely from exposure to high temperatures by vertical movements in response to diurnal temperature fluctuations.

Further studies in New Zealand in additional localities and using different soil types showed that reduction in summer pasture cover to a height of 5 cm or less can result in a 50–80% reduction in grass grub populations in autumn compared with less severely defoliated pasture. However, for high summer mortalities to occur, periods of 4 weeks or more were required when soil moisture was low enough to suppress pasture regrowth and to allow soil temperature to reach levels that were lethal to second-instar and third-instar grass grubs. Summer rainfall, irrigation and shading prevented soil temperatures from reaching the lethal range, resulting in little or no effect of summer defoliation on grass grub numbers.

In considering the minimum period of short-pasture maintenance required for significant grass grub mortality, East and Willoughby (1980) concluded that a single grazing to 5 cm early in a prolonged dry spell might be as effective in reducing populations as weekly cutting. It also appeared that a single grazing would reduce grass grub more if the pasture had been allowed to grow for a long spell (e.g. 5 weeks). Under those conditions defoliation produced a more open sward with more bare ground that could markedly influence soil temperature. Also, since suppression of grass grub numbers by hard summer grazing carries over into the next generation, it may be necessary to use this grazing strategy only in alternate years or less frequently in infested fields.

Control of grass grub by hard summer grazing in New Zealand is heavily dependent on suitable conditions; it may be possible only in dry years and depends on accurate weather forecasting. The feasibility of this approach is also affected by factors such as topography and pasture species: northern slopes have lower average soil moisture and higher maximum soil temperatures than southern slopes, while soil temperatures reach higher levels under pasture species such as ryegrass, which are dormant during dry summer weather, than under summer-active species such as *Paspalum dilatatum* and tall fescue *Festuca arundinacea* (East and Willoughby, 1980 and references therein).

The commonest way in which grazing has been manipulated to control pasture pests in New Zealand has been by the use of high stocking rates ('mob stocking') to produce a heavy trampling effect for the control of grass grub in late autumn and winter when larvae are actively feeding in the top 2.5–5 cm of the soil layer. Increases in grass grub mortality of up to 40% in mob-stocked pasture over that in unstocked areas have been recorded, but only in situations where populations were high enough to cause serious damage (East and Pottinger, 1975). Heavy grazing has also been recommended for the control of cockchafer (*Melolontha melolontha*) larvae in European pastures. Murbach, Keller and Bourqu (1952) reported that larval numbers were reduced by 44–60% in pasture strip-grazed by dairy cows compared with pasture where set stocking was practised. Although other scarabs such as *Sericesthis geminata* appear to be favoured by moderate levels of grazing, they may nevertheless be amenable to control by varying grazing management. Roberts (1979b) reported a progressive reduction in numbers of scarabs in Australian pastures subjected to alternating high and moderate sheep-stocking densities in spring.

There also appears to be some potential for controlling porina, soldier fly and slugs in New Zealand by manipulating pasture management. Intensive stocking for short periods can cause high levels of mortality among surface-dwelling larvae of porina (Holmes, 1981). Hard grazing or cutting during the adult flight periods in spring and autumn reduced numbers of soldier fly larvae (Kain and Burton, 1975), while high stocking rates reduced reinfestation (Robertson *et al.*, 1979). High sheep stocking rates (14–20 sheep equivalents ha^{-1}) can result in significant slug mortality in lowland pastures in moderate to high rainfall areas of New Zealand (Barker, 1989). Stock treading can cause mortality through physical injury, and rapid pasture defoliation under rotational grazing exposes surface populations to periodic threat of desiccation. Slug mortality due to the combined effects of treading and habitat alteration can be as high as 90% in extreme cases.

There are some serious constraints on the use of grazing animals for the control of pasture pests, which relate to factors such as varying efficacy depending on weather and topography and vegetation composition, critical dependence upon time of application, and possible detrimental effects on animal and pasture production (East and Pottinger, 1983). High stocking rates of the kind needed to control pests could lead to lower feed intake and depressed animal production, while both hard summer grazing and winter treading can severely reduce subsequent pasture production. Thus, pest control by grazing animals is appropriate only where adverse effects of the treatment are likely to be outweighed by the benefits associated with controlling the pest. However, there is scope for using grazing management in combination with other approaches, such as plant resistance or tolerance, use of natural enemies, and judicious use of insecticides.

(d) Altering cultural practices

Newly sown grass leys are particularly vulnerable to pest damage, but the degree of risk can be very much influenced by the method of establishment. The risk of infestation is greater in leys established by minimum cultivation techniques than in leys established by conventional ploughing and cultivation. In direct-drilled grass leys, a large proportion of shoot-mining larvae can migrate successfully from the desiccated sward into the newly emerging sward and cause severe damage (Bentley, 1984). The timing of sowing can also be important; for example, in the case of autumn sown leys in northern Europe peak times of egg laying by frit fly can be avoided by delaying sowing until after mid-September (Clements, Bentley and Jackson, 1990). Slugs are very susceptible to damage associated with soil cultivation; they are most likely to be troublesome in new leys established by minimum cultivation or in poorly prepared conventional seedbeds where the soil is unconsolidated and cloddy. Slug numbers can be reduced by intensive grazing of pasture prior to direct seeding. Ferguson, Barratt and Jones (1988) found that mob stocking with 1500 sheep d^{-1} ha^{-1} reduced *Deroceras reticulatum* numbers by 90% in New Zealand pasture. Open furrows formed by drill coulters provide very suitable habitats for slugs; covering of these by press wheels, by harrowing or by stock treading improves seedling emergence and survival and reduces exposure to slugs (Barker, 1989). Herbicidal desiccation of the old sward prior to seeding suppresses competition and enhances the prospects for successful establishment, but it does appear to increase risk of pest damage through aggregation of insects and nematodes on seedling rows. Band application of herbicides may be an attractive option in some situations. Although it does not suppress competition as well as complete treatment, it reduces the risk of pest infestation by leaving some shelter and food for the indigenous fauna (Barker, 1989).

Heavy rolling may be an effective alternative to mob stocking for the control of grass grub and perhaps other pests in established pastures in some circumstances (Stewart and van Toor, 1983), while well-consolidated seedbeds are less susceptible to damage from pests such as leatherjackets and slugs.

(e) Pesticides

Routine application of insecticides to established grassland is rare, although insecticides might occasionally be used to control potentially damaging pest outbreaks. While losses in grass production attributable to non-outbreak levels of a range of phytophagous species are probably common in northern European grasslands (Table 6.13), repeated insecticidal treatments to prevent such losses would generally be uneconomic as well as being environmentally damaging. A stronger economic argument has been made for leatherjacket control based on population monitoring in wetter grassland (Blackshaw, 1984, 1985). However, view of the ecological importance of grassland as the dominant seminatural habitat in many countries, more extensive use of pesticides would appear ill-advised. While insecticides in current use are less persistent than the organochlorines, many widely used materials such as chlorpyrifos and synthetic pyrethroids have a broad spectrum of activity and can have at least transient effects on non-target organisms (Chapter 5). There is no evidence that occasional use of such materials has any serious consequence, but accumulating evidence indicating progressive depletion of polyphagous predators in arable land receiving routine pesticide treatment (e.g. Burn, 1988) points to the need to reduce insecticide use in agricultural land generally.

Newly sown leys are particularly vulnerable to pest damage and sometimes insecticidal treatment may be warranted to ensure the development of a good stand. Henderson and Clements (1980) reported that a single application of phorate (10% granules) to spring-sown Italian ryegrass at Rothamsted Experimental Station, England, increased seedling emergence, dry-matter yield and ryegrass contribution to the sward in the first year. Insecticidal applications increased the number of seedlings produced by 24–35% and increased the mean annual yield of herbage dry matter from 8.2 to 9.3 t ha^{-1}. Subsequently, Clements, Bentley and Jackson (1986) investigated the influence on Italian ryegrass swards of a range of insecticides applied as granules either broadcast or direct drilled with the seed and in most cases grass yields were enhanced by a sufficient amount for the treatments to be economically viable. Similar responses have also been reported for perennial ryegrass and white clover leys, with clover establishment benefiting particularly from insecticidal treatment (Mowat and Shakeel, 1988; Lewis, Cook and van der Ende, 1991). Some encouraging results have been obtained from the use of seed dressings (Clements, 1980; Mowat and Jess, 1985), and there may be scope for the development of slow-release formulations to protect seedlings from emergence to tillering, the stage most vulnerable to pest attack. A desirable feature of seed dressings compared with other forms of insecticidal application is the small amount of active ingredient used and hence the greatly reduced environmental risk.

Insecticides continue to play an important role in the control of recurring pasture pests, although it is now widely recognized that they should be used more selectively than in the past. Ideally, insecticides should be integrated into an overall pest-management strategy as in the case of the use of selective aphicides for the control of lucerne pests in California (Flint and van den Bosch, 1981). In the past, organochlorine insecticides have provided a relatively cheap and effective means of controlling soil pests in New Zealand pastures but this option is no longer available. The absence of satisfactory alternatives coupled with a generally low level of profitability associated with pastoral agriculture has led to a sharp decline in the use of insecticides and a search for more effective approaches.

Current recommendations for the chemical control of grass grub are directed against larvae feeding close to the soil surface in the autumn. This treatment has variable results and is very inefficient in terms of chemical use since less than 1% of the chemical applied to the soil actually comes in contact with the target organism (Henzell and Lauren, 1977c). An attractive alternative would be to control the adult stage in the spring before pasture damage occurs, but this requires careful timing of application to coincide with male flight activity and insecticides with rapid knockdown properties, since effective control can be achieved only if a high percentage kill is obtained before the females can mate and lay eggs. Reduction in adult numbers would appear to have the additional merit of complementing rather than disrupting natural larval mortality factors. Sex-attractant traps using controlled-release formulations of phenols have provided an effective means of monitoring grass grub flight so that the optimum time to spray can be established (Henzell and Lauren, 1977b; Lauren, 1979; Henzell, Lauren and du Toit, 1979). The synthetic pyrethroids cypermethrin and permethrin are effective at low concentrations and provide rapid adult knockdown, reducing subsequent third-instar larval numbers by up to 96% (Henzell and Lauren, 1977a; Pottinger *et al.*, 1979); however, repeated applications are required to achieve these results. Three applications at 6-day intervals are required to maintain toxic cover in the foliage over the major grass grub mating period of 14–18 days or longer in hilly areas. Improved formulation (e.g. microencapsulation), which would extend the chemical and biological field life of pyrethroids, might offer improved prospects for practical control of adult grass grub (Lauren *et al.*, 1979; Lauren Henzell and du Toit, 1980).

As in the case of arable crops, there is likely to be a continuing need for pesticides to control damaging pests in grassland, but improved formulations and delivery systems are required to increase their effectiveness and to minimize environmentally adverse effects. There is a need to integrate chemicals with other approaches – for example, biological-control agents, cutting and grazing management, irrigation, fertilizer application, cultivation and reseeding, and use of pest-resistant cultivars – insofar as possible insecticides should complement rather than replace other aspects of pasture management. A good example of a possible complementary effect of chemical and grazing manipulation relates to soldier fly control in New Zealand. Chemical control of this pest by itself is uneconomical but can be cost-effective when used in combination with hard grazing over the adult flight period (Robertson, 1979).

6.7 CONCLUSIONS

Invertebrate decomposers have only a limited capacity to digest plant structural compounds, and their role in primary litter decomposition is minor compared with that of microorganisms. Available data on community energetics, which are inevitably tentative given the complexity of the grassland community, indicate that invertebrate respiration typically accounts for 5–15% of heterotrophic soil respiration in temperate grasslands. However, low assimilation efficiencies are accompanied by high ingestion rates and invertebrates promote decomposition by ingesting and fragmenting litter, by incorporating it into the soil and by stimulating microbial activity in a variety of ways. Earthworms are capable of consuming most of the litter produced in base-rich temperate grasslands, and when their activities are suppressed a surface layer of undecayed organic material rapidly develops and soil fertility declines. Termites are generally the main consumers of litter in tropical grasslands. However, they tend to concentrate plant material in their mounds, resulting in a depletion of soil organic matter in the surrounding area. They also have a high assimilation efficiency and their faeces are a poor substrate for other decomposers.

Coprophagous invertebrates such as dung beetles, dipterous larvae and earthworms accelerate dung decomposition mainly by promoting microbial activity and by incorporating it into the soil. In the absence of an effective coprophagous community, dung accumulates on the soil surface and pasture fouling occurs.

In mull soils, earthworms may make a significant contribution to nutrient cycling, directly through the mineralization of assimilated plant litter and indirectly through the production of casts, which form a rich substrate for microbial activity. Microcosm studies indicate that microbial grazing by invertebrates may be an important mechanism for the release of mineral elements immobilized in microbial biomass; however, the importance of this mechanism under field conditions is difficult to assess.

Large invertebrates modify the physical and chemical properties of soil through their burrowing and mixing activities. The role of earthworms in the development of mull soils, with their granular structure and intimately mixed organic and mineral constituents, has received particular attention. Large channels created by burrowing influence water infiltration, aeration and root penetration, while medium-sized pores created by earthworm casting in the soil influence water-holding capacity. These effects are apparent in situations where earthworms have been introduced into reclaimed land, while dramatic deterioration in soil structure after the elimination of earthworms has also been reported. The potential impact of the recently established New Zealand planarian *Artioposthia triangulata* on earthworm populations in maritime areas of northern

Europe is a matter for particular concern in this context (Blackshaw, 1990). The beneficial effects of earthworms on soil structure and fertility have been reflected in enhanced plant growth in many experiments. Termites and ants can also modify soil properties significantly through nest building and foraging.

Herbage consumption by invertebrates at non-outbreak levels normally amounts to 1–10% of net primary production in energy terms, but there are no simple correlations between the quantities consumed and the effects on plant productivity. Moderate levels of grazing can increase grass growth by stimulating the production of new roots and shoots, but wilting associated with sap-sucking insects may cause severe check in plant growth while the damage associated with pests that sever plant shoots at the base is out of all proportion to the amount of tissue consumed. The effects of pest attack are likely to be more severe in swards that are already under stress due, for example, to drought or to overgrazing. Selective feeding on preferred plant species can alter the botanical composition to the detriment of sward stability and productivity. Insect-transmitted disease can further depress productivity, as in the case of aphid-transmitted barley yellow dwarf virus (BYDV) in ryegrass swards.

On the positive side, pollination by flower-visiting bees is vitally important for legume seed production, while weed-feeding species may limit the distribution of some pasture weeds. Herbivorous invertebrates also contribute to the turnover of organic matter and nutrient cycling in grassland, via mineralization of their excreta and corpses and of the dead plant material that results from their destructive feeding.

Problems associated with the control of recurring grassland pests by chemical means have stimulated an interest in the pest-management approach. Spotted alfalfa aphid, for example, has been successfully controlled by a combination of biological, cultural and selective insecticidal means in California. The development of pest-management systems requires detailed life-table data for the major pests and reliable economic injury threshold data which take account of how pests affect animal production under a range of grazing pressures. There are very few instances in which adequate information is available, but considerable progress has been made in the cases of damaging pasture pests, such as grass grub, porina, black beetle and Australian soldier fly in New Zealand. Strategies that appear to have potential in pest-management programmes for the control of these pests include classical biological control, altering botanical composition, adjustment of stocking densities or cutting regimes, altering reseeding practices, and more accurate timing of insecticidal applications.

An important question for the future is the degree to which global change will influence the composition and functioning of grassland invertebrate communities. Clearly, this is an area of great uncertainty, but some recent 'best guess' studies based on the output of current climate models provide a basis for some speculation. One such study conducted in Ireland on the assumptions that mean annual temperature will increase by 2 °C, winter precipitation will increase by 5–10% and summer precipitation will decrease by 5–10% between now and the year 2030 suggests that grass production could increase by about 20% due to elevated levels of atmospheric carbon dioxide (McWilliams, 1991). This, in turn, would result in a greatly increased input of plant litter, with higher C:N ratios, to the soil. This poorer quality litter would probably decompose more slowly than at present, leading to an increase in soil organic matter levels (Coûteaux *et al.*, 1991). Phytophagous invertebrates such as aphids could be expected to benefit from the milder winters, warmer summers and extended growing season, but this effect could be countered to a degree by poorer resource quality and by changes in natural enemy activity. Earthworms in Irish grasslands are probably more often limited by food quality than by quantity; a deterioration in litter quality coupled with higher levels of summer mortality could have adverse consequences for this group. Litter-dwelling macroinvertebrates and microarthropods are likely to benefit from the increased litter supply but may also be adversely affected by the increased incidence of summer drought. In summary, it appears unlikely that climatic change of the order envisaged *per se* will have a very pronounced effect on the invertebrate fauna of Irish grassland; any changes in management practices that may occur in response to climatic change will probably be of much greater consequence.

References

Adams, E.C.G. (1971) Ecological studies of microarthropods in a New Zealand pasture soil with special reference to Collembola (1.). *Pedobiologia*, **11**, 321–37.

Addison, J.A. (1977) Population dynamics and biology of Collembola on True-love Lowland, Devon Island, N.W.T., in *Truelove Lowland, Devon Island, Canada: A High Arctic Ecosystem*, (ed. L.C. Bliss), University of Alberta Press, Edmonton, pp. 363–82.

Albert, A.M. (1982) Species spectrum and dispersion patterns of chilopods in 3 Solling habitats. *Pedobiologia*, **23**, 337–47.

Al Dabbagh, K.Y. and Block, W. (1981) The population ecology of a terrestrial isopod in two breckland grass heaths. *Journal of Animal Ecology*, **50**, 61–77.

Allan, J.D., Alexander, H.J. and Greenberg, R. (1975) Foliage arthropod communities of crop and fallow fields. *Oecologia*, **22**, 49–56.

Allison, L.E. (1964) Salinity in relation to irrigation. *Advances in Agronomy*, **16**, 139–80.

Alvarado, A., Berish, C.W., and Peralta, F. (1981) Leaf-cutter ant (*Atta cephalotes*) influence on the morphology of Andepts in Costa Rica. *Soil Science Society of America Journal*, **45**, 790–4.

Amante, E. (1962) Ensaio de campo para controlar o cupim de monticulo *Cornitermes cumulans* (Kollar, 1832) (Isoptera, Termitidae). *Arquivos do Instituto Biológico São Paulo*, **29**, 133–8.

Amante, E. (1967) A formiga saúva *Atta capiguara*, praga das pastagens. *Biológica*, **33**, 113–20.

Amante, E. (1972) Influência de alguns factores microclimáticos sobre a formiga saúva *Atta laevigata* (F. Smith, 1858), *Atta sexdens rubropilosa* (Forel, 1908), *Atta bisphaerica* (Forel, 1980), e *Atta capiquara* (Gonçalves, 1944) (Hymenoptera, Formicidae), em Formigueiros Localizados no Estado de São Paulo. Tese, Doutor em Agronomia, Escola Superior de Agricultura 'Luiz de Queiróz', Universidade de Sao Paulo, em Piracicaba.

Andersen, C. (1983) Nitrogen turnover by earthworms in animal manured plots, in *Earthworm Ecology – from Darwin to Vermiculture*, (ed. J.E. Satchell), Chapman & Hall, London, pp. 139–50.

Anderson, J.M. (1988) Invertebrate-mediated transport processes in soils. *Agriculture, Ecosystems and Environment*, **24**, 5–19.

Anderson, J.M. and Bignell, O.E. (1980) Bacteria in the food, gut contents and faeces of the litter-feeding millipede *Glomeris marginata* (Villers). *Soil Biology and Biochemistry*, **12**, 251–4.

Anderson, J.M., Huish, S.A., Ineson, P. *et al.* (1985) Interactions of invertebrates, micro-organisms and tree roots in nitrogen and mineral element fluxes in deciduous woodland soils, in *Ecological Interactions in Soil, Plants, Microbes and Animals*, (eds A.H. Fitter, D. Atkinson, D.J. Read and M.B. Usher), Blackwell, Oxford, pp. 377–92.

Anderson, J.M. and Ineson, P. (1983) Interactions between soil arthropods and microorganisms in carbon, nitrogen and mineral element fluxes from decomposing leaf litter, in *Nitrogen as an Ecological Factor*, (eds J.A. Lee, S. McNeill and I.H. Rorison), Blackwell, Oxford, pp. 413–22.

Anderson, J.M., Ineson, P. and Huish, S.A. (1983a) The effects of animal feeding activities on element release from deciduous forest litter and soil organic matter, in *New Trends in Soil Biology*, (eds Ph. Lebrun, H.M. André, A. de Medts, C. Grégòire-Wibo and G. Wauthy), Imprimeur Dieu-Brichart, Ottignies, Louvain-la-Neuve, pp. 87–100.

Anderson, J.M., Ineson, P. and Huish, S.A. (1983b) Nitrogen and cation mobilization by soil fauna feeding on leaf litter and soil organic matter from deciduous woodlands. *Soil Biology and Biochemistry*, **15**, 463–7.

Anderson, R.C. (1982) An evolutionary model summarizing the roles of fire, climate and grazing animals in the origin and maintenance of grasslands: an end paper, in *Grasses and Grasslands*, (eds J.R. Estes, R.J. Tyrl and J.N. Brunken), University of Oklahoma Press, Norman, Oklahoma, pp. 297–308.

Anderson, R.V., Coleman, D.C. and Cole, C.V. (1981) Effects of saprotrophic grazing on net mineralization, in *Terrestrial Nitrogen Cycles*, (eds F.E. Clark and T. Rosswall), *Ecological Bulletins* (Stockholm), **33**, 201–16.

Anderson, R.V., Elliott, E.T., Coleman, D.C. and Cole, C.V. (1981) Effect of the nematodes *Acrobeloides* sp. and *Mesodiplogaster Iheritieri* on substrate utilization and nitrogen and phosphorous mineralization in soil. *Ecology*, **62**, 549–55.

Andow, D.A. (1991) Vegetational diversity and arthropod population response. *Annual Review of Entomology*, **36**, 561–86.

Andrén, O. and Lagerlöf, J. (1983) Soil fauna (Microarthropods, Enchytraeids, Nematodes) in Swedish agricultural cropping systems. *Acta Agriculturae Scandinavica*, **33**, 33–52.

Andrén, O., Lindberg, T., Paustian, K. and Rosswall, T. (eds) (1990) Ecology of arable land – organisms, carbon and nitrogen cycling. *Ecological Bulletins* (Copenhagan), **40**, 1–222.

Andrewartha, H.G. and Birch, L.C. (1954) *The Distribution and Abundance of Animals*, University of Chicago Press, Chicago.

Andrzejewska, L. (1962) *Macrosteles laevis* Rib. as an unsettlement index of natural meadow associations of Homoptera. *Bulletin de l'Académie Polonaise des Sciences Classe 11 Série des Sciences Biologiques*, **10**, 231–6.

Andrzejewska, L. (1965) Stratification and its dynamics in meadow communities of Auchenorrhyncha (Homoptera). *Ekologia Polska*, **13A**, 685–715.

Andrzejewska, L. (1967) Estimation of the effects of feeding of the sucking insect *Cicadella viridis* L. (Homoptera-Auchenorrhyncha) on plants, in *Secondary Productivity of Terrestrial Ecosystems*, Vol. 2, (ed. K. Petrusewicz), Polish Academy of Sciences, Warsaw, pp. 791–805.

Andrzejewska, L. (1976a) The influence of mineral fertilization on the meadow phytophagous fauna. *Polish Ecological Studies*, **2**, 93–109.

Andrzejewska, L. (1976b) The effect of mineral fertilization on the Auchenorrhyncha (Homoptera) fauna. *Polish Ecological Studies*, **2**, 111–28.

Andrzejewska, L. (1979a) Herbivorous fauna and its role in the economy of grassland ecosystems I. Herbivores in natural and managed meadows. *Polish Ecological Studies*, **5**, 5–44.

Andrzejewska, L. (1979b) Herbivorous fauna and its role in the economy of grassland ecosystems II. The role of herbivores in trophic relationships. *Polish Ecological Studies*, **5**, 45–76.

Andrzejewska, L. and Gyllenberg, G. (1980) Small herbivore subsystem, in *Grasslands, Systems Analysis and Man*, International Biological Programme, 19, (eds A.I. Breymeyer and G.M. Van Dyne), Cambridge University Press, Cambridge, pp. 201–67.

Andrzejewska, L. and Wojcik, Z. (1970) The influence of Acridoidea on the primary production of a meadow (field experiment). *Ekologia Polska*, **18**, 89–109.

Andrzejewska, L. and Wojcik, Z. (1971) Productivity investigation of two types of meadows in the Vistula Valley, VII. Estimation of the effect of phytophagous insects on the vascular plant biomass of the meadow. *Ekologia Polska*, **19**, 173–82.

Anglade, P. (1967) Étude de populations de Symphyles en sol cultivé et d'influence de traitements du sol, in *Progress in Soil Biology*, (eds O. Graff and J.E. Satchell), North Holland, Amsterdam, pp. 372–81.

Athias, F. (1976) Recherche sur les microarthropodes du sol de la savane de Lamto (Côte d'Ivoire). *Annales de l'Université d'Abidjan, série E (Ecologie)*, **9**, 193–271.

Athias, F., Josens, G. and Lavelle, P. (1974) Analyse d'un écosystème tropical humide: la savane de Lamto (Côte d'Ivoire). Traits généraux du peuplement endogé: le peuplement animal. *Bulletin de Liaison des Chercheurs de Lamto* (Paris), 5, (n° spécial), pp. 45–54.

Atlavinyté, O. (1976) The effect of land reclamation and field management on the change in specific composition and densities of Lumbricidae. *Polish Ecological Studies*, **2**, 147–52.

Ausmus, B.S., Edwards, N.T. and Witkamp, M. (1976) Microbial immobilization of carbon, nitrogen, phosphorus and potassium: implications for forest ecosystem processes, in *The Role of Terrestrial and Aquatic Organisms in Decomposition Processes*, (eds J.M. Anderson and A. Macfadyen), Blackwell, Oxford, pp. 387–416.

Baker, D.G. (1965) Factors affecting soil temperature. *Minnesota Farm and Home Science*, **22**, 11–13.

Baker, E.W. and Wharton, G.W. (1952) *An Introduction to Acarology*, Macmillan, New York.

Baker, G.H. (1978a) The distribution and dispersal of the introduced millipede, *Ommatoiulus moreletii* (Diplopoda: Iulidae), in Australia. *Journal of Zoology, London*, **185**, 1–11.

Baker, G.H. (1978b) The population dynamics of the millipede *Ommatoiulus moreletii* (Diplopoda: Iulidae). *Journal of Zoology, London*, **186**, 229–42.

Baker, G.H. (1980) The water and temperature relationships of *Ommatoiulus moreletii* (Diplopoda: Iulidae). *Journal of Zoology, London*, **190**, 97–108.

Baker, G.H. (1981) Notes on the ecology of the beetle *Dascillus cervinus* (Coleoptera: Dascillidae). *Irish Naturalists Journal*, **20**, 336–8.

Baker, G.H. (1983) The distribution, abundance and species associations of earthworms (Lumbricidae) in a reclaimed peat soil in Ireland. *Holarctic Ecology*, **6**, 74–80.

Baker, G.H. (1986) The biology and control of white snails (Mollusca: Helicidae), introduced pests in Australia. *CSIRO Australia Division of Entomology Technical Paper*, No. 25, Canberra.

Baker, G.H. and Vogelzang, B.K. (1988) Life history, population dynamics and polymorphism of *Theba pisana* (Mollusca: Helicidae) in Australia. *Journal of Applied Ecology*, **25**, 867–88.

Banage, W.B. (1963) The ecological importance of free-living nematodes with special reference to those of moorland soil. *Journal of Animal Ecology*, **32**, 133–40.

Banage, W.B. and Visser, S.A. (1967) Micro-organisms and nematodes from a virgin bush site in Uganda, in *Progress in Soil Biology*, (eds. O. Graff and J.E. Satchell), North Holland, Amsterdam, pp. 93–101.

Barker, G.M. (1989) Slug problems in New Zealand pastoral agriculture, in *Slugs and Snails in World Agriculture*, (ed. I. Henderson), *British Crop Protection Council Monograph*, **41**, 59–68.

Barley, K.P. (1959a) The influence of earthworms on soil fertility I. Earthworm populations found in agricultural land near Adelaide. *Australian Journal of Agricultural Research*, **10**, 171–8.

Barley, K.P. (1959b) The influence of earthworms on soil fertility II. Consumption of soil and organic matter by the earthworm *Allolobophora caliginosa* (Savigny). *Australian Journal of Agricultural Research*, **10**, 179–85.

Barley, K.P. (1959c) Earthworms and soil fertility IV. The influence of earthworms on the physical properties of a red-brown earth. *Australian Journal of Agricultural Research*, **10**, 371–6.

Barley, K.P. (1964) Earthworms and the decay of plant litter and dung – a review. *Proceedings of the Australian Society of Animal Production*, **5**, 236–40.

Barley, K.P. and Jennings, A.C. (1959) Earthworms and soil fertility III. The influence of earthworms on the availability of nitrogen. *Australian Journal of Agricultural Research*, **10**, 364–70.

Barley, K.P. and Kleinig, C.R. (1964) The occupation of newly irrigated lands by earthworms. *Australian Journal of Science*, **26**, 290–1.

Barlow, N.D. (1985a) Laboratory studies on porina (*Wiseana* spp.) feeding behaviour: the 'functional response' to herbage mass, in *Proceedings of the 14th Australasian Conference on Grassland Invertebrate Ecology*, Lincoln College, Canterbury, 13–17 May 1985, (ed. R.B. Chapman), Caxton Press, Christchurch, New Zealand, pp. 1–6.

Barlow, N.D. (1985b) A model for the impact and control of porina on sheep farms: development and field application, in *Proceedings of the 14th Australasian Conference on Grassland Invertebrate Ecology*, Lincoln college, Canterbury, 13–17 May 1985, (ed. R.B. Chapman), Caxton Press, Christchurch, New Zealand, pp. 152–9.

Barlow, N.D. (1987) Pastures, pests and productivity: simple grazing models with two herbivores. *New Zealand Journal of Ecology*, **10**, 43–55.

Barlow, N.D., French, R.A. and Pearson, J.F. (1986) Population ecology of *Wiseana cervinata*, a pasture pest in New Zealand. *Journal of Applied Ecology*, **23**, 415–31.

Barnes, H.F. (1946) *Gall Midges of Economic Importance*, Vol. II. *Gall Midges of Fodder Crops*, Crosby, Lockwood, London.

Baroni-Urbani, C. (1968) Studie sulla mirmecofauna d'Italia. V. Aspettilecologici della riviera del M. Conero. *Bollettino di Zoologia, Pubblicato dall 'Unione Zoologica Italiana*, **35**, 39–76.

Baroni-Urbani, C. (1969) Ant communities of the high-altitude Appenine grasslands. *Ecology*, **50**, 488–92.

Baylis, J.P., Cherrett, J.M. and Ford, J.B. (1986) A survey of the invertebrates feeding on living clover roots (*Trifolium repens* L.) using ^{32}P as a radiotracer. *Pedobiologia*, **29**, 201–8.

Beard, J.S. (1978) The physiognomic approach, in *Classification of Plant Communities*, (ed. R.H. Whittaker), Junk, The Hague, pp. 35–64.

Behan, V.M. and Hill, S.B. (1980) Distribution and diversity of North American Arctic soil Acari, in *Soil Biology as Related to Land Use Practices*, (ed. D.L. Dindal), Environmental Protection Agency (EPA), Washington DC, pp. 717–40.

Behan-Pelletier, V.M. and Hill, S.B. (1983) Feeding habits of sixteen species of

Oribatei (Acari) from an acid peat bog, Glenamoy, Ireland. *Revue d'Écologie et de Biologie du Sol*, **20**, 221–67.

Belfield, W. (1956) The arthropods of the soil in a West African pasture. *Journal of Animal Ecology*, **25**, 275–87.

Bengtson, S.-A., Nilsson, A., Nordström, S. and Rundgren, S. (1975) Habitat selection of lumbricids in Iceland. *Oikos*, **26**, 253–63.

Bentley, B.R. (1984) Stem-boring dipterous larvae (e.g. *Oscinella frit*) as pests of direct-drilled grass re-seeds with particular reference to the mode of infestation and seasonality of damage inflicted. M. Phil. thesis, University of Reading.

Berger, H., Foissner, W. and Adam, H. (1986) Field experiments on the effects of fertilizers and lime on the soil microfauna of an alpine pasture. *Pedobiologia*, **29**, 261–72.

Bernays, E.A. and Chapman, R.F. (1970) Food selection by *Chortippus parallelus* (Zetterstedt) (Orthoptera: Acrididae) in the field. *Journal of Animal Ecology*, **39**, 383–94.

Bernays, E.A. and Chapman, R.F. (1974) The regulation of food intake by Acridids, in *Experimental Analysis of Insect Behaviour*, (ed. B. Browne), Springer-Verlag, Berlin, pp. 47–59.

Bernays, E.A. and Chapman, R.F. (1978) Plant chemistry and acridoid feeding behaviour, in *Biochemical Aspects of Plant and Animal Coevolution*, (ed. J.B. Harborne), Academic Press, London, pp. 99–141.

Bernays, E.A., Chapman, R.F., Horsey, J. and Leather, E.M. (1974) The inhibitory effect of seedling grasses on feeding and survival of acridids (Orthoptera). *Bulletin of Entomological Research*, **64**, 413–20.

Bernays, E.A., Cooper Driver, G. and Bilgener, M. (1989) Herbivores and plant tannins. *Advances in Ecological Research*, **19**, 263–302.

Berthet, P. (1963) Mesure de la consommation d'oxygène des Oribatides (Acarines) de la littière des forêts, in *Soil Organisms*, (eds J. Doeksen and J. van der Drift), North-Holland, Amsterdam, pp. 18–31.

Biel, E.R. (1961) Microclimate, bioclimatology and notes on comparative dynamic climatology. *American Scientist*, **49**, 326–57.

Blackshaw, R.P. (1984) The impact of low numbers of leatherjackets on grass yield. *Grass and Forage Science*, **39**, 339–43.

Blackshaw, R.P. (1985) A preliminary comparison of some management options for reducing grass losses created by leatherjackets in Northern Ireland. *Annals of Applied Biology*, **107**, 279–85.

Blackshaw, R.P. (1990) Studies on *Artioposthia triangulata* (Dendy) (Tricladida: Terricola), a predator of earthworms. *Annals of Applied Biology*, **116**, 169–76.

Blackshaw, R.P. and Newbold, J.W. (1987) Interactions between fertilizer use and leatherjacket control in grassland. *Grass and Forage Science*, **42**, 343–6.

Block, W.C. (1965) Distribution of soil mites (Acarina) on the Moor House National Nature Reserve, Westmorland, with notes on numerical abundance. *Pedobiologia*, **5**, 244–51.

Block, W. (1966) Seasonal fluctuations and distribution of mite populations in moorland soils, with a note on biomass. *Journal of Animal Ecology*, **35**, 487–503.

Block, W. (1970) Micro-arthropods in some Uganda soils, in *Methods of Study in Soil Ecology*, (ed. J. Phillipson), Unesco, Paris, pp. 195–202.

Block, W. and Banage, W.B. (1968) Population density and biomass of earthworms in some Uganda soils. *Revue d'Écologie et de Biologie du Sol*, **5**, 515–21.

Blower, J.G. (1955) Millipedes and centipedes as soil animals, in *Soil Zoology*, (ed. D.K. McE. Kevan), pp. 138–51.

Blower, J.G. (1956) Some relations between millipedes and the soil. *Proceedings VIe Congrès International de la Science du Sol, Paris,* III, 169–76.

Bock, C.E., Bock, J.H. and Grant, M.C. (1992) Effects of bird predation on grasshopper densities in an Arizona grassland. *Ecology,* **73**, 1706–17.

Bodine, M.C. and Ueckert, D.N. (1975) Effect of desert termites on herbage and litter in a shortgrass ecosystem in West Texas. *Journal of Range Management,* **28**, 353–8.

Bolger, T. and Curry, J.P. (1980) Effects of cattle slurry on soil arthropods in grassland. *Pedobiologia,* **20**, 246–53.

Bolger, T. and Curry, J.P. (1984) Influences of pig slurry on soil microarthropods in grassland. *Revue d'Écologie et de Biologie du Sol,* **21**, 269–81.

Bolton, P.J. and Phillipson, J. (1976) Burrowing, feeding, egestion and energy budgets of *Allolobophora rosea* (Savigny) (Lumbricidae). *Oecologia,* **23**, 225–45.

Booth, R.G. and Anderson, J.M. (1979) The influence of fungal food quality on the growth and fecundity of *Folsomia candida* (Collembola: Isotomidae). *Oecologia,* **38**, 317–23.

Booth, R.G. and Usher, M.B. (1986) Arthropod communities in a maritime Antarctic moss-turf habitat: life history strategies of the prostigmatid mites. *Pedobiologia,* **29**, 209–18.

Bornemissza, G.F. (1970) Insectary studies on the control of dung breeding flies by the activity of the dung beetle, *Onthophagus gazella* F. (Coleoptera, Scarabaeidae). *Journal of the Australian Entomological Society,* **9**, 31–41.

Bossenbrock, P., Kessler, A., Liem, A.S.N. and Vlijm, L. (1977) The significance of plant growth forms as 'shelter' for terrestrial animals. *Journal of Zoology, London,* **182**, 1–6.

Boström, U. (1988a) *Ecology of Earthworms in Arable Land: Population Dynamics and Activity in Four Cropping Systems,* Department of Ecology and Environmental Research, Report No. 34, Swedish University of Agricultural Sciences, Uppsala.

Boström, U. (1988b) Growth and cocoon production by the earthworm *Aporrectodea caliginosa* in soil mixed with various plant materials. *Pedobiologia,* **32**, 77–80.

Boström, U. and Lofs-Holmin, A. (1986) Growth of earthworms (*Allolobophora caliginosa*) fed shoots and roots of barley, meadow fescue and lucerne. Studies in relation to particle size, protein, crude fibre content and toxicity. *Pedobiologia,* **29**, 1–12.

Bouché, M.B. (1972) *Lombriciens de France: Écologie et Systématique,* Institut National de la Recherche Agronomique, Annales de Zoologie – Écologie animale (N° hors-série), Paris.

Bouché, M.B. (1976) Étude de l'activité des invertébrés épigés prairaux I. Resultats generaux et geodrilogiques (Lumbricidae: Oligochaeta). *Revue d'Écologie et de Biologie du Sol,* **13**, 261–81.

Bouché, M.B. (1977) Fauna in the grassland system: role of earthworms, in *Department of Zoology: Orientations and Means; Presentation of Some Research Projects,* INRA publication, Ministry of Agriculture, France, pp. 21–3.

Bouché, M.B. and Ferrière, G. (1986) Cinétique de l'assimilation de l'azote d'origine lombricienne par une végétation prairiale non perturbée. *Comptes Rendus de l'Academie des Sciences, Paris,* **302**, Série III, 75–80.

Bouché, M.B., Ferrière, G. and Soto, P. (1987) The role of earthworms in the decomposition and nitrogen nutrition of plants in a grassland, in *On Earthworms,* (eds A.M. Bonvicini Pagliai and P. Omodeo), *Selected Symposia and Monographs,* U.Z.I., 2, Mucchi, Modena, pp. 113–29.

Bouillon, A. (1970) Termites of the Ethiopian region, in *Biology of Termites*, (eds K. Krishna and F.M. Weesner), Academic Press, London, pp. 153–280.

Boyd, J.M. (1956) The Lumbricidae in the Hebrides, II. Geographical distribution. *Scottish Naturalist*, **68**, 165–72.

Boyd, J.M. (1957) Ecological distribution of the Lumbricidae in the Hebrides. *Proceedings of the Royal Society of Edinburgh*, **66**, 311–38.

Boyle, K.E. (1990) The ecology of earthworms (Lumbricidae) in grasslands on reclaimed cutover peatland and their impact on soil physical properties and grass yield. Ph.D. thesis, National University of Ireland, Dublin.

Braun-Blanquet, J. (1932) *Plant Sociology: the Study of Communities*, McGraw Hill, New York.

Bray, J.R. and Curtis, J.T. (1957) An ordination of upland forest communities of southern Wisconsin. *Ecological Monographs*, **27**, 325–49.

Breymeyer, A. (1971) Productivity investigation of two types of meadows in the Vistula Valley. XIII. Some regularities in structure and function of the ecosystem. *Ekologia Polska*, **19**, 249–61.

Breymeyer, A. (1974) Analysis of a sheep pasture in the Pieniny Mountains (the Carpathians). XI. The role of coprophagous beetles (Coleoptera, Scarabaeidae) in the utilization of sheep dung. *Ekologia Polska*, **22**, 617–34.

Breymeyer, A. (1978) An analysis of the trophic structure of some grassland ecosystems. *Polish Ecological Studies*, **4**, 55–128.

Breymeyer, A. (1980) Trophic structure and relationships, in *Grasslands, Systems Analysis and Man*, International Biological Programme 19, (eds A.J. Breymeyer and G.M. Van Dyne), Cambridge University Press, Cambridge, pp. 799–819.

Breymeyer, A.J. and Van Dyne, G.M. (eds) (1980) *Grasslands, Systems Analysis and Man*, International Biological Programme 19, Cambridge University Press, Cambridge.

Brian, M.V. (1965) *Social Insect Populations*, Academic Press, London.

Bro-Larsen, E. (1949) The influence of the severe winters of 1939–42 on the soil fauna of Tipperne. *Oikos*, **1**, 184–207.

Brown, A.W.A. (1977) *Ecology of Pesticides*, John Wiley, New York.

Brown, V.K. and Southwood, T.R.E. (1983) Trophic diversity, niche breadth and generation times of exopterygote insects in a secondary succession. *Oecologia*, **56**, 220–5.

Brown, W.L. Jr (1973) A comparison of hylaean and Congo-West African rainforest ant faunas, in *Tropical Forest Ecosystems in Africa and South America: a Comparative Review*, (eds B.J. Meggers, E.S. Ayensu and W.D. Duckworth), Smithsonian Institute, Washington DC, pp. 161–85.

Bulanova-Zachvatkina, E.M. (1967) Pancirnye klešči-Oribatidy. *Izdatel'stvo 'Vysšaja Škola', Moskva*, 1–254.

Bunnell, F.L., MacLean, S.F. and Brown, J. (1975) Barrow, Alaska, USA, in *Structure and Function of Tundra Ecosystems*, (eds T. Rosswall and O.W. Heal). *Ecological Bulletins* (Stockholm), **20**, 73–124.

Buol, S.W., Hole, F.D. and McCracken, R.J. (1973) *Soil Genesis and Classification*, Iowa State University Press, Ames, Iowa.

Burford, J.R. (1976) Effects of the application of cow slurry to grassland on the composition of the soil atmosphere. *Journal of Science, Food and Agriculture*, **27**, 115–26.

Burn, A.J. (1988) Cereal pest/predator interactions, in *The Boxworth Project 1988 Report*, (ed. P. Greig-Smith), Ministry of Agriculture, Fisheries and Food, Tolworth, Surrey, pp. 56–73.

Byers, R.A. (1967) Increased yields of coastal bermudagrass after application of insecticides to control insect complex. *Journal of Economic Entomology*, **60**, 315–18.

Carne, P.B. (1956) An ecological study of the pasture scarab *Aphodius howitti* Hope. *Australian Journal of Zoology*, **4**, 259–314.

Carpenter, A. (1985) Studies on invertebrates in a grassland soil with particular reference to root herbivores. Ph.D. thesis, University of Wales.

Carpenter, A. (1988) The biology of *Campodea staphylinus* (Campodeidae: Diplura) in a grassland soil. *Pedobiologia*, **32**, 31–8.

Carter, J.B. (1976) A survey of microbial, insect and nematode parasites of Tipulidae (Diptera) larvae in north-east England. *Journal of Applied Ecology*, **13**, 103–22.

Carter, J.B. (1978) Field trials with Tipula iridescent virus against *Tipula* sp. larvae in grassland. *Entomophaga*, **23**, 169–74.

Carter, N., McLean, I.F.G., Watt, A. and Dixon, A.F.G. (1980) Cereal aphids: a case study and review. *Applied Biology*, **5**, 272–348.

CAST (Council for Agricultural Science and Technology) (1979) *Application of Sewage Sludge to Cropland: Appraisal of Potential Hazards of the Heavy Metals to Plants and Animals*, Report No. 64, CAST, Ames, Iowa.

Catherall, P.L. (1971) Virus diseases of cereals and grasses, in *Diseases of Crop Plants*, (ed. J.H. Western), Macmillan, London, pp. 308–22.

Chang, A.C., Page, A.L., Warneke, J.E. *et al.* (1983) Accumulation of cadmium and zinc in barley grown on sludge-treated soils: a longterm field study. *Journal of Environmental Quality*, **12**, 391–7.

Chant, D.A. (1959) Phytoseiid mites (Acarina: Phytoseiidae). *The Canadian Entomologist*, **91**, Suppl. 12, 1–166.

Chapman, R.F. (1974) The chemical inhibition of feeding by phytophagous insects: a review. *Bulletin of Entomological Research*, **64**, 339–63.

Chapman, R.F. (1976) *A Biology of Locusts*, Edward Arnold, London.

Chernova, Yu. I., Dorogostaiskaya, E.V., Gerasimenko, T.V. *et al.* (1975) Tareya, USSR, in *Structure and Function of Tundra Ecosystems*, (eds T. Rosswall and O.W. Heal), *Ecological Bulletins* (Stockholm), **20**, 159–81.

Cherrett, J.M. (1964) The distribution of spiders on the Moor House National Nature Reserve, Westmorland. *Journal of Animal Ecology*, **33**, 27–48.

Cherrett, J.M. (1986) The biology, pest status and control of leaf-cutting ants. *Agricultural Zoology Reviews*, **1**, 1–37.

Cherrett, J.M., Pollard, G.V. and Turner, J.A. (1974) Preliminary observations on *Acromyrmex landolti* (For.) and *Atta laevigata* (Fr. Smith) as pasture pests in Guyana. *Tropical Agriculture*, **51**, 69–74.

Chew, R.M. (1974) Consumers as regulators of ecosystems: an alternative to energetics. *Ohio Journal of Science*, **74**, 359–70.

Chiverton, P.A. (1986) Predator density manipulation and its effects on populations of *Rhopalosiphum padi* (Hom: Aphididae) in spring barley. *Annals of Applied Biology*, **109**, 49–60.

Christensen, O. (1988) The direct effects of earthworms on nitrogen turnover in cultivated soils. *Ecological Bulletins* (Copenhagen), **39**, 41–44.

Christiansen, K. (1964) Bionomics of Collembola. *Annual Review of Entomology*, **9**, 147–78.

Clarholm, M. (1981) Protozoan grazing of bacteria in soil – impact and importance. *Microbial Ecology*, **7**, 343–50.

Clarholm, M. (1984) Microbes as predators or prey. Heterotrophic, free-living protozoa: neglected microorganisms with an important task in regulating bacterial populations, in *Current Perspectives on Microbial Ecology*, (eds M.J. Klug and C.A. Reddy), American Society for Microbiology, Washington DC, pp. 321–6.

Clarholm, M. (1985) Interactions of bacteria, protozoa and plants leading to mineralization of soil nitrogen. *Soil Biology and Biochemistry*, 17, 181–7.

Clear Hill, B.H., van Emden, H.F. and Clements, R.O. (1990) Control of frit fly (*Oscinella* spp.) in newly sown grass using a combination of low doses of pesticide, resistant grass cultivars and indigenous parasitoids. *Crop Protection*, 9, 97–100.

Clements, R.O. (1980) Grassland pests – an unseen enemy. *Outlook on Agriculture*, 10, 219–23.

Clements, R.O. (1982) Some consequences of large and frequent pesticide applications to grassland, in *Proceedings of the 3rd Australasian Conference on Grassland Invertebrate Ecology*, (ed. K.E. Lee), South Australian Government Printer, Adelaide, pp. 393–6.

Clements, R.O., Bentley, B.R. and Jackson, C.A. (1986) The impact of granular formulations of phorate, terbufos, carbofuran, carbosulfan and thiofanox on newly sown Italian ryegrass (*Lolium multiflorum*). *Crop Protection*, 5, 389–94.

Clements, R.O., Bentley, B.R. and Jackson, C.A. (1990) Influence of date of sowing on frit-fly damage to newly sown Italian ryegrass. *Crop Protection*, 9, 101–4.

Clements, R.O., Bentley, B.R., Murray, P.J. and Henderson, I.F. (1991) Improvement of persistence and yield of Italian ryegrass (*Lolium multiflorum*) through pesticide use. *Annals of Applied Biology*, 119, 513–19.

Clements, R.O., Chapman, P.F. and Henderson I.F. (1983). Seasonal distribution of stem-boring larvae including frit fly (Diptera: *Oscinella frit* L.) and pest damage in perennial ryegrass. *Grass and Forage Science*, 38, 283–6.

Clements, R.O., French, N. and Henderson, I.F. (1981) Pest damage to grassland in Northern England. *Proceedings of the Conference on Crop Protection in Northern Britain*, Dundee University, 17–19 March 1981, pp. 213–18.

Clements, R.O. and Henderson, I.F. (1979) Insects as a cause of botanical changes in swards, in *Changes in Sward Composition and Productivity*, Occasional Symposium No. 10, British Grassland Society, pp. 157–60.

Clements, R.O. and Henderson, I.F. (1980) The importance of frit-fly in grassland. *ADAS Quarterly Review*, 36, 14–26.

Clements, R.O. and Murray, P.J. (1991) Incidence and severity of pest damage to white clover. *Aspects of Applied Biology*, 27, 369–72.

Clements, R.O., Murray, P.J., Bentley, B.R. and Henderson, J.F. (1990a) Herbage yield and botanical composition over 20 years of a predominantly ryegrass sward treated frequently with phorate pesticide and three rates of nitrogen fertilizer. *Journal of Agricultural Science*, Cambridge, 115, 23–7.

Clements, R.O., Murray, P.J., Bentley, B.R., Lewis, G.C. and French, N. (1990b) The impact of pests and diseases on the herbage yield of permanent grassland at eight sites in England and Wales. *Annals of Applied Biology* 117, 349–57.

Clements, R.O., Murray, P.J. and Sturdy, R.G. (1991) The impact of 20 years' absence of earthworms and three levels of N fertilizer on a grassland soil environment. *Agriculture, Ecosystems and Environment*, 36, 75–85.

Cole, L. (1946) A study of the Cryptozoa of an Illinois woodland. *Ecological Monographs*, 16, 49–86.

Cole, L.C. (1949) The measurement of interspecific association. *Ecology*, 30, 411–24.

Coleman, D.C. (1985) Through a ped darkly: an ecological assessment of root–soil–microbial–faunal interactions, in *Ecological Interactions in Soils: Plants, Microbes and Animals*, (eds A.H. Fitter, D. Atkinson, D.J. Read and M.B. Usher), Blackwell, Oxford, pp. 1–21.

Coleman, D.C., Anderson, R.V., Cole, C.V. *et al.* (1978) Trophic interactions in soils as they affect energy and nutrient dynamics. IV. Flows of metabolic and biomass carbon. *Microbial Ecology*, **4**, 373–80.

Coleman, D.C., Andrews, R., Ellis, J.E. and Singh, J.S. (1976) Energy flow and partitioning in selected man-managed and natural ecosystems. *Agro-Ecosystems*, **3**, 45–54.

Coleman, D.C., Cole, C.V., Anderson, R.V. *et al.* (1977) An analysis of rhizosphere-saprophage interactions in terrestrial ecosystems, in *Soil Organisms as Components of Ecosystems*, (eds U. Lohm and T. Persson), *Ecological Bulletins* (Stockholm), **25**, 299–309.

Coleman, D.C., Reid, C.P.P. and Cole, C.V. (1983) Biological strategies of nutrient cycling in soil systems. *Advances in Ecological Research*, **13**, 1–55.

Coleman, D.C. and Sasson, A. (1980) Decomposer subsystem, in *Grasslands, Systems Analysis and Man*, International Biological Programme 19, (eds A.J. Breymeyer and G.M. Van Dyne), Cambridge University Press, Cambridge, pp. 609–55.

Collins, N.M. (1981) The role of termites in the decomposition of woodland leaf litter in the Southern Guinea Savanna of Nigeria. *Oecologia*, **51**, 389–99.

Connell, J.H. and Slatyer, R.O. (1977) Mechanisms of succession in natural communities and their role in community stability and organisation. *American Naturalist*, **111**, 1119–44.

Conrady, D. (1986) Ökologische Untersuchungen über die Wirkung von Umweltchemikalien auf die Tiergemeinschaft eines Grünlandes. *Pedobiologia*, **29**, 273–284.

Cooke, A. and Luxton, M. (1980) Effect of microbes on food selection by *Lumbricus terrestris*. *Revue d'Écologie et de Biologie du Sol*, **17**, 365–70.

Cotton, D.C.F. and Curry, J.P. (1980a) The effects of cattle and pig slurry fertilizers on earthworms (Oligochaeta, Lumbricidae) in grassland managed for silage production. *Pedobiologia*, **20**, 181–8.

Cotton, D.C.F. and Curry, J.P. (1980b) The response of earthworm populations (Oligochaeta, Lumbricidae) to high applications of pig slurry. *Pedobiologia*, **20**, 189–96.

Coulson, J.C. (1962) The biology of *Tipula subnodicornis* Zetterstedt, with comparative observations on *Tipula paludosa* Meigen. *Journal of Animal Ecology*, **31**, 1–21.

Coulson, J.C. and Whittaker, J.B. (1978) Ecology of moorland animals, in *The Ecology of Some British Moors and Montane Grasslands*, (eds O.W. Heal and D.F. Perkins), Springer-Verlag, Berlin, pp. 52–93.

Coupland, R.T. (ed.) (1979) *Grassland Ecosystems of the World: Analysis of Grasslands and their uses*, International Biological Programme 18, Cambridge University Press, Cambridge.

Coûteaux, M.-M. (1980) Effects of the annual burnings on testacea of two kinds of savannah in Ivory coast, in *Soil Biology as Related to Land Use Practices*, (ed. D.L. Dindal), Environmental Protection Agency. Washington DC, pp. 466–71.

Coûteaux, M.-M., Mousseau, M., Célérier, M.L. and Bottner, P. (1991) Increased atmospheric CO_2 and litter quality: decomposition of sweet chestnut leaf litter with animal food webs of different complexities. *Oikos*, **61**, 54–64.

Coûteaux, M.-M. and Pussard, M. (1982) Nature du régime alimentaire des protozaires du sol, in *New Trends in Soil Biology*, (eds P.H. Lebrun, M. André, A. de Medts *et al.*), Dieu-Brichart, Ottignies-Louvain-la-Neuve, pp. 179–194.

Cox, C.B., Healey, I.N. and Moore, P.D. (1976) *Biogeography*, 2nd edn, Blackwell, Oxford.

Cowan, F.T. and Shipman, H.J. (1947) Quantities of food consumed by Mormon crickets. *Journal of Economic Entomology*, **40**, 825–8.

Cragg, J.B. (1961) Some aspects of the ecology of moorland animals. *Journal of Ecology*, **49**, 477–506.

Crawley, M.J. (1983) *Herbivory: the Dynamics of Animal–Plant Interactions*. Blackwell, Oxford.

Crawley, M.J. and Gillman, M.P. (1989) Population dynamics of cinnabar moth and ragworth in grassland. *Journal of Animal Ecology*, **58**, 1035–50.

Crosby, D.G. (1973) The fate of pesticides in the environment. *Annual Review of Plant Physiology*, **24**, 467–92.

Crossley, D.A., Proctor, C.W. and Gist, C. (1975) Summer biomass of soil microarthropods of the Pawnee National Grassland, Colorado. *American Midland Naturalist*, **93**, 491–5.

Crossley, D.A. Jr and Witkamp, M. (1964) Forest soil mites and mineral cycling. *Acarologia*, **6**, 137–46.

Curry, J.P. (1968) A study of some aspects of the ecology and biological activity of grassland soil fauna. Ph.D. thesis, National University of Ireland, Dublin.

Curry, J.P. (1969a) The qualitative and quantitative composition of the fauna of an old grassland site at Celbridge, Co. Kildare. *Soil Biology and Biochemistry*, **1**, 219–27.

Curry, J.P. (1969b) Studies on the decomposition of organic matter. I. The role of the fauna in the decomposition of decaying grassland herbage. *Soil Biology and Biochemistry*, **1**, 253–8.

Curry, J.P. (1969c) Studies on the decomposition of organic matter. II. The fauna of decaying grassland herbage. *Soil Biology and Biochemistry*, **1**, 259–66.

Curry, J.P. (1970) The effects of different methods of new sward establishment and the effects of the herbicides paraquat and dalapon on the soil fauna. *Pedobiologia*, **10**, 329–61.

Curry, J.P. (1971) Seasonal and vertical distribution of the arthropod fauna of an old grassland soil. *Scientific Proceedings of the Royal Dublin Society*, **3B**, 49–71.

Curry, J.P. (1973) The arthropods associated with the decomposition of some common grass and weed species in the soil. *Soil Biology and Biochemistry*, **5**, 645–57.

Curry, J.P. (1976a) The arthropod fauna of some common grass and weed species of pasture near Dublin. *Proceedings of the Royal Irish Academy*, **76B**, 1–35.

Curry, J.P. (1976b) Some effects of animal manures on earthworms in grassland. *Pedobiologia*, **16**, 425–38.

Curry, J.P. (1979) The arthropod fauna associated with cattle manure applied as slurry to grassland. *Proceedings of the Royal Irish Academy*, **79B**, 15–27.

Curry, J.P. (1986) Above-ground arthropod fauna of four Swedish cropping systems and its role in carbon and nitrogen cycling. *Journal of Applied Ecology*, **23**, 853–70.

Curry, J.P. and Boyle, K.E. (1987) Growth rates, establishment and effects on herbage yield of introduced earthworms in grassland on reclaimed cutover peat. *Biology and Fertility of Soils*, **3**, 95–8.

Curry, J.P. and Cotton D.C.F. (1980) Effects of heavy pig slurry contamination on earthworms in grassland, in *Soil Biology as Related to Land Use Practices*, (ed. D.L. Dindal), EPA, Washington DC, pp. 336–43.

Curry, J.P. and Cotton, D.C.F. (1983) Earthworms and land reclamation, in *Earthworm Ecology – From Darwin to Vermiculture*, (ed. J.E. Satchell), Chapman & Hall, London, pp. 215–28.

Curry, J.P., Cotton, D.C.F., Bolger, T. and O'Brien, V. (1980) Effects of land-spread animal manures on the fauna of grassland, in *Effluents from Livestock*, (ed. J.K.R. Gasser), Applied Science, Barking, Essex, pp. 314–25.

Curry, J.P., Kelly, M. and Bolger, T. (1985) Role of invertebrates in the decomposition of *Salix* litter in reclaimed cutover peat, in *Ecological Interactions in Soil*, (eds A.H. Fitter, D. Atkinson, D.J. Read and M.B. Usher), Blackwell, Oxford, pp. 393–7.

Curry, J.P. and Momen, F.M. (1988) The arthropod fauna of grassland on reclaimed cutaway peat in Central Ireland. *Pedobiologia*, 32, 99–109.

Curry, J.P. and O'Neill, N. (1979) A comparative study of the arthropod communities of various swards using the D-vac suction sampling technique. *Proceedings of the Royal Irish Academy*, 79B, 247–58.

Curry, J.P. and Tuohy, C.F. (1978) Studies on the epigeal microarthropod fauna of grassland swards managed for silage production. *Journal of Applied Ecology*, 15, 727–41.

Curtis, J.T. (1959) *Vegetation of Wisconsin: an Ordination of Plant communities*. University of Wisconsin Press, Madison, Wisconsin.

Czerwínski, Z., Jakubczyk, H. and Petal, J. (1971) Influence of ant hills on the meadow soils. *Pedobiologia*, 11, 277–85.

Czerwínski, Z., Jakubczyk, H. and Nowak, E. (1974) Analysis of a sheep pasture ecosystem in the Pieniny mountains (the Carpathians). XII. The effect of earthworms on the pasture soil. *Ekologia Polska*, 22, 635–50.

Darbyshire, J.F. and Greaves, M.P. (1967) Protozoa and bacteria in the rhizosphere of *Sinapis alba* L., *Trifolium repens* L. and *Lolium perenne* L. *Canadian Journal of Microbiology*, 13, 1057–68.

Darbyshire, J.F., Wheatley, R.F., Greaves, M.P. and Inkson, R.H.E. (1974) A rapid micromethod for estimating bacterial and protozoan populations in soil. *Révue d'Écologie et de Biologie du Sol*, 11, 465–75.

D'Arcy-Burt, S. and Blackshaw, R.P. (1991) Bibionids (Diptera: Bibionidae) in agricultural land: review of damage, benefits, natural enemies and control. *Annals of Applied Biology*, 118, 695–708.

Darwin, C.R. (1881) *The Formation of Vegetable Mould through the Action of Worms, with Observations on their Habits*. Murray, London.

Dash, M.C. (1975) Numbers and distribution of testacea (Protozoa) in tropical grassland soils from Southern Orissa, India. *Indian Biologist*, 7(1), 53–6.

Dash, M.C. and Cragg, J.B. (1972) Selection of microfungi by Enchytraeidae (Oligochaeta) and other members of the soil fauna. *Pedobiologia*, 12, 282–6.

Dash, M.C. and Guru, B.C. (1980) Distribution and seasonal variation in numbers of Testacea (Protozoa) in some Indian soils. *Pedobiologia*, 20, 325–42.

Dash, M.C., Nandra, B. and Mishra, P.C. (1981) Digestive enzymes in three species of Enchytraeidae. *Oikos*, 36, 316–18.

Dash, M.C. and Patra, U.C. (1977) Density, biomass and energy budget of a tropical earthworm population from a grassland site in Orissa, India. *Revue d'Écologie et de Biologie du Sol*, 14, 461–71.

Daubenmire, R. (1968) Ecology of fire in grasslands. *Advances in Ecological Research*, 5, 209–66.

Davidson, J. (1932) Factors affecting oviposition of *Smynthurus viridis* L. (Collembola). *Australian Journal of Experimental Biology and Medical Science*, 10, 1–16.

Davidson, J. (1933) The distribution of *Smynthurus viridis* L. (Collembola) in South Australia, based on rainfall, evaporation and temperature. *Australian Journal of Experimental Biology and Medical Science*, 11, 59–66.

Davidson, J. (1934) The 'lucerne flea' *Smynthurus viridis* L. (Collembola) in Australia. *Bulletin of the Council for Scientific and Industrial Research, Australia*, **79**, 1–66.

Davidson, R.L. (1969) Influence of soil moisture and organic matter on scarab damage to grasses and clover. *Journal of Applied Ecology*, **6**, 237–46.

Davidson, R.L. (1979) Effects of root feeding on foliage yield, in *Proceedings of the 2nd Australasian Conference on Grassland Invertebrate Ecology*, (eds T.K. Crosby and R.P. Pottinger), Government Printer, Wellington, pp. 117–20.

Davidson, R.L., Hilditch, J.A., Wiseman, J.R. and Wolfe, V.J. (1979) Growth, fecundity and mortality responses of Scarabaeidae (Coleoptera) contributing to population increases in improved pastures, in *Proceedings of the 2nd Australasian Conference on Grassland Invertebrate Ecology*, (eds T.K. Crosby and R.P. Pottinger), Government Printer, Wellington, pp. 66–70.

Davidson, R.L. and Roberts, R.J. (1968) Influence of plants, manure and soil moisture on survival and liveweight gain of two scarabaeid larvae. *Entomologia Experimentalis et Applicata*, **11**, 305–14.

Davidson, R.L., Shackley, A., Wolfe, V.J. and Donelan, M.J. (1979) Anomalous increases in pasture yield after use of insecticides on soil, in *Proceedings of the 2nd Australasian Conference on Grassland Invertebrate Ecology*, (eds T.K. Crosby and R.P. Pottinger), Government Printer, Wellington, pp. 30–2.

Davidson, R.L., Wiseman, J.R. and Wolfe, V.J. (1970) A systems approach to pasture scarab problems in Australia, in *Proceedings of the XIth International Grassland Congress*, (ed. M.J.T. Norman), pp. 681–4.

Davidson, R.L., Wiseman, J.R. and Wolfe, V.J. (1972a) Environmental stress in the pasture scarab *Sericesthis nigrolineata* Boisd. 1. Mortality in larvae caused by high temperature. *Journal of Applied Ecology*, **9**, 783–97.

Davidson, R.L., Wiseman, J.R. and Wolfe, V.J. (1972b) Environmental stress in the pasture scarab *Sericesthis nigrolineata* Boisd. II. Effects of soil moisture and temperature on survival of first-instar larvae. *Journal of Applied Ecology*, **9**, 799–806.

Davidson, S.J. (1981) Environmental gradients of cattle droppings and corresponding population densities of microarthropods. *Pedobiologia*, **21**, 236–41.

Davis, B.N.K. (1963) A study of microarthropod communities in mineral soils near Corby, Northants. *Journal of Animal Ecology*, **32**, 49–71.

Davis, B.N.K. (1968) The soil macrofauna and organochlorine insecticide residues at twelve agricultural sites near Huntingdon. *Annals of Applied Biology*, **61**, 29–45.

Davis, B.N.K. (1973) The Hemiptera and Coleoptera of stinging nettle (*Urtica dioica* L.) in East Anglia. *Journal of Applied Ecology*, **10**, 213–37.

Davis, R.C. and Sutton, S.L. (1978) A comparative study of the changes in biomass of isopods inhabiting dune grassland. *Scientific Proceedings of the Royal Dublin Society*, **6A**, 223–33.

Dean, G.J. (1974) Effects of parasites and predators on the cereal aphids *Metopolophium dirhodum* (Wlk.) and *Macrosiphum avenae* (F.) (Hem., Aphididae). *Bulletin of Entomological Research*, **63**, 411–22.

Dean, G.J. (1975) The natural enemies of cereal aphids. *Annals of Applied Biology*, **80**, 130–2.

Debauche, H.R. (1962) The structural analysis of animal communities of the soil, in *Progress in Soil Zoology* (ed. P.W. Murphy), Butterworths, London, pp. 10–25.

Delchev, Kh. and Kajak, A. (1974) Analysis of a sheep pasture ecosystem in the Pieniny Mountans (the Carpathians). XVI. Effect of pasture management on the number and biomass of spiders (Aranaea) in two climatic regions (The Pieniny and the Sredna Gora Mountains). *Ekologia Polska*, **22**, 693–710.

Delettre, Y.R. and Lagerlöf, J. (1992) Abundance and life history of terrestrial Chironomidae (Diptera) in four Swedish agricultural cropping systems. *Pedobiologia*, **36**, 69–78.

Dempster, J.P. (1963) The population dynamics of grasshoppers and locusts. *Biological Reviews*, **38**, 490–529.

Dempster, J.P. (1967) A study of the effects of DDT applications against *Pieris rapae* on the crop fauna. *Proceedings of the 4th British Insecticide and Fungicide Conference 1967*, **1**, 19–25.

Dempster, J.P. (1971) Some effects of grazing on the population ecology of the Cinnabar Moth (*Tyria jacobaeae* L.), in *The Scientific Management of Animal and Plant Communities for Conservation*, (eds E. Duffey and A.S. Watt), Blackwell, Oxford, pp. 517–26.

Dempster, J.P. (1975) *Animal Population Ecology*. Academic Press, London.

Dempster, J.P. and Lakhani, K.H. (1979) A population model for cinnabar moth and its food plant, ragwort. *Journal of Animal Ecology*, **48**, 143–64.

den Boer, P.J. (1981) On the survival of populations in a heterogenous and variable environment. *Oecologia*, **50**, 39–53.

den Boer, P.J. (1985) Fluctuations of density and survival of carabid populations. *Oecologia*, **67**, 322–30.

den Boer, P.J. (1986) Facts, hypotheses and models on the part played by food in the dynamics of carabid populations. *Report of the Fifth Meeting of European Carabidologists, Poland, September 1982*, Warsaw Agricultural University Press, Warsaw, pp. 81–96.

Denno, R.F. (1977) Comparison of the assemblages of sap-feeding insects (Homoptera-Hemiptera) inhabiting two structurally different salt marsh grasses in the genus *Spartina*. *Environmental Entomology*, **6**, 359–72.

Denno, R.F. (1983) Tracking variable host plants in space and time, in *Variable Plants and Herbivores in Natural and Managed Systems*, (eds R.F. Denno and M.S. McClure), Academic Press, New York, pp. 291–341.

Desender, K. (1982) Ecological and faunal studies on Coleoptera in agricultural land. II. Hibernation of Carabidae in agro-ecosystems. *Pedobiologia*, **23**, 295–303.

Desender, K., Maelfait, J.-P., D'Hulster, M. and Vanherche, L. (1981) Ecological and faunal studies on Coleoptera in agricultural land. I. Seasonal occurrence of Carabidae in the grassy edges of a pasture. *Pedobiologia*, **22**, 379–84.

Dhillon, B.S. and Gibson, N.H.E. (1962) A study of the Acarina and Collembola of agricultural soils. I. Numbers and distribution in undisturbed grassland. *Pedobiologia*, **1**, 189–209.

Dickinson, N.M. (1983) Decomposition of grass litter in a successional grassland. *Pedobiologia*, **25**, 117–26.

Dietz, S.M. and Harwood, R.F. (1960) Host range and damage by the grass mealy bug *Heterococcus graminicola*. *Journal of Economic Entomology*, **53**, 737–40.

Dindal, D.L., Newell, L.T. and Moreau, J.-P. (1979) Municipal wastewater irrigation: effects on community ecology of soil invertebrates, in *Utilization of Municipal Sewage Effluent and Sludge on Forest and Disturbed Land*, (eds W.E. Sopper and S.N. Kerr), Pennsylvania State University Press, University Park, Pennsylvania, pp. 197–205.

Dindal, D.L., Schwert, D.P., Moreau, J.-P. and Theoret, L. (1977). Earthworm communities and soil nutrient levels as affected by municipal wastewater irrigation, in *Soil Organisms as Components of Ecosystems*, (eds U. Lohm and T. Persson), *Ecological Bulletins* (Stockholm), **25**, 284–90.

Dixon, G.M. and Campbell, A.G. (1978) Relationships between grazing animals and populations of the pasture insects *Costelytra zealandica* (White) and *Inopus rubriceps* (Macquart). *New Zealand Journal of Agrcultural Research*, **21**, 301–5.

Dmowska, E. and Koztowska, J. (1988) Communities of nematodes in soil treated with semi-liquid manure. *Pedobiologia*, **32**, 323–30.

Dodd, J.L. and Lauenroth, U.K. (1979) Analysis of the response of a grassland ecosystem to stress, in *Perspectives in Grassland Ecology*, (ed. N. French), Springer-Verlag, New York, pp. 43–58.

Dondale, C.D. (1971) Spiders of Heasman's field, a mown meadow near Belleville, Ontario. *Proceedings of the Entomological Society of Ontario*, **101**, 62–9.

Doube, B.M. (1987) Spatial and temporal organization in communities associated with dung pads and carcasses, in *Organization of Communities Past and Present*, (eds J.H.R. Gee and P.S. Giller), Blackwell, Oxford, pp. 255–80.

Doull, K.M. (1956) Thrips infesting cocksfoot in New Zealand. II. The biology and economic importance of the cocksfoot thrips *Chirothrips manicatus* Haliday. *New Zealand Journal of Science and Technology*, **38A**, 56–65.

Dowdy, W.W. (1944) The influence of temperature on vertical migration of invertebrates inhabiting different soil types. *Ecology*, **25**, 449–60.

Drury, W.H. and Nisbet, I.C.T. (1973) Succession. *Journal of the Arnold Arboretum*, **54**, 331–68.

Duffey, E. (1962) A population study of spiders in limestone grassland: description of study area, sampling methods and population characteristics. *Journal of Animal Ecology*, **31**, 571–99.

Duffey, E., Morris, M.G., Sheail, J. *et al.* (1974) *Grassland Ecology and Wildlife Management*. Chapman & Hall, London.

Dunbier, M.W., Palmer, T.P., Ellis, T.J. and Bennett, P.P. (1979) Field evaluation of lucerne cultivars for *Ditylenchus dipsaci* (Nematoda: Tylenchidae) and *Acyrthosiphon kondoi* (Hemiptera: Aphididae), in *Proceedings of the 2nd Australasian Conference on Grassland Invertebrate Ecology*, (eds T.K. Crosby and R.P. Pottinger), Government Printer, Wellington, pp. 99–102.

Dunger, W. (1956) Untersuchungen über Laubzersetzung durch Collembolen. *Zoologische Jahrbücher. Abteilung für Systematik, Ökologie und Geographie der Tiere*, **84**, 75–98.

Dunger, W. (1958) Über die Zersetzung der Laubstreu durch die Boden Makrofauna in Auenwald. *Zoologische Jahrbücher. Abteilung für Systematik, Ökologie und Geographie der Tiere*, **86**, 139–80.

Dunger, W. (1969a) Fragen der naturlichen und experimentellen Besiedlung kulturfeindlicher Boden durch Lumbriciden. *Pedobiologia*, **9**, 146–51.

Dunger, W. (1969b) Über den Anteil der Arthropoden an der Umsetzung des Bestandesabfalles in Anfangs-Bodenbildungen. *Pedobiologia*, **9**, 366–71.

Dunger, W. (1989) The return of soil fauna to coal mined areas in the German Democratic Republic, in *Animals in Primary Succession: the Role of Fauna in Reclaimed Lands*, (ed. J.D. Majer), Cambridge University Press, Cambridge, pp. 307–37.

Dwivedi, R.S. (1979) Microorganisms, in *Grassland Ecosystems of the World: Analysis of Grasslands and their Uses*, International Biological Programme 19, (ed. R.T. Coupland), Cambridge Universty Press, Cambridge, pp. 227–30.

Dyer, M.I. .(1975) The effects of red-winged blackbirds (*Agelaius phoeniceus* L.) on biomass production of corn grains (*Zea mays* L.). *Journal of Applied Ecology*, **12**, 719–26.

Dyer, M.I. (1980) Mammalian epidermal growth factor promotes plant growth. *Proceedings of the National Academy of Science, USA.*, **77**, 4836–7.

Dyer, M.I. and Bokhari, U.G. (1976) Plant-animal interactions: studies of the effects of grasshopper grazing on blue grama grass. *Ecology*, **57**, 562–72.

Dyer, M.I., Detling, J.K., Coleman, D.C. and Hilbert, D.W. (1982) The role of herbivores in grassland, in *Grasses and Grasslands*, (eds T.R. Estes, R.J. Tyrl and J.N. Brunken), University of Oklahoma Press, Norman, Oklahoma, pp. 255–95.

East, R. (1972) Starling (*Sturnus vulgaris* L.) Predation on grass grub (*Costelytra zealandica* (White), Melolonthinae) populations in Canterbury. Ph.D. thesis, University of Canterbury, New Zealand.

East, R. (1976) The effects of pasture species on white-fringed weevil populations. *Proceedings of the 29th New Zealand Weed and Pest Control Conference*, pp. 165–7.

East, R. (1977) The effects of pasture and forage crop species on longevity, fecundity and oviposition rate of adult white-fringed weevils *Graphognathus leucoloma* (Boheman). *New Zealand Journal of Experimental Agriculture*, 5, 177–81.

East, R. (1979a) Population studies of Australasian pasture Scarabaeidae (Coleoptera), in *Proceedings of the 2nd Australasian Conference on Grassland Invertebrate Ecology*, (eds T.K. Crosby and R.P. Pottinger), Government Printer, Wellington, pp. 54–62.

East, R. (1979b) Effects of grazing management on *Costelytra zealandica* populations (Coleoptera: Scarabaeidae), in *Proceedings of the 2nd Australasian Conference on Grassland Invertebrate Ecology*, (eds T.K. Crosby and R.P. Pottinger), Government Printer, Wellington, pp. 180–4.

East, R., King, P.D. and Watson, R.N. (1981) Population studies of grass grub (*Costelytra zealandica*) and black beetle (*Heteronychus arator*) (Coleoptera: Scarabaeidae). *New Zealand Journal of Ecology*, 4, 56–64.

East, R. and Pottinger, R.P. (1975) Starling (*Sturnus vulgaris* L.) predation on grass grub (*Costelytra zealandica* (White), Melolonthinae) populations in Canterbury. *New Zealand Journal of Agricultural Research*, 18, 417–52.

East, R. and Pottinger, R.P. (1983) Use of grazing animals to control insect pests of pasture. *New Zealand Entomologist*, 7, 352–9.

East, R. and Willoughby, B.E. (1980) Effects of pasture defoliation in summer on grass grub (*Costelytra zealandica*) populations. *New Zealand Journal of Agricultural Research*, 23, 547–62.

Edwards, C.A. (1955) Soil sampling for symphylids and a note on populations, in *Soil Zoology*, (ed. D.K. McE. Kevan), Butterworths, London, pp. 152–6.

Edwards, C.A. (1958) The ecology of Symphyla. Part I. Populations. *Entomologia Experimentalis et Applicata*, 1, 308–19.

Edwards, C.A. (1959) The ecology of Symphyla. Part II. Seasonal soil migrations. *Entomologia Experimentalis et Applicata*, 2, 257–67.

Edwards, C.A. (1965) Effects of pesticide residues on soil invertebrates and plants, in *Ecology and the Industrial Society*, (eds G.T. Goodman, R.W. Edwards and J.M. Lambert), Blackwell, Oxford, pp. 239–61.

Edwards, C.A. (1973) *Persistent Pesticides in the Environment*. CRC Press, Cleveland, Ohio.

Edwards, C.A. (1974a) Macroarthropods, in *Biology of Plant Litter Decomposition*, (eds C.H. Dickinson and G.J.F. Pugh), Academic Press, London, pp. 533–54.

Edwards, C.A. (1974b) Some effects of insecticides on myriapod populations, in *Myriapoda*, (ed. J.G. Blower), *Symposia of the Zoological Society of London*, 32, 645–55.

Edwards, C.A. (1975) Effects of direct drilling on the soil fauna. *Outlook on Agriculture*, 8, 243–4.

Edwards, C.A. (1977) Investigations into the influence of agricultural practice on soil invertebrates. *Annals of Applied Biology*, 87, 515–20.

Edwards, C.A. (1980) Interactions between agricultural practice and earthworms, in *Soil Biology as Related to Land Use Practices*, (ed. D.L. Dindal), EPA, Washington DC, pp. 3–12.

Edwards, C.A. (1983) Earthworm ecology in cultivated soils, in *Earthworm Ecology – From Darwin to Vermiculture*, (ed. J.E. Satchell), Chapman & Hall, London, pp. 123–37.

Edwards, C.A. (1988) Breakdown of animal, vegetable and industrial organic wastes by earthworms, in *Earthworms in Waste and Environmental Management*, (eds C.A. Edwards and E.F. Neuhauser), SPB Academic Publishing, The Hague, pp. 21–31.

Edwards, C.A., Butler, C.G. and Lofty, J.R. (1976) The invertebrate fauna of the Park Grass Plots. II. Surface fauna. *Rothamsted Experimental Station Report for 1975, Part 2*, pp. 63–89.

Edwards, C.A., Dennis, E.B. and Empson, D.W. (1967) Pesticides and the soil fauna: effects of aldrin and DDT in an arable field. *Annals of Applied Biology*, **60**, 11–22.

Edwards, C.A. and Heath, G.W. (1963) The role of soil animals in breakdown of leaf material, in *Soil Organisms*, (eds J. Doeksen and J. van der Drift), North Holland, Amsterdam, pp. 76–85.

Edwards, C.A. and Lofty, J.R. (1975a) The invertebrate fauna of the Park Grass Plots. I. Soil fauna. *Rothamsted Experimental Station Report for 1974, Part 2*, pp. 133–54.

Edwards, C.A. and Lofty, J.R. (1975b) The influence of cultivation on soil animal populations, in *Progress in Soil Zoology*, (ed. J. Vanek), Academia, Prague, pp. 399–407.

Edwards, C.A. and Lofty, J.R. (1978) The influence of arthropods and earthworms upon root growth of direct drilled cereals. *Journal of Applied Ecology*, **15**, 789–95.

Edwards, C.A. and Lofty, J.R. (1980) Effects of earthworm inoculation upon the root growth of direct drilled cereals. *Journal of Applied Ecology*, **17**, 533–43.

Edwards, C.A. and Lofty, J.R. (1982) Nitrogenous fertilizers and earthworm populations in agricultural soils. *Soil Biology and Biochemistry*, **14**, 515–21.

Edwards, C.A., Sunderland, K.D. and George, K.S. (1979) Studies on polyphagous predators of cereal aphids. *Journal of Applied Ecology*, **16**, 811–23.

Edwards, C.A. and Thompson, A.R. (1973) Pesticides and the soil fauna. *Residue Reviews*, **45**, 1–79.

Edwards, C.A. and Thompson, A.R. (1975) Some effects of insecticides on predatory beetles. *Annals of Applied Biology*, **80**, 132–5.

Edwards, C.A., Thompson, A.R. and Beynon, K.I. (1968) Some effects of chlorfenvinphos, an organophosphorous insecticide, on populations of soil animals. *Revue d'Écologie et de Biologie du Sol*, **5**, 199–224.

Edwards, P.J. and Wratten, S.D. (1980) *Ecology of Insect–Plant Interactions*, Edward Arnold, London.

Ehrlich, P.R. and Raven, P.H. (1964) Butterflies and plants: a study in coevolution. *Evolution*, **18**, 586–608.

Eijsackers, H. (1987) The impact of heavy metals on terrestrial ecosystems: biological adaptation through behavioural and physiological avoidance, in *Ecological Assessment of Environmental Degradation, Pollution and Recovery*, (ed. O. Ravera), Elsevier, Amsterdam, pp. 245–59.

Eijsackers, H. and van der Drift, J. (1976) Effects on the soil fauna, in *Herbicides: Physiology, Biochemistry, Ecology*, Vol. 2, 2nd edn, (ed. L.J. Audus), Academic Press, London, pp. 149–74.

Elliott, E.T. and Coleman, D.C. (1977) Soil protozoan dynamics of a Colorado shortgrass prairie. *Soil Biology and Biochemistry*, **9**, 113–18.

Elliott, P.W., Knight, D. and Anderson, J.M. (1990) Denitrification in earthworm casts and soil from pastures under different fertilizer and drainage regimes. *Soil Biology and Biochemistry*, **22**, 601–5.

Ellis, S.A., Clements, R.O. and Bale, J.S. (1990) A comparison of the effects of sward improvement on invertebrates, sward establishment and herbage yield. *Annals of Applied Biology*, **116**, 343–56.

Emmanuel, N., Curry, J.P. and Evans, G.O. (1985) The soil acari of barley plots with different cultural treatments. *Experimental and Applied Acarology*, **1**, 101–13.

Engelmann, M.D. (1961) The role of soil arthropods in the energetics of an old field community. *Ecological Monographs*, **31**, 221–38.

Evans, A.C. (1948) Studies on the relationships between earthworms and soil fertility. II. Some effects of earthworms on soil structure. *Annals of Applied Biology*, **35**, 1–13.

Evans, A.C. and Guild, W.J.McL. (1947) Studies on the relationships between earthworms and soil fertility. I. Biological studies in the field. *Annals of Applied Biology*, **34**, 307–30.

Evans, A.C. and Guild, J.W.McL. (1948a) Studies on the relationships between earthworms and soil fertility. IV. On the life cycles of some British Lumbricidae. *Annals of Applied Biology*, **35**, 471–84.

Evans, A.C. and Guild, J.W.McL. (1948b) Studies on the relationships between earthworms and soil fertility. V. Field populations. *Annals of Applied Biology*, **35**, 485–93.

Evans, F.C. and Murdoch, W.W. (1968) Taxonomic composition, trophic structure and seasonal occurrence in a grassland insect community. *Journal of Animal Ecology*, **37**, 259–73.

Eyre, M.D., Luff, M.L. and Rushton, S.P. (1990) The ground beetle (Coleoptera, Carabidae) fauna of intensively managed agricultural grasslands in northern England and southern Scotland. *Pedobiologia*, **34**, 11–18.

Fagan, E.B. and Kuitert, L.C. (1970) Evaluation of insecticides for control of the two-lined spittlebug of Florida pastures. *Journal of Economic Entomology*, **63**, 716–19.

Fager, E.W. (1957) Determination and analysis of recurrent groups. *Ecology*, **38**, 586–95.

FAO–UNESCO (1974) *Soil Map of the World, Vol. I. Legend*. Unesco, Paris.

Feeny, P. (1976) Plant apparency and chemical defence. *Recent Advances in Phytochemistry*, **10**, 1–40.

Fenemore, P.G. (1966) Effects of grass grub infestations on pasture. *Proceedings of the New Zealand Ecological Society*, **13**, 137–45.

Ferguson, C.M., Barratt, B.I.P. and Jones, P.A. (1988) Control of the grey field slug (*Deroceras reticulatum* (Muller)) by stock management prior to direct-drilled pasture establishment. *Journal of Agricultural Science, Cambridge*, **111**, 443–9.

Ferguson, C.M., Barratt, B.P. and Jones, P.H. (1989) A new technique for estimating the density of the grey field slug *Deroceras reticulatum*, in *Slugs and Snails in World Agriculture*, (ed. I.F. Henderson), British Crop Protection Council Monograph No. 41, pp. 331–6.

Ferrar, P. (1973) CSIRO dung beetle project. *Wool Technology and Sheep Breeding*, **20**, 73–5.

Ferrar, P. and Watson, J.A.L. (1970) Termites (Isoptera) associated with dung in Australia. *Journal of the Australian Entomological Society*, **9**, 100–2.

Ferriére, G. and Bouché, M.B. (1985) Première mesure écophysiologique d'un débit d'élément dans un animal endogé: le débit d'azote de *Nicodrilus longus longus* (Ude) (Lumbricidae, Oligochaeta) dans la prairie de Cîteaux. *Comptes Rendus de l'Académie des Sciences*, Paris, **301**, Série III, 789–94.

FitzPatrick, E.A. (1980) *Soils: their Formation, Classification and Distribution*, Longman, London.

FitzPatrick, E.A. (1984) *Micromorphology of Soils*, Chapman & Hall, London.

Fletcher, L.R., Popay, A.J. and Tapper, B.A. (1991) Evaluation of several loli-trem-free endophyte/perennial ryegrass combinations. *Proceedings of the New Zealand Grassland Association*, **53**, 215–19.

Flint, M.L. and van den Bosch, R. (1981) *Introduction to Integrated Pest Management*, Plenum, New York.

Forschler, B.T. and Gardner, W.A. (1991) Field efficacy and persistence of entomogenous nematodes in the management of white grubs (Coleoptera, Scarabaeidae) in turf and pasture. *Journal of Economic Entomology*, **84**, 1454–9.

Fortuner, R. and Jacq, V.A. (1976) *In vitro* study of toxicity of soluble sulfides to three nematodes parasitic on rice in Senegal. *Nematologia*, **22**, 343–51.

Foth, H.D. (1978) *Fundamentals of Soil Science*, 6th edn, Wiley, New York.

Fox, C.J.S. (1967) Effects of several chlorinated hydrocarbon insecticides on the springtails and mites of grassland soil. *Journal of Economic Entomology*, **60**, 77–9.

Freckman, D.W., Duncan, D.A. and Larson, J.R. (1979) Nematode density and biomass in an annual grassland ecosystem. *Journal of Range Management*, **32**, 418–22.

Freckman, D.W. and Mankau, R. (1977) Distribution and trophic structure of nematodes in desert soils. *Ecological Bulletins* (Stockholm), **25**, 511–14.

French, N. (1969) Assessment of leatherjacket damage to grassland and economic aspects of control. *Proceedings of the Fifth British Insecticide and Fungicide Conference*, **2**, 511–21.

Fungo, N.K. and Curry, J.P. (1983) The effects of eight fungicides on the glasshouse spider mite, *Tetranychus urticae* (Koch). *Journal of Life Sciences, Royal Dublin Society*, **4**, 175–81.

Gaspar, Ch. (1971) Les formicides de la Famenne. I. Une étude zoosociologique. *Revue d'Écologie et de Biologie du Sol*, **8**, 553–607.

Gauch, H.G. Jr (1982) *Multivariate Analysis in Community Ecology*, Cambridge University Press, Cambridge.

Gauch, H.G. and Whittaker, R.H. (1972) Comparison of ordination techniques. *Ecology*, **53**, 868–75.

Gauch, H.G. Jr, Whittaker, R.H. and Wentworth, T.R. (1977) A comparative study of reciprocal averaging and other ordination techniques. *Journal of Ecology*, **65**, 157–74.

Gaugler, R. (1988) Ecological considerations in the biological control of soil-inhabiting insects with entomopathogenic nematodes. *Agriculture, Ecosystems and Environment*, **24**, 351–60.

Gelernter, W.D. (1990) *Bacillus thuringiensis*, bioengineering and the future of bioinsecticides, in *Proceedings of the 1990 British Crop Protection Conference – Pests and Diseases*, pp. 617–24.

Georgiadis, N.J., Ruess, R.W., McNaughton, S.J. and Western, D. (1989) Ecological conditions that determine when grazing stimulates grass production. *Oecologia*, **81**, 316–22.

Georgis, R. and Gaugler, R. (1991) Predictability in biological control using entomopathogenic nematodes. *Journal of Economic Entomology*, **84**, 713–20.

Gerard, B.M. (1967) Factors affecting earthworms in pastures. *Journal of Animal Ecology*, **36**, 235–52.

Gerard, P.J. (1979) The effects of pasture species on *Inopus rubriceps* (Diptera: Stratiomyidae), in *Proceedings of the 2nd Australasian Conference on Grassland Invertebrate Ecology*, (eds T.K. Crosby and R.P. Pottinger), Government Printer, Wellington, pp. 267–70.

Ghilarov, M.S. and Mamajev, B.M. (1966) Über die Ansiedlung von Regenwürmern in den artesisch bewässerten Oasen der Wüste Kysyl-Kum. *Pedobiologia*, **6**, 197–218.

Gibbs, G.W. (1976) The role of insects in natural terrestrial ecosystems. *New Zealand Entomologist*, **6**, 113–21.

Gibson, C.W.D. (1976) The importance of food plants for the distribution and abundance of some Stenodemini (Heteroptera: Miridae) of limestone grassland. *Oecologia*, **25**, 55–76.

Gibson, C.W.D. (1980) Niche use patterns among some Stenodemini (Heteroptera: Miridae) of limestone grassland, and an investigation of the possibility of interspecific competition between *Notostira elongata* Geoffroy and *Megaloceraea recticornis* Geoffroy. *Oecologia*, **47**, 352–64.

Gibson, C. and Visser, M. (1982) Interspecific competition between two field populations of grass-feeding bugs. *Ecological Entomology*, **7**, 61–7.

Gillard, O. (1967) Coprophagous beetles in pasture ecosystems. *Journal of the Australian Institute of Agricultural Science*, **33**, 30–4.

Giller, P.S. and Doube, B.M. (1989) Experimental analysis of inter- and intraspecific competition in dung beetle communities. *Journal of Animal Ecology*, **58**, 129–42.

Gillon, D. (1971) The effect of bush fire on the principal pentatomid bugs (Hemiptera) of an Ivory Coast savanna, in *Proceedings of the Annual Tall Timbers Fire Ecology Conference*, No. 11, pp. 377–417.

Gillon, Y. (1971) The effect of bush fire on the principal Acridid species of an Ivory Coast savanna, in *Proceedings of the Annual Tall Timbers Fire Ecology Conference*, No. 11, pp. 419–71.

Gillon, Y. and Gillon, D. (1973) Recherches écologiques sur une savanna Sahelienne du Ferlo Septentrionel, Sénégal: donnêes quantitatives sur les Arthropodes. *Terre et Vie*, **27**, 297–323.

Gisiger, L. (1961) Neue Erkentnisse Über die Bereitung der Gülle und ihre zweckmässige Anwendung, in *Bundesversuchsanstalt für Alpenländische Landwirtschaft, Gumpenstein: Berichte 3. Arbeitstagun 'Fragen der Güllerei' 1960*, pp. 103–22.

Glasgow, J.P. (1939) A population study of subterranean soil Collembola. *Journal of Animal Ecology*, **8**, 323–54.

Godan, D. (1983) *Pest Slugs and Snails: Biology and Control*, Springer-Verlag, Berlin.

Goffinet, G. (1975) Écologie édaphique des milieux naturels du Haut-Shaba (Zaïre). 1. Caractéristiques, écotopiques et synécologie comparée des zoocénoses intercaliques. *Revue d'Écologie et de Biologie du Sol*, **12**, 691–722.

Goffinet, G. (1976) Écologie édaphique des écosystemes naturels de haut-Shaba (Zaïre) III. Les peuplements en termites épigés au niveau des latosols. *Revue d'Écologie et de Biologie du Sol*, **13**, 459–76.

Golley, F.B. and Gentry, J.B. (1964) Bioenergetics of the southern harvester ant *Pogonomyrmex badius*. *Ecology*, **45**, 217–25.

Good, J.A. and Giller, P.S. (1990) Staphylinid beetles (Coleoptera) from cereal and grass fields in south-west Ireland. *Bulletin of the Irish Biogeographical Society*, **13**, 2–22.

Gordon, A.J. and Kluge, R.L. (1991) Biological control of St. John's Wort, *Hypericum perforatum* (Clusiaceae) in South Africa. *Agriculture, Ecosystems and Environment*, **37**, 77–90.

Goring, C.A.I. and Laskowski, D.A. (1982) The effects of pesticides on nitrogen transformations in soils, in *Nitrogen in Agricultural Soils*, (ed. F.J. Stevenson), ASA-CSSA-SSSA, Madison, Wisconsin, pp. 689–720.

Gould, H.J. (1961) Observations on slug damage to winter wheat in East Anglia 1957–1959. *Plant Pathology*, **11**, 147–52.

Gower, J.C. (1966) Some distance properties of latent root and vector methods used in multivariate analysis. *Biometrika*, **53**, 325–38.

Graff, O. (1971) Stickstoff, Phosphor und Kalium in der Regenwurmlosung auf der Wiesenversuchsflache des Sollingprojekektes, in *IV Colloquium Pedobiologiae*, (ed. J. d'Aguilar), Institut National des Recherches Agriculturelles, Publ. 71-7, Paris, pp. 503–11.

Grant, W.C. (1955) Studies on moisture relationships in earthworms. *Ecology*, **36**, 400–7.

Grassé, P.P. (1949) Ordre des Isoptères on termites, in *Traité de Zoologie*, Vol. IX, (ed. P.P. Grassé), Masson, Paris, pp. 408–544.

Gray, J.S. (1987) Species-abundance patterns, in *Organization of Communities Past and Present*, (eds J.H.R. Gee and P.S. Giller), Blackwell, Oxford, pp. 53–67.

Greenslade, P.J.M. (1983) Adversity selection and the habitat templet. *The American Naturalist*, **122**, 352–65.

Griffiths, B.S., Wood, S. and Cheshire, M.V. (1989) Mineralisation of ^{14}C-labelled plant material by *Porcellio scaber* (Crustacea, Isopoda). *Pedobiologia*, **33**, 350–60.

Guild, W.J.McL. (1948) Effect of soil type on populations. *Annals of Applied Biology*, **35**, 181–92.

Guild, W.J.McL. (1955) Earthworms and soil structure, in *Soil Zoology*, (ed. D.K.McE. Kevan), Butterworths, London, pp. 83–98.

Guiran, G. de, Bonnel, L. and Abirached, M. (1980) Landspreading of pig manures. IV. Effect on soil nematodes, in *Effluents from Livestock*, (ed. J.K.R. Gasser), Applied Science Publishers, London, pp. 109–19.

Gupta, R.K. and Abrol, L.P. (1990) Salt affected soils: their reclamation and management for crop production, in *Soil Degradation*, (eds R. Lal and B.A. Stewart), *Advances in Soil Science*, **11**, 223–88.

Gupta, S.R., Rajvanshi, R. and Singh, J.S. (1981) The role of the termite *Odontotermes gurdaspurensis* (Isopoda: Termitidae) in plant decomposition in a tropical grassland. *Pedobiologia*, **22**, 254–61.

Gyllenberg, G. (1969) The energy flow through a *Chorthippus parallelus* (Zett.) (Orthoptera) population in a meadow in Tvärminne, Finland. *Acta Zoologica Fennica*, **123**, 4–74.

Haarløv, N. (1960) Microarthropods from Danish soils: ecology, phenology. *Oikos*, Suppl. **3**, 176 pp.

Hågvar, S. and Abrahamsen, G. (1980) Colonisation by Enchytraeidae, Collembola and Acari in sterile soil samples with adjusted pH levels. *Oikos*, **34**, 245–58.

Hågvar, S. and Östbye, E. (1972) Quantitative and qualitative investigations of the invertebrate fauna under stones (hypolithion) in some alpine habitats at Finse, South Norway. *Norsk Entomologisk Tidsskrift*, **19**, 1–10.

Hale, W.G. (1965) Observations on the breeding biology of Collembola (I). *Pedobiologia*, **5**, 146–152; (II) *Pedobiologia*, **5**, 161–177.

Hale, W.G. (1966) A population study of moorland Collembola. *Pedobiologia*, **6**, 65–99.

Hale, W.G. (1967) Collembola, in *Soil Biology*, (eds A. Burgess and F. Raw), Academic Press, London, pp. 397–411.

Hale, W.G. (1980) Production and energy flow in two species of *Onychiurus* (Collembola, Insecta Apterygota). *Pedobiologia*, **20**, 274–87.

Hammond, R.F. (1979) *The Peatlands of Ireland*, Soil Survey Bulletin, No. 35, An Foras Taluntais, Dublin.

Hanlon, R.D.G. (1981) Some factors influencing microbial growth on soil animal faeces. II. Bacterial and fungal growth on soil animal faeces. *Pedobiologia*, **21**, 264–70.

Hanlon, R.D.G. and Anderson, J.M. (1979) The effects of Collembola grazing on microbial activity in decomposing leaf litter. *Oecologia*, **38**, 93–9.

Harberd, D.J. (1962) Some observations on natural clones in *Festuca ovina. New Phytologist*, **61**, 85–100.

Harding, D.J.L. and Stuttard, R.A. (1974) Microarthropods, in *Biology of Plant Litter Decomposition*, (eds C.H. Dickinson and G.J.F. Pugh), Academic Press, London, pp. 489–532.

Harper, J.L. (1969) The role of predation in vegetational diversity. *Brookhaven Symposia in Biology*, **22**, 48–62.

Harris, P. (1973) Insects in the population dynamics of plants, in *Insect/Plant Relationships*, Royal Entomological Society of London Symposium No. 6, (ed. H.F. van Emden), Blackwell, Oxford, pp. 201–9.

Harris, P. (1974) A possible explanation of plant yield increases following insect damage. *Agro-Ecosystems*, **1**, 219–25.

Harris, W. (1969) Some effects of a porina caterpillar (*Wiseana* spp.) infestation on perennial ryegrass, cocksfoot and white clover. *New Zealand Journal of Agricultural Research*, **12**, 543–52.

Harris, W. and Brock, J.L. (1972) Effect of porina caterpillar (*Wiseana* spp.) infestation on yield and competitive interactions of ryegrass and white clover varieties. *New Zealand Journal of Agricultural Research*, **15**, 723–40.

Harris, W.V. (1963) *Exploration du Parc National de la Garamba, Part 42, Isoptera*, Hayez, Brussels.

Hartenstein, R. (1982) Soil macroinvertebrates, aldehyde oxidase, catalase, cellulase and peroxidase. *Soil Biology and Biochemistry*, **14**, 387–91.

Hartenstein, R. (1986) Earthworm biotechnology and global biogeochemistry. *Advances in Ecological Research*, **15**, 379–409.

Hartenstein, R. and Bisesi, M.S. (1989) Use of earthworm biotechnology for the management of effluents from intensively housed livestock. *Outlook on Agriculture*, **18**, 72–6.

Hartenstein, R., Neuhauser, E.F. and Collier, J. (1980) Accumulation of heavy metals in the earthworm *Eisenia foetida. Journal of Environmental Quality*, **9**, 23–6.

Hartwig, E.K. (1955) Control of snouted harvester termites. *Farming in South Africa*, **30**, 361–6.

Hassall, M. (1983) Population metabolism of the terrestrial isopod *Philoscia muscorum* in a dune grassland ecosystem. *Oikos*, **41**, 17–26.

Hassall, M. and Dangerfield, J.M. (1989) Interspecific competition and the relative abundance of grassland isopods. *Monitore Zoologico Italiano (N.S.) Monografie*, **4**, 379–97.

Hassall, M. and Rushton S.P. (1982) The role of coprophagy in the feeding strategies of terrestrial isopods. *Oecologia*, **53**, 374–81.

Hassall, M. and Rushton, S.P. (1985) The adaptive significance of coprophagous behaviour in the terrestrial isopod *Porcellio scaber. Pedobiologia*, **28**, 169–75.

Hassall, M. and Sutton, S.L. (1978) The role of isopods as decomposers in a dune grassland ecosystem. *Scientific Proceedings of the Royal Dublin Society*, **6A**, 235–45.

Hassall, M. and Sutton, S.L. (1985) Immobilization of mineral nutrients by *Philoscia muscorum* (Isopoda, Oniscoidea) in a dune grassland ecosystem, in *Pro-*

ceedings of the 9th International Colloquium of Soil Zoology, Moscow, 1985, (ed. B.R. Striganova), pp. 29–37.

Hassall, M., Turner, T.G. and Rands, M.R.W. (1987) Effects of terrestrial isopods on the decomposition of woodland leaf litter. *Oecologia*, **72**, 597–604.

Hassell, M.P., Latto, J. and May, R.M. (1989) Seeing the wood for the trees: detecting density dependence from existing life table studies. *Journal of Animal Ecology*, **58**, 883–92.

Hawkins, J.A., Wilson, B.H., Mondart, C.L. *et al.* (1979) Leafhoppers and plant-hoppers in coastal Bermuda grass: effect on yield and quality and control by harvest frequency. *Journal of Economic Entomology*, **72**, 101–4.

Heal, O.W. (1965) Observations on testate amoebae (Protozoa: Rhizopoda) from Signy Island, South Orkney Islands. *British Antarctic Survey Bulletin*, **6**, 43–7.

Heal, O.W. and Dighton, J. (1985) Resource quality and trophic structure in the soil system, in *Ecological Interactions in Soil*, (eds A.H. Fitter, D. Atkinson, D.J. Read and M.B. Usher), Blackwell, Oxford, pp. 339–54.

Heal, O.W. and MacLean, S.F. (1975) Comparative productivity in ecosystems: secondary productivity, in *Unifying Concepts in Ecology*, (eds W.H. van Dobben and R.H. Lowe-McConnell), Junk, The Hague, pp. 89–108.

Heal, O.W. and Perkins, D.F. (1976) IPB studies on montane grasslands and moorlands. *Philosophical Transactions of the Royal Society of London*, **274B**, 295–314.

Healy, B. and Bolger, T. (1984) The occurrence of species of semi-aquatic Enchytraeidae (Oligochaeta) in Ireland. *Hydrobiologia*, **115**, 159–170.

Heard, A.J. and Hopper, M.J. (1963) Populations of stem-boring Diptera in leys and subsequent winter wheat and the effects of aldrin and phorate. *Annals of Applied Biology*, **51**, 301–11

Hébrant, F. (1970) Étude du flux énergétique chez deux espèces du genre *Cubitermes* Wasmann (Isoptera, Termitinae), termites humivores des savanes tropicales de la région éthiopienne. Thèse de Docteur en Sciences, Université Catholique de Louvain.

Henderson, I.F. and Clements, R.O. (1977a) Grass growth in different parts of England in relation to invertebrate numbers and pesticide treatment. *Journal of the British Grassland Society*, **32**, 89–93.

Henderson, I.F. and Clements, R.O. (1977b) Stem-boring Diptera in grassland in relation to management practice. *Annals of Applied Biology*, **87**, 524–7.

Henderson, I.F. and Clements, R.O. (1979) Differential susceptibility to pest damage in agricultural grasses. *Journal of Agricultural Science, Cambridge*, **73**, 465–72.

Henderson, I.F. and Clements, R.O. (1980) The effect of insecticide treatment on the establishment and growth of Italian ryegrass under different sowing conditions. *Grass and Forage Science*, **35**, 235–41.

Hendrix, P.F. and Parmelee, R.W. (1985) Decomposition, nutrient loss and microarthropod densities in herbicide-treated grass litter in a Georgia Piedmont agroecosystem. *Soil Biology and Biochemistry*, **17**, 421–8.

Henzell, R.F. and Lauren, D.R. (1977a) Control of grass grub adults with synthetic pyrethroids. *Proceedings of the 30th New Zealand Weed and Pest control Conference*, 211–16.

Henzell, R.F. and Lauren, D.R. (1977b) Use of sex attractant traps to estimate the development stage of grass grub, *Costelytra zealandica* (White) (Coleoptera; Scarabaeidae), in the soil. *New Zealand Journal of Agricultural Research*, **20**, 75–8.

Henzell, R.F. and Lauren, D.R. (1977c) Grass grub control through sex attractant and insecticide combinations. *Proceedings of the Ruakura Farmers Conference*, 20–3,

Henzell, R.F., Lauren, D.R. and du Toit, G.D.G. (1979) Insecticides for the control of adult grass grub. II. Further tests with various synthetic pyrethroid formulations. *Proceedings of the 32nd New Zealand Weed and Pest Control Conference*, 96–100.

Hill, D.S. (1987) *Agricultural Insect Pests of Temperate Regions and Their Control*, Cambridge University Press, Cambridge.

Hill, M.G. (1976) The population and feeding ecology of five species of leafhoppers (Homoptera) on *Holcus mollis* L. Ph.D. thesis, University of London.

Hill, M.O. (1979a) *DECORANA: a FORTRAN Program for Detrended Correspondence Analysis and Reciprocal Averaging*, Ecology and Systematics, Cornell University, Ithaca, New York.

Hill, M.O. (1979b) *TWINSPAN: a FORTRAN Program for Arranging Multivariate Data in an Ordered Two-way Table by Classification of the Individuals and Attributes*, Ecology and Systematics, Cornell University, Ithaca, New York.

Hinton, J.M. (1971) Energy flow in a natural population of *Neophilaenus lineatus* (Homoptera). *Oikos*, 22, 155–71.

Hodges, L.R. (1973) *Nematodes and Their Control*, Union Carbide Corporation, Salinas, California.

Hofsvang, T. (1972) *Tipula excisa* Schum. (Diptera, Tipulidae) life cycle and population dynamics. *Norsk Entomologisk Tidsskrift*, 19, 43–8.

Hofsvang, T. (1973) Energy flow in *Tipula excisa* Schum (Diptera, Tipulidae) in a high mountain area, Finse, South Norway. *Norwegian Journal of Zoology*, 21, 7–16.

Hölldobler, B. and Wilson, E.O. (1990) *The Ants*, Belknap Press, Harvard University, Cambridge, Mass.

Holmes, N. (1981) Mob stocking beats porina. *New Zealand Journal of Agricultural Research*, 143, 38–9.

Holter, P. (1973) A chromic oxide method for measuring consumption in dung-eating *Aphodius* larvae. *Oikos*, 24, 117–22.

Holter, P. (1974) Food utilization of dung-eating *Aphodius* larvae (Scarabaeidae). *Oikos*, 25, 71–9.

Holter, P. (1975) Energy budget of a natural population of *Aphodius rufipes* larvae (Scarabaeidae). *Oikos*, 26, 177–86.

Holter, P. (1979) Effect of dung-beetle (Aphodius spp.) and earthworms on the disappearance of cattle dung. *Oikos*, 32, 393–402.

Hoogerkamp, M., Rogäar, H. and Eijsackers, H.J.P. (1983) Effect of earthworms on grassland on recently reclaimed polder soils in the Netherlands, in *Earthworm Ecology – From Darwin to Vermiculture*, (ed. J.E. Satchell), Chapman & Hall, London, pp. 85–104.

Hopp, H. and Slater, C.S. (1948) Influence of earthworms on soil productivity. *Soil Science*, 66, 421–8.

Horn, H.S. (1974) The ecology of secondary succession. *Annual Review of Ecology and Systematics*, 5, 25–37.

Hoy, J.B. (1980) Effects of lindane, carbaryl and chlorpyrifos on non-target soil arthropod communities, in *Soil Biology as Related to Land Use Practices*, (ed. D.L. Dindal), EPA, Washington DC, pp. 71–81.

Hughes, A.M. (1976) *The Mites of Stored Food and Houses*, 2nd edn, Ministry of Agriculture, Fisheries and Food Technical Bulletin 9, HMSO, London.

Hughes, M.K., Lepp, N.W. and Phipps, D.A. (1980) Aerial heavy metal pollution and terrestrial ecosystems. *Advances in Ecological Research*, 11, 218–327.

Hughes, R.D. and Walker, J. (1970) The role of food in the population dynamics of the Australian Bush Fly *Musca vetustissima*, in *Animal Populations in Relation to their Food Supply*, (ed. A. Watson), Blackwell, Oxford, pp. 255–69.

Huhta, V., Ikonen, E. and Vilkamaa, P. (1979) Succession of invertebrate populations in artificial soil made of sewage sludge and crushed bark. *Annales Zoologica Fennici*, **16**, 223–70.

Hukkinen, Y. (1936) Tutkimuksia nurmipuntarpään (*Alopecurus pratensis* L.) siementutiolaisista. I. *Chirothrips hamatus* Tryb., puntarpääripsiänen. *Maatalouskoetoiminnan Julkaisuja*, **81**, 1–132.

Humphreys, W.F. (1979) Production and respiration in animal populations. *Journal of Animal Ecology*, **48**, 427–53.

Hunt, H.W., Coleman, D.C., Ingham, E.R. *et al.* (1987) The detrital food web in a shortgrass prairie. *Biology and Fertility of Soils*, **3**, 57–68.

Hunter, B.A. and Johnson, M.S. (1982) Food chain relationships of copper and cadmium in contaminated grassland ecosystems. *Oikos*, **38**, 108–17.

Hunter, B.A., Johnson, M.S. and Thompson, D.J. (1987) Ecotoxicology of copper and cadmium in a contaminated grassland ecosystem. II. invertebrates. *Journal of Applied Ecology*, **24**, 587–99.

Hunter, P.J. (1966) Distribution and abundance of slugs on an arable plot in Northumberland. *Journal of Animal Ecology*, **35**, 543–7.

Hunter, P.J. (1967) The effect of cultivations on slugs of arable ground. *Plant Pathology*, **16**, 153–6.

Hurd, L.E. and Wolf, L.L. (1974) Stability in relation to nutrient enrichment in arthropod consumers of old-field successional ecosystems. *Ecological Monographs*, **44**, 465–82.

Hurlbert, S.H. (1971) The nonconcept of species diversity: a critique and alternative parameters. *Ecology*, **54**, 577–86.

Hussey, N.W. (1984) Biological control – a commercial evaluation. *Proceedings of the 1984 British Crop Protection Conference – Pests and Diseases*, 379–86.

Hutchinson, K.J. and King, K.L. (1980) The effects of sheep stocking level on invertebrate abundance, biomass and energy utilization in a temperate, sown grassland. *Journal of Applied Ecology*, **17**, 369–87.

Hutchinson, K.J. and King, K.L. (1982) Invertebrates and nutrient cycling, in *Proceedings of the 3rd Australasian Conference on Grassland Invertebrate Ecology*, (ed. K.E. Lee), South Australian Government Printer, Adelaide, pp. 331–8.

Hutson, B.R. (1980) Colonization of industrial reclamation sites by Acari, Collembola and other invertebrates. *Journal of Applied Ecology*, **17**, 255–75.

Hutson, B.R. (1981) Age distribution and the annual reproductive cycle of some Collembola colonizing reclaimed land in Northumberland, England. *Pedobiologia*, **21**, 410–16.

Ibarra, E.L., Wallwork, J.A. and Rodriquez, J.C. (1965) Ecological studies of mites found in sheep and cattle pastures. I. Distribution patterns of oribatid mites. *Annals of the Entomological Society of America*, **58**, 153–9.

Imms, A.D. (1977) *A General Textbook of Entomology*, 10th edn, revised by O.W. Richards and R.G. Davies, Chapman & Hall, London.

Ingham, E.R., Trofymow, J.A., Ames, R.N. *et al.*, Trophic interactions and nitrogen cycling in a semi-arid grassland soil. II. System responses to removal of different groups of soil microbes or fauna. *Journal of Applied Ecology*, **23**, 615–30.

Ingham, R.E. (1985) Review of the effects of 12 selected biocides on target and non-target soil organisms. *Crop Protection*, **4**, 3–32.

Ingham, R.E., Trofymow, J.A., Ingham, E.R. and Coleman, D.C. (1985) Interactions of bacteria, fungi and their nematode grazers: effects on nutrient cycling and plant growth. *Ecological Monographs*, **55**, 119–40.

Ireland, M.P. (1983) Heavy metal uptake and tissue distribution in earthworms, in *Earthworm Ecology – From Darwin to Vermiculture*, (ed. J.E. Satchell), Chapman & Hall, London, pp. 247–65.

Jackson, T.A. and Trought, T.E.T. (1982) Progress with the use of nematodes and bacteria for the control of grass grub, in *Proceedings of the 35th New Zealand Weed and Pest Control Conference*, (ed. M.J. Hartley), pp. 103–6.

James, M. (1985) *Classification Algorithms*, Collins, London.

James, S.W. (1982) Effects of fire and soil type on earthworm populations in a tallgrass prairie. *Pedobiologia*, **24**, 37–40.

Janzen, D.H. (1973) Sweep samples of tropical foliage insects: description of study sites, with data on species abundances and size distributions. *Ecology*, **54**, 659–86.

Jepson, O.L.R., Keifer, H.H. and Baker, E.W. (1975) *Mites Injurious to Economic Plants*, University of California Press, Berkeley, California.

Johansson, E. (1946) Studier och försök rörande de på gräs och sädesslag levande tripsarnas biologi och skadegörelse. II. Tripsarnas frekvens och spridning i jämförelse med andra sugande insekters samt deras fröskadegörande betydelse. *Meddelanden från Växtskyddsanstalt*, **46**, 1–59.

Jones, C.G. (1983) Phytochemical variation, colonization and insect communities: the case of bracken fern (*Pteridium aquilinum*), in *Variable Plants and Herbivores in Natural and Managed Systems*, (eds R.F. Denno and M.S. McClure), Academic Press, New York, pp. 513–58.

Jones, F.G.W. and Jones, M.G. (1984) *Pests of Field Crops*, 2nd edn, Edward Arnold, London.

Jones, M.G. and Dean, G.J. (1975) Observations on cereal aphids and their natural enemies in 1972. *Entomologists' Monthly Magazine*, **111**, 69–78.

Jonkman, J.C.M. (1977) Biology and ecology of *Atta vollenweideri*, Forel 1893, and its impact in Paraquayan pastures. Thesis, Universiteitsbibliotheck, Leiden.

Jonkman, J.C.M. (1978) Nests of the leaf-cutting ant *Atta vollenweideri* as accelerators of succession in pastures. *Zeitschrift für Angewandte Entomologie*, **86**, 25–34.

Joosse, E.N.G. and Groen, J.B. (1970) Relationship between saturation deficit and survival and locomotory activity of surface dwelling Collembola. *Entomologia Expermentalis et Applicata*, **13**, 229–335.

Josens, G. (1972) Études biologiques et écologiques des termites (Isoptera) de la savane de Lamto-Pakobo (Côte d'Ivoire). Doctoral thesis, University of Brussels, Brussels.

Josens, G. (1973) Observations sur les bilans énergétiques dans deux populations de termites à Lamto (Côte d'Ivoire). *Annales de la Société Royale Zoologique de Belgique*, **103**, 169–76.

Kain, W.M. (1975) Population dynamics and pest assessment studies of grass grub (*Costelytra zealandica* (White) Melolonthinae) in the North Island of New Zealand. Ph.D. thesis, Lincoln College, University of Canterbury, New Zealand.

Kain, W.M. (1979) Pest management systems for control of pasture insects in New Zealand, in *Proceedings of the 2nd Australasian Conference on Grassland Invertebrate Ecology*, (eds T.K. Crosby and R.P. Pottinger), Government Printer, Wellington, pp. 172–9.

Kain, W.M. and Atkinson, D.S. (1970) A rational approach to grass grub control. *Proceedings of the 23rd New Zealand Weed and Pest Control Conference*, pp. 180–3.

Kain, W.M. and Atkinson, D.S. (1975) Problems of insect pest assessment. *New Zealand Entomologist*, **6**, 9–13.

Kain, W.M. and Burton, G. (1975) Effect of ground cover and pasture height on oviposition and establishment of soldier fly larvae. *Proceedings of the 28th New Zealand Weed and Pest Control Conference*, pp. 237–41.

Kain, W.M., East, R. and Douglas, J.A. (1979) *Costelytra zealandica* – pasture species relationships on the pumice soils of the central North Island of New Zealand (Coleoptera: Scarabaeidae), in *Proceedings of the 2nd Australasian Conference on Grassland Invertebrate Ecology*, (eds T.K. Crosby and R.P. Pottinger), Government Printer, Wellington, pp. 88–91.

Kajak, A. (1965) An analysis of food relations between the spiders *Araneus cornutus* Clerck and *Araneus quadratus* Clerck and their prey in a meadow. *Ekologia Polska*, **13**, 717–64.

Kajak, A. (1967) Productivity of some populations of web spiders, in *Secondary Productivity of Terrestrial Ecosystems*, Vol. 2, (ed K. Petrusewicz), Polish Scientific Publishers, Warsaw, pp. 807–20.

Kajak, A. (1971) Productivity investigation of two types of meadows in the Vistula Valley. IX. Production and consumption of field layer spiders. *Ekologia Polska*, **19**, 197–211.

Kajak, A. (1974) Analysis of a sheep pasture ecosystem in the Pieniny Mountains (the Carpathians). XVIII. Analysis of the transfer of carbon. *Ekologia Polska*, **22**, 711–32.

Kajak, A. (1975) Energy flow through a mountain pasture ecosystem, in *Progress in Soil Zoology*, (ed. J. Vanek), Junk, The Hague, pp. 95–102.

Kajak, A. (1980) Invertebrate predator subsystem, in *Grasslands, Systems Analysis and Man*, International Biological Programme 19, (eds A.J. Breymeyer and G.M. Van Dyne), Cambridge University Press, Cambridge, pp. 539–89.

Kajak, A., Breymeyer, A., Petal, J. and Olechowicz, E. (1972) The influence of ants on the meadow invertebrates. *Ekologia Polska*, **20**, 163–71.

Kajak, A., Olechowicz, E. and Petal, J. (1972) The influence of ants and spiders on the elimination of Diptera on meadows. *Proceedings of the XIII International Congress of Entomology*, Vol. 3, Nauka, Leningrad, pp. 364–6.

Kalmakoff, J. (1979) Enzootic virus control of *Wiseana* spp. (Lepidoptera: Hepialidae) in the pasture environment, in *Proceedings of the 2nd Australasian Conference on Grassland Invertebrate Ecology*, (eds T.K. Crosby and R.P. Pottinger), Government Printer, Wellington, pp. 202–4.

Kamm, J.A. (1972) Thrips that affect production of grass seed in Oregon. *Journal of Economic Entomology*, **65**, 1050–5.

Karieva, P. (1983) Influence of vegetation structure on herbivore populations: resource concentration and herbivore movement, in *Variable Plants and Herbivores in Natural and Managed Systems*, (eds R.F. Denno and M.S. McClure), Academic Press, New York, pp. 259–89.

Kauri, H., Holdung, T. and Solhöy, T. (1969) Turnbull and Nicholls' 'quick trap' for acquiring standing crop of evertebrates in high mountain grassland communities. 1. Report from the grazing project of the Norwegian IBP Committee. *Norsk Entomologisk Tidsskrift*, **16**, 133–6.

Kelsey, J.M. (1958) Damage in ryegrass by *Hyperodes griseus* Hust. *New Zealand Journal of Agricultural Research*, **1**, 790–5.

Keogh, R.G. (1979) Lumbricid earthworm activities and nutrient cycling in pasture ecosystems, in *Proceedings of the 2nd Australasian Conference on Grassland Invertebrate Ecology*, (eds T.K. Crosby and R.P. Pottinger), Government Printer, Wellington, pp. 49–51.

Keogh, R.G. and Whitehead, P.H. (1975) Observations on some effects of pasture spraying with benomyl and carbendazim on earthworm activity and litter removal from pasture. *New Zealand Journal of Experimental Agriculture*, **3**, 103–4.

Khalaf El-Duweini, A. and Ghabbour, S.I. (1965) Population density and biomass of earthworms in different types of Egyptian soils. *Journal of Applied Ecology*, **2**, 271–87.

Khalaf El-Duweini, A. and Ghabbour, S.I. (1971) Nitrogen contribution by live earthworms to the soil, in *IV Colloquium Pedobiologia*, (ed. J. d'Aquilar), Institut National des Recherches Agriculturelles, Publ. 71–7, Paris, pp. 495–501.

King, K.L. and Hutchinson, K.J. (1976) The effects of sheep stocking intensity on the abundance and distribution of mesofauna in pastures. *Journal of Applied Ecology*, **13**, 41–55.

King, P.D. and East, R. (1979) Effects of pasture composition on the dynamics of *Heteronychus arator and Graphagnathus leucoloma* populations (Coleoptera: Scarabaeidae and Curculionidae), in *Proceedings of the 2nd Australasian Conference on Grassland Invertebrate Ecology*, (eds T.K. Crosby and R.P. Pottinger), Government Printer, Wellington, pp. 79–82.

King, P.D. and Mercer, C.F. (1979) Effect of *Adelina* sp. (Protozoa: Coccidia) on *Heteronychus arator* fecundity and populations (Coleoptera: Scarabaeidae), in *Proceedings of the 2nd Australasian Conference on Grassland Invertebrate Ecology*, (eds T.K. Crosby and R.P. Pottinger), Government Printer, Wellington, pp. 200–2.

King, P.D., Mercer, C.F. and Meekings, J.S. (1981a) Ecology of black beetle *Heteronychus arator*. Influence of plant species on larval consumption, utilization and growth. *Entomologia Experimentalis et Applicata*, **29**, 109–16.

King, P.D., Mercer, C.F. and Meekings, J.S. (1981b) Ecology of black beetle *Heteronychus arator* (Coleoptera: Scarabaeidae) – population modelling. *New Zealand Journal of Agricultural Research*, **24**, 99–105.

King, P.D., Mercer, C.F., Stirling, J. and Meeking, J.S. (1975) Resistance of lucerne to black beetle. *Proceedings of the 28th New Zealand Weed and Pest Control Conference*, pp. 262–3.

Kirchner, T.B. (1977) The effects of resource enrichment on the diversity of plants and arthropods in a shortgrass prairie. *Ecology*, **58**, 1334–44.

Knülle, W. (1957) Die Verteilung der Acari, Oribatei im Boden. *Zeitschrift für Morphologie und Ökologie der Tiere*, **46**, 397–432.

Knutson, H. and Campbell, J.B. (1976) Relationships of grasshoppers (Acrididae) to burning, grazing and range sites of native tallgrass prairie in Kansas, in *Proceedings of the Tall Timber Conference on Animal Control by Habitat Management*, No. 6, 1974, Gainsville, Florida, pp. 107–20.

Kontkanen, P. (1950) Quantitative and seasonal studies on the leafhopper fauna of the field stratum on open areas in North Karelia. *Suomalaisen Eläinja Kasvitieteellisen Seuran Vanamon Kasvitieteellisia Julkaiswa*, **13**, 1–91.

Kontkanen, P. (1954) Studies on insect populations. 1. The number of generations of some leafhopper species in Finland and Germany. *Archivum Societatis Zoologicae Botanicae Fennicae*, **8**, 150–6.

Krantz, G.W. and Lindquist, E.E. (1979) Evolution of phytophagous mites (Acari). *Annual Review of Entomology*, **24**, 121–58.

Krischik, V.A. and Denno, R.F. (1983) Individual, population and geographic patterns in plant defense, in *Variable Plants and Herbivores in Natural and Managed Systems*, (eds R.F. Denno and M.S. McClure), Academic Press, New York, pp. 463–512.

Krull, W.H. (1939) Observations on the distribution and ecology of the oribatid mites. *Journal of the Washington Academy of Science*, **29**, 519–28.

Kubiena, W.L. (1955) Animal activity in soils as a decisive factor in establishment of humus forms, in *Soil Zoology*, (ed. D.K.McE. Kevan), Butterworths, London, pp. 73–82.

Kühnelt, W. (1961) *Soil Biology with Special Reference to the Animal Kingdom* (English edn), Faber & Faber, London.

Kuikman, P.J., Jansen, A.G., van Veen, J.A. and Zehnder, A.J.B. (1990) Protozoan predation and the turnover of soil organic carbon and nitrogen in the presence of plants. *Biology and Fertility of Soils*, **10**, 22–8.

Labrador, J.R., Martinez, I.J.Q. and Mora, A. (1972) *Acromyrmex landolti* Forel, Plaga del Pasto Guinea (*Panicum maximum*) en el Estado Zulia. *Jornadas Agronomicas, Universidad del Zulia, Venezuela*, **8**, 1–12.

Lacy, H. (1977) Putting new life in wormless soil. *New Zealand Farmer*, **98**, 20–2.

Lagerlöf, J., Andrén, O. and Paustian, K. (1989) Dynamics and contribution to carbon flows of Enchytraeidae (Oligochaeta) under four cropping systems. *Journal of Applied Ecology*, **26**, 183–9.

Lagerlöf, J. and Scheller, U. (1989) Abundance and activity of Pauropoda and Symphyla (Myriapoda) in four cropping systems. *Pedobiologia*, **33**, 315–21.

Laker, M.C., Hewitt, P.H., Nel, A. and Hunt, R.P. (1982) Effects of the termite *Trinervitermes trinervoides* Sjöstedt on the organic carbon and nitrogen contents and particle size distribution of soils. *Revue d'Écologie et de Biologie du Sol*, **19**, 27–39.

Lakhani, K.H. and Satchell, J.E. (1970) Production by *Lumbricus terrestris* (L.). *Journal of Animal Ecology*, **39**, 473–92.

Lal, R. (1987) *Tropical Ecology and Edaphology*, Wiley, New York.

Lal, R. (1988) Effects of macrofauna on soil properties in tropical ecosystems. *Agriculture, Ecosystems and Environment*, **24**, 101–116.

Lal, V.B. (1981) An Ecological study of Enchytraeidae (Oligochaeta) in tropical grassland soil, Varanasi, India. Ph.D. thesis, Banaras Hindu University.

Lamotte, M. (1947) Recherches écologiques sur le cycle saisonnier d'une savane guinéene. *Bulletin de la Société Zoologique de France, Paris*, **72**, 88–90.

Lamotte, M. (1975) The structure and function of a tropical savannah ecosystem, in *Tropical Ecological Systems: Analysis and Synthesis*, (eds F.B. Golley and E. Medina), Springer-Verlag, New York, pp. 179–222.

Lamotte, M. (1979) Structure and functioning of the savanna ecosystem of Lamto (Ivory Coast), in *Tropical Grazing Land Ecosystems*, Unesco, Paris, pp. 511–61.

Lamotte, M., Bourlière, F., Borbault, R. *et al.* (1979). Secondary production: consumption and decomposition, in *Tropical Grazing Land Ecosystems*, Unesco, Paris, pp. 146–206.

Latteur, G. (1973) Étude de la dynamique des populations des pucerons des céréales: Premières données relatives aux organismes aphidiphages en trois localités différentes. *Parasitica*, **29**, 134–51.

Lauren, D.R. (1979) Controlled release formulations for phenols: use as sex attractant lures for the grass grub beetle. *Environmental Entomology*, **8**, 914–16.

Lauren, D.R., Henzell, R.F. and du Toit, G.D.G. (1980) Insecticides for control of adult grass grub, *Costelytra zealandica* (White). 1. Chemical and biological field life of various synthetic pyrethroid formulations. *New Zealand Journal of Agricultural Research*, **23**, 111–15.

Lauren, D.R., Henzell, R.F., Smith, B.N. and Graham, D.P.F. (1979) Alternative formulations of synthetic pyrethroids for the control of adult grass grub. *Proceedings of the 32nd New Zealand Weed and Pest Control Conference*, pp. 101–6.

Laurence, B.R. (1954) The larval inhabitants of cow pats. *Journal of Animal Ecology*, **23**, 234–60.

Lavelle, P. (1973) Peuplement et production des vers de terre dans la savane de Lamto (Côte d'Ivoire). *Annales de l'Université d'Abidjan, série E*, **6**, 79–98.

Lavelle, P. (1974) Les vers de terre de la savane de Lamto, in *Analyse d'un Éco-*

système Tropicale Humide: la Savane de Lamto (Côte d'Ivoire). Les Organismes Endogés de la Savane de Lamto, (eds F. Athias, G. Josens, P. Lavelle and R. Schaefer), *Bulletin de Liaison des Chercheurs de Lamto (Paris), n° spécial*, **5**, 133–66.

Lavelle, P. (1979) Relations entre types écologiques et profils démographiques chez les vers de terre de la savane de Lamto (Côte d'Ivoire). *Revue d'Écologie et de Biologie du Sol*, **16**, 85–101.

Lavelle, P. (1988) Earthworm activities and the soil system. *Biology and Fertility of Soils*, **6**, 237–51.

Lavelle, P., Barois, I., Martin, A., Zaidi, Z. and Schaefer, R. (1989) Management of earthworm populations in agro-ecosystems: a possible way to maintain soil quality, in *Ecology of Arable Land*, (eds M. Clarholm and L. Bergström), Kluwer Academic Publishers, Dordrecht, pp. 109–22.

Lavelle, P. and Pashanasi, B. (1989) Soil macrofauna and land management in Peruvian Amazonia (Yurimaguas, Loreto). *Pedobiologia*, **33**, 283–91.

Lavelle, P., Schaefer, R. and Zadi, Z. (1989) Soil ingestion and growth in *Millsonia anomala*, a tropical earthworm, as influenced by the quality of the organic matter ingested. *Pedobiologia*, **33**, 379–88.

Lavelle, P., Sow, B. and Schaefer, R. (1980) The geophagous earthworm community in the Lamto savanna (Ivory Coast); niche partitioning and utilization of soil nutrient resources, in *Soil Biology as Related to Land Use Practices*, (ed. D.L. Dindal), EPA, Washington DC, pp. 653–72.

Laverack, M.S. (1961) Tactile and chemical perception in earthworms. II. Responses to acid pH solutions. *Comparative Biochemistry and Physiology*, **2**, 22–34.

Laverack, M.S. (1963) *The Physiology of Earthworms*, Pergamon Press, Oxford.

Lawton, J.H. (1982) Vacant niches and unsaturated communities: a comparison of bracken herbivores at sites on two continents. *Journal of Animal Ecology*, **51**, 573–95.

Lawton, J.H. (1983) Plant architecture and the diversity of phytophagous insects. *Annual Review of Entomology*, **28**, 23–9.

Lebrun, P. (1965) Contribution a l'étude écologique des oribates de la litière dans une forêt de moyenne-Belgique. *Mémoires de L'Institut Royale des Sciences Naturelle de Belgique*, **153**, 1–96.

Lee, K.E. (1979) The role of invertebrates in nutrient cycling and energy flow in grasslands, in *Proceedings of the 2nd Australasian Conference on Grassland Invertebrate Ecology*, (eds T.K. Crosby and R.P. Pottinger), Government Printer, Wellington, pp. 26–9.

Lee, K.E. (1983a) The influence of earthworms and termites on soil nitrogen cycling, in *New Trends in Soil Biology*, (eds Ph. Lebrun, H.M. André, A. de Meds *et al.*), Imprimeur Dieu-Brichart, Ottignies-Louvain-la-Neuve, pp. 35–48.

Lee, K.E. (1983b) Earthworms of tropical regions – some aspects of their ecology and relationships with soils, in *Earthworm Ecology – From Darwin to Vermiculture*, (ed. J.E. Satchell), Chapman & Hall, London, pp. 179–93.

Lee, K.E. (1985) *Earthworms: Their Ecology and Relationships with Soils and Land Use*, Academic Press, Sydney.

Lee, K.E. and Wood, T.G. (1971a) Physical and chemical effects on soils of some Australian termites and their pedological significance. *Pedobiologia*, **11**, 376–409.

Lee, K.E. and Wood, T.G. (1971b) *Termites and Soils*, Academic Press, London.

Lepage, M. (1972) Recherches écologiques sur une savane sahélienne du Ferlo septentrional, Sénégal: données preliminaires sur l'écologie des termites. *La Terre et la Vie*, **26**, 383–409.

Lepage, M. (1974) Les termites d'une savane sahélienne (Ferlo septentrional, Sénégal): peuplement, populations, consommation, rôle dans l'écosystème. Doctoral Thesis, University of Dijon, Dijon.

Lewis, G.C. and Clements, R.O. (1986) A survey of ryegrass endophyte (*Acremonium loliae*) in the UK and its apparent ineffectuality on a seedling pest. *Journal of Agricultural Science, Cambridge*, **107**, 633–8.

Lewis, G.C., Cook, R. and van der Ende, A. (1991) Effect of agrochemical applied at sowing on seedling emergence and herbage yield of perennial ryegrass and white clover. *Grass and Forage Science*, **46**, 121–9.

Lewis, G.C. and Thomas, B.J. (1991) Incidence and severity of pest and disease damage to white clover foliage at 16 sites in England and Wales. *Annals of Applied Biology*, **118**, 1–8.

Lewis, J.K. (1971) The grassland biome: a synthesis of structure and function, 1970, in *Preliminary Analysis of Structure and Function in Grasslands*, (ed. N.J. French), Colorado State University, Fort Collins, Colorado, pp. 317–87.

Lewis, T. (1969) The diversity of the insect fauna in a hedgerow and neighbouring fields. *Journal of Applied Ecology*, **6**, 453–8.

Lewis, T. (1973) *Thrips: Their Biology, Ecology and Economic Importance*, Academic Press, London.

Lewis, T. and Dibley, G.C. (1970) Air movement near windbreaks and a hypothesis of the mechanism of the accumulation of airborne insects. *Annals of Applied Biology*, **66**, 477–84.

Lienhard, C. (1980) Zurkenntnis der Collembolen eines alpinen Caricetum firmae im Schweizerischen Nationalpark. *Pedobiologia*, **20**, 369–86.

Lindemann, R.L. (1942) The trophic-dynamic aspect of ecology. *Ecology*, **23**, 399–418.

Lindsten, K. and Gerhardson, B. (1969) Investigations on barley yellow dwarf virus (BYDV) in leys in Sweden. *Meddelanden från Statens Växtskyddsanstalt*, **14**, 261–80.

Lloyd, J.E., Kumar, R., Grow, R.R., Leetham, J.W. and Keith, V. (1973) *Abundance and biomass of soil macroinvertebrates of the Pawnee site collected from pasture subjected to different grazing pressures, irrigation and/or nitrogen fertilization 1970–71*, US/IBP Grassland Biome Technical Report, No. 239, Colorado State University, Fort Collins, Colorado.

Lofs-Holmin, A. (1981) Influence in field experiments of benomyl and carbendazim on earthworms (Lumbricidae) in relation to soil texture. *Swedish Journal of Agricultural Research*, **11**, 141–7.

Lofs-Holmin, A. (1983a) Earthworm population dynamics in different agricultural rotations, in *Earthworm Ecology – From Darwin to Vermiculture*, (ed. J.E. Satchell), Chapman & Hall, London, pp. 151–60.

Lofs-Holmin, A. (1983b) Reproduction and growth of common arable land and pasture species of earthworms (Lumbricidae) in laboratory cultures. *Swedish Journal of Agricultural Research*, **13**, 31–7.

Loots, G.C. and Ryke, P.A.J. (1966) A comparative, quantitative study of the micro-arthropods in different types of pasture soil. *Zoologica Africana*, **2**, 167–92.

Loquet, M. (1978) The study of respiratory and enzymatic activities of earthworm-made pedological structures in a grassland soil at Citeaux, France. *Scientific Proceedings of the Royal Dublin Society*, **6A**, 207–14.

Loquet, M., Bhatnagar, T., Bouché, M.B. and Rouelle, J. (1977) Essai d'estimation de l'influence écologique des lombriciens sur les microorganismes. *Pedobiologia*, **17**, 400–17.

Lozek, V. (1962) Soil conditions and their influence on terrestrial Gasteropoda in central Europe, in *Progress in Soil Zology*, (ed. P.W. Murphy), Butterworths, London, pp. 334–42.

Luff, M.L. (1966) The abundance and diversity of the beetle fauna of grass tussocks. *Journal of Animal Ecology*, **35**, 189–208.

Luff, M.L. (1987) Biology of polyphagous ground beetles in agriculture. *Agricultural Zoology Reviews*, **2**, 237–78.

Lutman, J. (1977) The role of slugs in an *Agrostis–Festuca* grassland, in *Production Ecology of British Moors and Montane Grassland*, (eds O.W. Heal and D.F. Perkins), Springer Verlag, Berlin, pp. 332–47.

Luxton, M. (1972) Studies on the oribatid mites of a Danish beechwood soil. 1. Nutritional biology. *Pedobiologia*, **12**, 434–63.

Luxton, M. (1981) Studies on the oribatid mites of a Danish beech wood soil. IV. Developmental biology. *Pedobiologia*, **21**, 312–40.

Luxton, M. (1982) Studies on the invertebrate fauna of New Zealand peat soils. IV. Pasture soils on Rukuhia peat. *Pedobiologia*, **24**, 297–308.

Luxton, M. (1983) Studies on the invertebrate fauna of New Zealand peat soils. V. Pasture soils on Kaipaki peat. *Pedobiologia*, **25**, 135–48.

Ma, Wei-chun (1988) Toxicity of copper to lumbricid earthworms in sandy agricultural soils amended with Cu-enriched organic waste materials. *Ecological Bulletins* (Copenhagen), **39**, 53–6.

MacArthur, R.H. and Wilson, E.O. (1963) An equilibrium theory of insular zoo-geography. *Evolution*, **17**, 373–87.

McBrayer, J.F. and Reichle, D.E. (1971) Trophic structure and feeding rates of forest soil invertebrate populations. *Oikos*, **22**, 381–8.

McBrayer, J.F., Reichle, D.E. and Witkamp, M. (1974) *Energy Flow and Nutrient Cycling in a Cryptozoan Food Web*. EDFB-IBP-73-8, Oakridge National Laboratory, Tennessee.

MacDonald, D.W. (1983) Predation on earthworms by terrestrial vertebrates, in *Earthworm Ecology – from Darwin to Vermiculture*, (ed. J.E. Satchell), Chapman & Hall, London, pp. 393–414.

Macfadyen, A. (1952) The small arthropods of a *Molinia* fen at Cothill. *Journal of Animal Ecology*, **21**, 87–117.

Macfadyen, A. (1963) The contribution of the microfauna to total soil metabolism, in *Soil Organisms*, (eds J. Doeksen and J. van der Drift), North Holland, Amsterdam, pp. 1–17.

Macfadyen, A. (1964) Relations between mites and microorganisms and their significance in soil biology. *Acarologia*, **6**, 147–49.

MacGregor, A.N. and Naylor, L.M. (1982) Effect of municipal sludge on the respiratory activity of a cropland soil. *Plant and Soil*, **65**, 149–52.

Mackay, A.D., Syers, J.K., Springett, J.A. and Gregg, P.E.F. (1982) Plant availability of phosphorus in superphosphate and a phosphate rock as influenced by earthworms. *Soil Biology and Biochemistry*, **14**, 281–7.

McLaren, G.F. and Crump, D.K. (1969) An economic study of the use of insecticides for the control of porina caterpillar. *Proceedings of the New Zealand Weed and Pest Control Conference*, **22**, 307–22.

MacLean, S.F. Jr (1973) Life cycle and growth energetics of the arctic crane fly *Pedicia hannai antenatta*. *Oikos*, **24**, 436–43.

McMillan, J.H. (1969) The ecology of the acarine and collembolan fauna of two New Zealand pastures. *Pedobiologia*, **9**, 372–404.

McNaughton, S.J. (1976). Serengeti migratory wildebeest: facilitation of energy flow by grazing. *Science* (Washington), **191**, 92–4.

McNaughton, S.J. (1979) Grazing as an optimization process: grass–ungulate relationships in the Serengeti. *American Naturalist*, **113**, 691–703.

McNeill, S. (1971) The energetics of a population of *Leptoterna dolabrata* (Heteroptera: Miridae). *Journal of Animal Ecology*, 40, 127–40.

McNeill, S. (1973) The dynamics of a population of *Leptoterna dolabrata* (Heteroptera: Miridae) in relation to its food resources. *Journal of Animal Ecology*, 42, 495–507.

McNeill, S. and Lawton, J.H. (1970) Annual production and respiration in animal populations. *Nature* (London), 225, 472–4.

McNeill, S. and Southwood, T.R.E. (1978) The role of nitrogen in the development of insect/plant relationships, in *Biochemical Aspects of Plant and Animal Coevolution*, (ed. J.B. Harborne), Academic Press, New York, pp. 77–98.

McWilliams, C.B. (ed.) (1991) *Climate Change: Studies on the Implications for Ireland*. Government Stationery Office, Dublin.

Madge, D.S. (1964a) The water relations of *Belba geniculosa* Oudms. and other species of oribatid mites. *Acarologia*, 6 199–223.

Madge, D.S. (1964b) The humidity reactions of oribatid mites. *Acarologia*, 6, 566–91.

Madge, D.S. (1964c) The longevity of fasting oribatid mites. *Acarologia*, 6, 718–29.

Madsen, M., Nielsen, B. Overgaard, Holter, P. *et al.* (1990) Treating cattle with ivermectin: effects on the fauna and decomposition of dung pats. *Journal of Applied Ecology*, 27, 1–15.

Majer, J.D. (ed.) (1989) *Animals in Primary Succession – the Role of Fauna in Reclaimed Lands*, Cambridge University Press, Cambridge.

Malecki, M.R., Neuhauser, E.F. and Loehr, R. (1982) The effect of metals on the growth and reproduction of *Eisenia foetida* (Oligochaeta, Lumbricidae). *Pedobiologia*, 24, 129–37.

Malone, C. (1969) Effects of diazinon contamination on an old-field ecosystem. *American Midland Naturalist*, 82, 1–27.

Manly, V.F.J. (1977) The determination of key factors from life-table data. *Oecologia*, 31, 111–17.

Manly, V.F.J. (1979) A note on key factor analysis. *Researches on Population Ecology*, 21, 30–9.

Mansell, G.P., Syers, J.K. and Gregg, P.E.H. (1981) Plant availability of phosphorous in dead herbage ingested by surface-casting earthworms. *Soil Biology and Biochemistry*, 13, 163–7.

Marcuzzi, G. (1970) Experimental observations on the role of *Glomeris* spp. (Myriapoda; Diplopoda) in the process of humification of litter. *Pedobiologia*, 10, 401–6.

Marshall, V.G. (1977) *Effects of Manures and Fertilizers on Soil Fauna: A Review*, Commonwealth Bureau of Soils Special Publication, No. 3, Commonwealth Agricultural Bureau, London.

Martin, N.A. (1975) Effect of four insecticides on the pasture ecosystem. II. The fauna collected in pit-traps. *New Zealand Journal of Agricultural Research*, 18, 179–82.

Martin, N.A. (1978) Earthworms in New Zealand Agriculture. *Proceedings of the 31st Weed and Pest Control Conference*, 176–80.

Martin, N.A. (1982) The interaction between organic matter in soil and the burrowing activity of three species of earthworms (Oligochaeta: Lumbricidae). *Pedobiologia*, 24, 185–90.

Martin, N.A. and Charles, J.C (1979) Lumbricid earthworms and cattle dung in New Zealand pastures, in *Proceedings of the 2nd Australasian Conference on Grassland Invertebrate Ecology*, (eds T.K. Crosby and R.P. Pottinger), Government Printer, Wellington, pp. 52–4.

Mason, C.F. (1974) Mollusca, in *Biology of Plant Litter Decomposition*, (eds C.H. Dickinson and G.J.F. Pugh), Academic Press, London, pp. 555–91.

Mathys, G. and Tencalla, Y. (1959) Note préliminaire sur la biologie et la valeur prédatrice de *Proctolaelaps hypudaei* Oudms. *Annuaire Agricole de la Suisse*, **60**, 645–54.

Mattson, W.J. (1980) Herbivory in relation to plant nitrogen content. *Annual Review of Ecology and Systematics*, **11**, 119–61.

Mattson, W.J. and Addy, N.D. (1975) Phytophagous insects as regulators of forest primary production. *Science*, **190**, 515–22.

May, R.M. (1973) *Stability and Complexity in Model Ecosystems*, Monograph in Population Biology, No. 6, Princeton University Press, Princeton.

May, R.M. (1975) Patterns of species abundance and diversity, in *Ecology and Evolution of Communities*, (eds M.L. Cody and J.M. Diamond), Belknap, Cambridge, Mass., pp. 81–120.

May, R.M. (1976) *Theoretical Ecology*, Blackwell, Oxford.

Mayer, H. (1957) Zur Biologie und Ethologie einheimischer Collembolen. *Zoologische Jahrbücher Abteilung für Systematik, Ökologie und Geographie der Tiere*, **85**, 501–70.

Meijer, J. (1972) An isolated earthworm population in the recently reclaimed Lauwerszeepolder. *Pedobiologia*, **12**, 409–11.

Meinhardt, U. (1973) Vergleichende Beobachtungen zur Laboratoriumsbiologie einheimischer Regenwurmarten. 1. Haltung und Zucht. *Zeitschrift für Angewandte Zoologie*, **60**, 233–55.

Meinhardt, U. (1974) Vergleichende Beobachtungen zur Laboratoriumsbiologie einheimischer Regenwurmarten. II. Biologie der gezüchteten Arten. *Zeitschrift für Angewandte Zoologie*, **61**, 137–82.

Menzie, C.M. (1972) Fate of pesticides in the environment. *Annual Review of Entomology*, **17**, 199–222.

Metcalf, C.L., Flint, W.P. and Metcalf, R.L. (1962) *Destructive and Useful Insects: Their Habits and Control*, 4th edn, McGraw Hill, New York.

Metcalf, R.L. (1975) Insecticides in pest management, in *Introduction to Integrated Pest Management*, (eds R.L. Metcalf and W.H. Luckmann), Wiley, New York, pp. 235–74.

Meyer, J.A. (1960) Résultats agronomiques d'un essai de nivellement des termitières réalisé dans la Cuvette centrale Congolaise. *Bulletin Agricole du Congo Belge*, **51**, 1047–59.

Michael, A.D. (1884–88) *British Oribatidae*, 2 vols, Ray Society, London.

Miln, A.J. (1979) The relationships between density, disease, and mortality in *Costelytra zealandica* populations (Coleoptera: Scarabaeidae), in *Proceedings of the 2nd Australasian Conference on Grassland Invertebrate Ecology*, (eds T.K. Crosby and R.P. Pottinger), Government Printer, Wellington, pp. 75–9.

Milne, A. (1984) Fluctuation and natural control of animal population, as exemplified in the garden chafer *Phyllopertha horticola* (L.). *Proceedings of the Royal Society of Edinburgh*, **82B**, 145–99.

Mitchell, B. (1963) Ecology of two carabid beetles, *Bembidion lampros* (Herbst.) and *Trechus quadristriatus* (Schrank). 1. Life cycles and feeding behaviour. *Journal of Animal Ecology*, **32**, 289–99.

Mitchell, J.E. and Pfadt, R.E. (1974) A role of grasshoppers in a shortgrass prairie ecosystem. *Environmental Entomology*, **3**, 358–60.

Mitchell, M.J., Hartenstein, R., Swift, B.L., Neuhauser, E.F. *et al* (1978) Effects of different sewage sludges on some chemical and biological characteristics of soil. *Journal of Environmental Quality*, **7**, 551–9.

Mohr, C.R. (1943) Cattle droppings as ecological units. *Ecological Monographs*, **13**, 276–97.

Möller, F. (1969) Ökologische Untersuchungen an terricolen Enchytraeiden-populationen. *Pedobiologia*, **9**, 114–19.

Momen, F.M. (1986) Mites associated with unsprayed orchards with particular reference to trophic relationships and feeding mechanisms. Ph.D. thesis, National University of Ireland, Dublin.

Moore, C.W.E. (1964) Distribution of grasslands, in *Grasses and Grasslands*, (ed. C. Barnard), Macmillan, London, pp. 182–205.

Moore, D. (1983) Hymenopterous parasitoids of stem-boring Diptera (e.g. *Oscinella frit* (L.)) in perennial ryegrass (*Lolium perenne*) in Britain. *Bulletin of Entomological Research*, **73**, 601–8.

Moore, D. and Clements, R.O. (1984) Stem boring Diptera in perennial ryegrass in relation to fertilizer. 1. Nitrogen level and form. *Annals of Applied Biology*, **105**, 1–6.

Moore, D., Clements, R.O. and Ridout, M.S. (1986) Effects of pasture establishment and renovation techniques on the hymenopterous parasitoids of *Oscinella frit* L. and other stem-boring Diptera in ryegrass. *Journal of Applied Ecology*, **23**, 871–81.

Morgan, C.V.G., Anderson, N.H. and Swales, J.E. (1958) Influence of some fungicides on orchard mites in British Columbia. *Canadian Journal of Plant Science*, **38**, 94–105.

Morris, M.G. (1967) Differences between the invertebrate faunas of grazed and ungrazed chalk grassland. 1. Responses of some phytophagous insects to cessation of grazing. *Journal of Applied Ecology*, **4**, 459–74.

Morris, M.G. (1968) Differences between the invertebrate faunas of grazed and ungrazed chalk grassland. II. The faunas of sample turves. *Journal of Applied Ecology*, **5**, 601–11.

Morris, M.G. (1969) Differences between the invertebrate faunas of grazed and ungrazed chalk grassland. III. The heteropterous fauna. *Journal of Applied Ecology*, **6**, 475–87.

Morris, M.G. (1971a) The management of grassland for the conservation of invertebrate animals, in *The Scientific Management of Animal and Plant Communities for Conservation*, (eds E. Duffey and A.S. Watt), Blackwell, Oxford, pp. 527–52.

Morris, M.G. (1971b) Differences between the invertebrate faunas of grazed and ungrazed chalk grassland. IV. Abundance and diversity of Homoptera-Auchenorrhyncha. *Journal of Applied Ecology*, **8**, 37–52.

Morris, M.G. (1973) The effects of seasonal grazing on the Heteroptera and Auchenorrhyncha (Hemiptera) of chalk grassland. *Journal of Applied Ecology*, **10**, 761–80.

Morris, M.G. (1979) Responses of grassland invertebrates to management by cutting. II. Heteroptera. *Journal of Applied Ecology*, **16**, 417–32.

Morris, M.G. (1981a) Responses of grassland invertebrates to management by cutting. III. Adverse effects on Auchenorrhyncha. *Journal of Applied Ecology*, **18**, 107–23.

Morris, M.G. (1981b) Responses of grassland invertebrates to management by cutting. IV. Positive responses to Auchenorrhyncha. *Journal of Applied Ecology*, **18**, 763–72.

Morris, M.G. and Lakhani, K.H. (1979) Responses of grassland invertebrates to management by cutting. I. Species diversity of Hemiptera. *Journal of Applied Ecology*, **16**, 77–98.

Morris, M.G. and Plant, R. (1983) Responses of grassland invertebrates to management by cutting. V. Changes in Hemiptera following cessation of management. *Journal of Applied Ecology,* 20, 157–77.

Morris, M.G. and Rispin, W.E. (1987) Abundance and diversity of the coleopterous fauna of a calcareous grassland under different cutting regimes. *Journal of Applied Ecology,* 24, 451–65.

Moursi, A. (1962) The lethal doses of CO_2, N_2 and H_2S for soil arthropods. *Pedobiologia,* 2, 9–14.

Mowat, D.J. (1974) Factors affecting the abundance of shoot-flies (Diptera) in grassland. *Journal of Applied Ecology,* 11, 951–62.

Mowat, D.J. and Jess, S. (1985) A comparison of insecticide treatments for the control of frit fly in seedling ryegrass. *Grass and Forage Science,* 40, 251–6.

Mowat, D.J. and Shakeel, M.A. (1988) The effect of pesticide application on the establishment of white clover in a newly-sown ryegrass/white clover sward. *Grass and Forage Science,* 43, 371–5.

Mueller, B.R., Beare, M.H. and Crossley, D.A. Jr (1990) Soil mites in detrital food webs of conventional and no-tillage agroecosystems. *Pedobiologia,* 34, 389–401.

Mulkern, G.B. (1967) Food selection by grasshoppers. *Annual Review of Entomology,* 12, 59–78.

Mulkern, G.B. (1972) The effects of preferred food plants on distribution and numbers of grasshoppers, in *Proceedings of the International Conference on Current and Future Problems in Acridology,* London, pp. 215–18.

Müller, H.T. (1978) Strukturanalyse der Zikadenfauna (Homoptera, Auchenorrhyncha) einer Rasenkatena Thüringens (Leutratal bei Jena). *Zoologische Jahrbücher Abteilung (für Systematik Ökologie und Geographie der Tiere,* 105, 258–334.

Müller, P.E. (1878) Studier over Skovjord, som Bidrag til Skovdyrkningens Theori. I. Om Bøgemuld, og Bøgemor paa Sand og Ler. *Tidsskrift for Skogbruk,* 3, 1–124.

Müller, P.E. (1884) Studier over Skovjord, som Bidrag til Skovdyrkningens Theori. II. Om Muld og Mori Egeskove og paa Heder. *Tidsskrift for Skogbruk,* 7, 1–232.

Mulligan, T.E. (1960) The transmission by mites, host range and properties of ryegrass mosaic virus. *Annals of Applied Biology,* 48, 575–9.

Munnelly, P.J. (1970) Survey of slug species and populations in farms and gardens. *Journal of the Irish Department of Agriculture,* 67, 1–8.

Murbach, R., Keller, Ch. and Bourqu, P. (1952) Eche au ver blanc. *Revue Romande d'Agriculture de Viticulture et d'Arboriculture,* 8, 19–21.

Murdoch, W.W. (1976) Diversity, complexity, stability and pest control. *Journal of Applied Ecology,* 13, 795–807.

Murdoch, W.W., Evans, F.C. and Peterson, C.H. (1972) Diversity and pattern in plants and insects. *Ecology,* 53, 819–29.

Murton, R.K. and Westwood, N.J. (1977) *Avian Breeding Cycles,* Clarendon Press, Oxford.

Nabialczyk-Karg, J. (1980) Density and biomass of soil inhabiting insect larvae in a rape field and in a meadow. *Polish Ecological Studies,* 6, 305–16.

Nakamura, M. (1965) Bio-economics of some larval populations of pleurostict Scarabaeidae on the flood plain of the river Tamagawa. *Japanese Journal of Ecology,* 15, 1–18.

Needham, A.E. (1957) Components of nitrogenous excreta in the earthworms *Lumbricus terrestris* L. and *Eisenia foetida* (Savigny). *Experimental Biology,* 34, 425–45.

Nel, J.J.C. and Hewitt, P.H. (1969) A study of the food eaten by a field population of the harvester termite, *Hodotermes mossambicus* (Hagen) and its relation to population density. *Journal of the Entomological Society of Southern Africa*, **32**, 123–31.

Nel, J.J.C., Hewitt, P.H. and Joubert, L. (1970) The collection and utilization of redgrass, *Themeda triandra* (Forsk.) by laboratory colonies of the harvester termite, *Hodotermes mossambicus* (Hagen) and its relation to population density. *Journal of the Entomological Society of Southern Africa*, **33**, 331–40.

Nerney, N.J. (1958) Grasshopper infestations in relation to range conditions. *Journal of Range Management*, **11**, 247.

Nerney, N.J. (1960) Grasshopper damage on short-grass rangeland of the San Carlos Apache Indian Reservation, Arizona. *Journal of Economic Entomology*, **53**, 640–6.

Neuhauser, E.F., Malecki, M.R. and Loehr, R.C. (1984) Growth and reproduction of the earthworm *Eisenia fetida* after exposure to sublethal concentrations of metals. *Pedobiologia*, **27**, 89–97.

Neuhauser, E.F., Loehr, R.C., Milligan, D.L. and Malecki, M.R. (1985) Toxicity of metals to the earthworm *Eisenia fetida*. *Biology and Fertility of Soils*, **1**, 149–57.

Newell, P.F. (1967) Mollusca, in *Soil Biology*, (eds A. Burges and F. Raw), Academic Press, London, pp. 413–33.

Newton, A.F. Jr (1984) Mycophagy in Staphylinoidea (Coleoptera), in *Fungus–Insect Relationships: Perspectives in Ecology and Evolution*, (eds Q. Wheeler and M. Blackwell), Columbia University Press, New York, pp. 302–53.

Nicholson, A.J. (1957) The self adjustment of populations to change. *Cold Spring Harbor Symposium of Quantitative Biology*, **22**, 153–72.

Nicholson, A.J. (1958) Dynamics of insect populations. *Annual Review of Entomology*, **3**, 107–36.

Nielsen, C.O. (1949) Studies on the soil microfauna. II. The soil inhabiting nematodes. *Natura Jutlandica*, **2**, 1–131.

Nielsen, C.O. (1955a) Studies on Enchytraeidae. 2. Field Studies. *Natura Jutlandica*, **4**, 1–58.

Nielsen, C.O. (1955b) Studies on Enchytraeidae. 5. Factors causing seasonal fluctuations in numbers. *Oikos*, **6**, 153–69.

Nielsen, C.O. (1961) Respiratory metabolism of some populations of enchytraeid worms and free-living nematodes. *Oikos*, **12**, 17–35.

Nielsen, C.O. (1962) Carbohydrases in soil and litter invertebrates. *Oikos*, **13**, 200–15.

Nielsen, C.O. (1967) Nematoda, in *soil Biology*, (eds A. Burges and F. Raw), Academic Press, London, pp. 197–211.

Nielsen, G.A. and Hole, F.D. (1963) A study of natural processes of incorporation of organic matter into soil in the University of Wisconsin Arboretum. *Wisconsin Academy of Science, Arts and Letters Transactions*, **52**, 213–28.

Nielsen, M.G. (1972) An attempt to estimate energy flow through a population of workers of *Lasius alienus* (Först.) (Hymenoptera: Formicidae). *Natura Jutlandica*, **16**, 97–107.

Nielsen, M.G. and Jensen, T.F. (1975) Okologiske studier over *Lasius alienus* (Först.) (Hymenoptera, Formicidae). *Entomologiske Meddelelser*, **43**, 5–16.

Nielson, R.L. (1951) Earthworms and soil fertility. *Proceedings of the New Zealand Grassland Association*, **13**, 158–67.

Nielson, R.L. (1965) Presence of plant growth substances in earthworms demonstrated by paper chromatography and the Went Pea test. *Nature* (London), **208**, 1113.

382 References

Nielson, M.W., Lehman, W.F. and Kodet, R.T. (1976) Resistance in alfalfa to *Acyrthosiphon kondoi*. *Journal of Economic Entomology*, **69**, 471–2.

Noble, J.C., Gordon, W.T. and Kleinig, C.R. (1970) The influence of earthworms on the development of mats of organic matter under irrigated pastures in southern Australia. *Proceedings of the XIth International Grassland Congress*, 465–8.

Nordström, S. (1975) Seasonal activity of lumbricids in southern Sweden. *Oikos*, **26**, 307–15.

Nordström, S. and Rundgren, S. (1973) Associations of lumbricids in southern Sweden. *Pedobiologia*, **13**, 301–26.

Nordström, S. and Rundgren, S. (1974) Environmental factors and lumbricid associations in southern Sweden. *Pedobiologia*, **14**, 1–27.

Norman, T.T., Kemp, A.W. and Tayler, J.E. (1957) Winter temperatures in long and short grass. *Meteorological Magazine*, **86**, 148–52.

Nowak, E. (1975) Population density of earthworms and some elements of their production in several grassland environments. *Ekologia Polska*, **23**, 459–91.

Nowak, E. (1976) The effect of fertilization on earthworms and other soil macrofauna. *Polish Ecological Studies*, **2**, 195–207.

Noy-Meir, I. (1975) Stability of grazing systems: an application of predator-prey graphs. *Journal of Ecology*, **63**, 459–81.

Noy-Meir, I. (1976) Rotational grazing in a continuously growing pasture: a simple model. *Agricultural Systems*, **1**, 87–112.

Nye, I.W.B. (1959) The distribution of shoot-fly larvae (Diptera, Acalypterae) within pasture grasses and cereals in England. *Bulletin of Entomological Research*, **50**, 53–62.

O'Connor, F.B. (1967) The Enchytraeidae, in *Soil Biology*, (eds A. Burges and F. Raw), Academic Press, London, pp. 213–57.

O'Connor, R.J. and Shrubb, M. (1987) *Farming and Birds*, Cambridge University Press, Cambridge.

Odum, E.P. (1969) The strategy of ecosystem development. *Science*, **164**, 262–70.

Odum, E.P. (1975) Diversity as a function of energy flow, in *Unifying Concepts in Ecology*, (eds W.H. van Dobben and R.H. Lowe-McConnell), Junk, The Hague, pp. 11–14.

Ohiagu, C.E. (1979) Nest and soil populations of *Trinervitermes* spp. with particular reference to *T. geminatus* (Wasmann) (Isoptera), in Southern Guinea Savanna near Mokwa, Nigeria. *Oecologia* (Berlin), **40**, 167–78.

Ohiagu, C.E. and Wood, T.G. (1976) A method for measuring rate of grass harvesting by *Trinervitermes geminatus* (Wasmann) (Isoptera, Nasutitermitinae) and observations on its foraging behaviour in Southern Guinea Savanna, Nigeria. *Journal of Applied Ecology*, **13**, 703–13.

Olechowicz, E. (1971) Productivity investigation of two types of meadows in the Vistula Valley. VIII. The number of emerged Diptera and their elimination. *Ekologia Polska*, **19**, 183–95.

Olechowicz, E. (1974) Analysis of sheep pasture ecosystem in the Pieniny Mountains (the Carpathians). X. Sheep dung and fauna colonizing it. *Ekologia Polska*, **22**, 589–616.

Olechowicz, E. (1976a) The effect of mineral fertilization on insect community of the herbage in a meadow. *Polish Ecological Studies*, **2**, 129–36.

Olechowicz, E. (1976b) The role of coprophagous dipterans in a mountain pasture ecosystem. *Ekologia Polska*, **24**, 125–65.

Olive, P.J.W. and Clark, R.B. (1978) Physiology of reproduction, in *Physiology of Annelids*, (ed. P.J. Mill), Academic Press, London, pp. 271–368.

O'Neill, N. (1981) Studies on the epigeal arthropod fauna of grassland at Grange, Co. Meath. M.Agr.Sc. thesis, National University of Ireland, Dublin.

Ossiannilsson, F. (1966) Insects in the epidemiology of plant viruses. *Annual Review of Entomology*, **11**, 213–32.

Østbye, E., Berg, A., Blehr, O., Espeland, M. *et al.* (1975) Hardangervidda, Norway, in *Structure and Function of Tundra Ecosystems*, (eds T. Rosswall and O.W. Heal). *Ecological Bulletins* (Stockholm), **20**, 225–64.

Owen, D.F. and Wiegert, R.G. (1976) Do consumers maximize plant fitness? *Oikos*, **27**, 488–92.

Pallant, D. (1972) The food of the grey field slug, *Agriolimax reticulatus* (Müller), in grassland. *Journal of Animal Ecology*, **41**, 761–9.

Palm, T. (1972) Kortvingar, fam. Staphylinidae, underfam. Aleocharinae (*Aleuronota–Tinotus*) H.7, in *Svensk Insektsfauna, Rekv. nr. 53, Entomologiska Foreningen, Stockholm.* Almqvist & Wiksells, Uppsala, pp. 297–467.

Palmgren, P. (1972) Studies on the spider populations of the surroundings of the Tvärminne Zoological Station, Finland. *Commentationes Biologicae, Societas Scientiarum Fennica, Helsinki*, **52**, 134.

Paris, O.H. (1963) The ecology of *Armadillidium vulgare* (Isopoda, Oniscoidea) in California grassland: food, enemies and weather. *Ecological Monographs*, **33**, 1–22.

Paris, O.H. and Pitelka, F.A. (1962) Population characteristics of the terrestrial isopod *Armadillidium vulgare* in California grassland. *Ecology*, **43**, 229–48.

Parker, F.D., Batra, S.W.T. and Tepedino, V.J. (1987) New pollinators for our crops. *Agricultural Zoology Reviews*, **2**, 279–304.

Parmelee, R.W., Beare, M.H., Cheng, W. *et al.* (1990) Earthworms and enchytraeids in conventional and no-tillage agroecosystems: a biocide approach to assess their role in organic matter breakdown. *Biology and Fertility of Soils*, **10**, 1–10.

Parr, T.W. (1978) An analysis of soil micro-arthropod succession. *Scientific Proceedings of the Royal Dublin Society*, **6A**, 185–96.

Pass, B.C. (1966) Control of the sod webworms *Crambus teterrellus, C. trisectus* and *C. mutabilis* in Kentucky. *Journal of Economic Entomology*, **59**, 19–21.

Paustian, K., Andrén, O., Clarholm, M. *et al.* (1990) Carbon and nitrogen budgets of four agro-ecosystems with annual and perennial crops, with and without N fertilization. *Journal of Applied Ecology*, **27**, 60–84.

Peachey, J.E. (1963) Studies on Enchytraeidae (Oligochaeta) of moorland soils. *Pedobiologia*, **2**, 81–95.

Peakin, G.J. and Josens, G. (1978) Respiration and energy flow, in *Production Ecology of Ants and Termites*, International Biological Programme 13, (ed. M.V. Brian), Cambridge University Press, Cambridge, pp. 111–63.

Peet, R.K. (1974) The measurement of species diversity. *Annual Review of Ecology and Systematics*, **5**, 285–307.

Perel, T.S. (1977) Differences in lumbricid organization connected with ecological properties, in *Soil Organisms as Components of Ecosystems*, (eds U. Lohm and T. Persson), *Ecological Bulletins* (Stockholm), **25**, 56–63.

Persson, T., Bååth, E., Clarholm, M. *et al.* (1980) Trophic structure, biomass dynamics and carbon metabolism of soil organisms in a Scots pine forest, in *Structure and Function of Northern Coniferous Forests – an Ecosystem Study*, (ed. T. Persson). *Ecological Bulletins* (Stockholm), **32**, 419–59.

Persson, T. and Lohm, U. (1977) Energetical significance of the annelids and arthropods in a Swedish grassland soil. *Ecological Bulletins* (Stockholm), **23**, 1–211.

Petal, J. (1967) Productivity and consumption of food in the *Myrmica laevinodis*

Nyl. population, in *Secondary Productivity of Terrestrial Ecosystems*, Vol. 2, (ed. K. Petrusewicz), Polish Academy of Sciences, Warsaw, pp. 841–57.

Petal, J. (1972) Methods of investigating the productivity of ants. *Ekologia Polska*, **20**, 9–22.

Petal, J. (1974) Analysis of a sheep pasture ecosystem in the Pieniny Mountains (the Carpathians). XV. The effect of pasture management on ant population. *Ekologia Polska*, **22**, 679–92.

Petal, J. (1976) The effect of mineral fertilization on ant populations in meadows. *Polish Ecological Studies*, **2**, 209–18.

Petal, J. (1978) The role of ants in ecosystems, in *Production Ecology of Ants and Termites*, International Biological Programme 13, (ed. M.V. Brian), Cambridge University Press, Cambridge, pp. 293–325.

Petal, J. (1980) Ant populations, their regulation and effect on soil in meadows. *Ekologia Polska*, **28**, 297–326.

Petal, J., Andrzejewska, L., Breymeyer, A. and Olechowicz, E. (1971) Productivity investigation of two types of meadows in the Vistula Valley. X. Role of ants as predators in a habitat. *Ekologia Polska*, **19**, 213–22.

Petal, J., Nowak, E., Jakubczyk, H. and Czerwiński, Z. (1977) Effects of ants and earthworms on soil habitat modification, in *Soil Organisms as Components of Ecosystems*, (eds U. Lohm and T. Persson), *Ecological Bulletins* (Stockholm), **25**, 501–3.

Petersen, H. (1978) Sex-ratios and the extent of parthenogenetic reproduction in some collembolan populations, in *First International Seminar on Apterygota, Siena, September 13–16, 1978*, (ed. R. Dallai), Academia delle Scienze di Siena dette de' Fisiocritici, pp. 19–35.

Petersen, H. and Krogh, P.H. (1987) Effects of perturbing microarthropod communities of a permanent pasture and a rye field by an insecticide and a fungicide, in *Soil Fauna and Soil Fertility*, (ed. B.R. Striganova), Nauka, Moscow, pp. 217–29.

Petersen, H. and Luxton, M. (1982) A comparative analysis of soil fauna populations and their role in decomposition processes. *Oikos*, **39**, 287–388.

Petrusewicz, K. (1967) Concepts in studies on the secondary productivity of terrestrial ecosystems, in *Secondary Productivity of Terrestrial Ecosystems*, Vol. 1, (ed. K. Petrusewicz), Polish Academy of Sciences, Warsaw, pp. 17–49.

Petrusewicz, K. and Macfadyen, A. (1970) *Productivity of Terrestrial Animals, Principles and Methods*, IBP Handbook No. 13, Blackwell, Oxford.

Phillipson, J. (1960) A contribution to the feeding biology of *Mitopus morio* (F.) (Phalangida). *Journal of Animal Ecology*, **29**, 35–43.

Piearce, T.G. (1978) Gut contents of some lumbricid earthworms. *Pedobiologia*, **18**, 153–7.

Piearce, T.G. (1981) Losses of surface fluids from lumbricid earthworms. *Pedobiologia*, **21**, 417–26.

Piearce, T.G. (1982) Recovery of earthworm populations following salt water flooding. *Pedobiologia*, **24**, 91–100.

Pielou, E.C. (1975) *Ecological Diversity*, Wiley, New York.

Pisarski, B. (1978) Comparison of various biomes, in *Production Ecology of Ants and Termites*, International Biological Programme 13, Cambridge University Press, Cambridge, pp. 326–31.

Pizl, V. (1988) Interactions between earthworms and herbicides. 1. Toxicity of some herbicides to earthworms in laboratory tests. *Pedobiologia*, **32**, 227–32.

Plumb, R.T. (1978) Invertebrates as vectors of grass viruses. *Scientific Proceedings of the Royal Dublin Society*, **6A**, 343–50.

Poinar, G.O. Jr (1979) *Nematodes for Biological Control of Insects*, CRC, Boca Raton, Florida.

Pollard, E. (1968) Hedges. III. The effect of removal of the bottom flora of a hawthorn hedgerow on the Carabidae of the hedge bottom. *Journal of Applied Ecology*, **5**, 125–39.

Poole, T.B. (1959) Studies on the food of Collembola in a Douglas fir plantation. *Proceedings of the Zoological Society of London*, **132**, 78–82.

Porres, M.A., McMurtry, J.A. and March, R.B. (1975) Investigations of leaf sap feeding by three species of phytoseiid mites by labelling with radioactive phosphoric acid (H_3^{32} PO_4). *Annals of the Entomological Society of America*, **68**, 871–2.

Port, C.M. and French, N. (1984) Damage to spring barley by larvae of *Dilophus febrilis* (L.) (Bibionidae: Diptera). *Plant Pathology*, **33**, 133–4.

Port, C.M. and Port, G.R. (1986) The biology and behaviour of slugs in relation to crop damage and control. *Agricultural Zoology Reviews*, **1**, 255–300.

Pottinger, R.P. (1976) The role of insects in modified terrestrial ecosystems. *New Zealand Entomologist*, **6**, 122–31.

Pottinger, R.P., Welsh, R.D., East, R. and Stratton, A.E. (1979) Control of *Costelytra zealandica* adults with cypermethrin (Coleoptera: Scarabaeidae), in *Proceedings of the 2nd Australasian Conference on Grassland Invertebrate Ecology*, (eds T.K. Crosby and R.P. Pottinger), Government Printer, Wellington, pp. 165–7.

Potts, G.R. and Vickerman, G.P. (1974) Studies on the cereal ecosystem. *Advances in Ecological Research*, **8**, 108–97.

Prestidge, R.A. (1982a) Instar duration, adult consumption, oviposition and nitrogen utilization efficiencies of leafhoppers feeding on different quality food (Auchenorrhyncha: Homoptera). *Ecological Entomology*, **7**, 91–101.

Prestidge, R.A. (1982b) The influence of nitrogenous fertilizer on the grassland Auchenorrhyncha (Homoptera). *Journal of Applied Ecology*, **19**, 735–49.

Prestidge, R.A., Barker, G.M. and Pottinger, R.P. (1991) A review of the effects of *Acremonium lolii* in perennial ryegrass on *Listronotus bonariensis* (Curevlionidae: Coleoptera). *Advances in Ecology*, **1**, 157–65.

Prestidge, R.A. and McNeill, S. (1983a) Auchenorrhyncha – a host plant interaction; leaf-hoppers and grasses. *Ecological Entomology*, **8**, 331–9.

Prestidge, R.A. and McNeill, S. (1983b) The role of nitrogen in the ecology of grassland Auchenorrhyncha, in *Nitrogen as an Ecological Factor*, (eds J.A. Lee, S. McNeill and I.H. Rorison), Blackwell, Oxford, pp. 257–81.

Preston, F.W. (1962) The canonical distribution of commonness and rarity, Part 1. *Ecology*, **43**, 185–215; Part II. *Ecology*, **43**, 410–32.

Pritchard, G. (1983) Biology of Tipulidae. *Annual Review of Entomology*, **28**, 1–22.

Pugh, G.J.F. and Williams, J.I. (1971) Effect of an organomercury fungicide on saprophytic fungi and on litter decomposition. *Transactions of the British Mycological Society*, **57**, 164–6.

Purvis, G. (1978) Studies on the effects of management on the arthropod communities of grassland. Ph.D. thesis, National University of Ireland, Dublin.

Purvis, G. and Bannon, J.W. (1992) Non-target effects of repeated methiocarb slug pellet application on carabid beetle (Coleoptera: Carabidae) activity in winter-sown cereals. *Annals of Applied Biology*, **121**, 401–22.

Purvis, G., Carter, N. and Powell, W. (1988) Observations on the effects of an autumn application of a pyrethroid insecticide on non-target predatory species in winter cereals, in *Integrated Crop Protection in Cereals*, (eds R. Cavalloro and K.D. Sunderland), A.A. Balkema, Rotterdam, pp. 153–66.

Purvis, G. and Curry, J.P. (1980) Successional changes in the arthropod fauna of a new ley pasture established on previously cultivated arable land. *Journal of Applied Ecology*, **17**, 309–21.

Purvis, G. and Curry, J.P. (1981) The influence of sward management on foliage arthropod communities in a ley grassland. *Journal of Applied Ecology*, **18**, 711–25.

Putman, R.J. and Wratten, S.D. (1984) *Principles of Ecology*, Croom Helm, London.

Raatikainen, M. (1967) Bionomics, enemies and population dynamics of *Javesella pellucida* (F.) (Hom. Delphacidae). *Annales Agriculturae Fenniae*, **6**, 1–149.

Rautapää, J. (1976) Population dynamics of cereal aphids and methods of predicting population trends. *Annales Agriculturae Fenniae*, **15**, 272–93.

Rautapää, J. (1977) Evaluation of predator-prey ratios using *Chrysopa carnea* Steph. in control of *Rhopalosiphum padi* (L.). *Annales Agriculturae Fenniae*, **16**, 103–9.

Raven, J.A. (1983) Phytophages of xylem and phloem: a comparison of animal and plant sap feeders. *Advances in Ecological Research*, **13**, 135–234.

Raw, F. (1956) The abundance and distribution of Protura in grassland. *Journal of Animal Ecology*, **25**, 15–21.

Raw, F. (1967) Arthropoda (except Acari and Collembola), in *Soil Biology*, (eds A. Burges and F. Raw), Academic Press, London, pp. 323–62.

Regier, H.A. and Cowell, E.B. (1972) Application of ecosystem theory, succession, diversity, stability, stress and conservation. *Biological Conservation*, **4**, 83–7.

Reichle, D.E. (1967) Radioisotope turnover and energy flow in terrestrial isopod populations. *Ecology*, **48**, 351–66.

Reichle, D.E. (1968) Relation of body size to food intake, oxygen consumption and trace element metabolism in forest floor arthropods. *Ecology*, **49**, 538–42.

Reichle, D.E. (1971) Energy and nutrient metabolism of soil and litter invertebrates, in *Productivity of Forest Ecosystems*, (ed. P. Duvigneaud), Unesco, Paris, pp. 465–77.

Reinecke, A.J. and Venter, J.M. (1985) Influence of dieldrin on the reproduction of the earthworm *Eisenia foetida* (Oligochaeta). *Biology and Fertility of Soils*, **1**, 39–44.

Reinecke, A.J. and Visser F.A. (1980) The influence of agricultural land use practices on the population densities of *Allolobophora trapezoides* and *Eisenia rosea* (Oligochaeta) in southern Africa, in *Soil Biology as Related to Land Use Practices*, (ed. D.L. Dindal), EPA, Washington, pp. 310–24.

Reynolds, J.W. (1973) Earthworm (Annelida: Oligochaeta) ecology and systematics, in *Proceedings of the 1st Soil Microcommunities Conference, Syracuse, New York, October 1971*, (ed. D.L. Dindal), Atomic Energy Commission, Office of Information Services, Technical Information Centre, Washington, DC, pp. 95–120.

Reynoldson, T.B. (1955) Observations on the earthworms of North Wales. *North Western Naturalist*, **3**, 291–304.

Rhoades, D.F. and Cates, R.G. (1976) Towards a general theory of plant antiherbivore chemistry. *Recent Advances in Phytochemistry*, **10**, 168–213.

Richards, O.W. and Waloff, N. (1954) Studies on the biology and population dynamics of British grasshoppers. *Anti-Locust Bulletin*, **17**, 182pp.

Richardson, A.M.M. (1975) Food, feeding rates and assimilation in the land snail *Cepaea nemoralis* L. *Oecologia*, **19**, 59–70.

Richter, G. (1958) Die Maikäferpopulationen im Gebiete der Deutschen Demokratischen Republik. *Nachrichtenblatt für den Deutschen Pflanzenschutz Dienst*, **12**, 21–35.

Ricou, G. (1967) Recherches sur les populations de tipules: action de certains facteurs écologiques sur *Tipula paludosa* Meig. *Annales de Épiphyties*, **18**, 451–81.

Ricou, G. (1976) La prairie permanente du nord-ouest Français le Pin-Au-Haras. *Polish Ecological Studies*, **2**, 51–66.

Ricou, G.A.E. (1979) Consumers in meadows and pastures. Pastures, in *Grassland Ecosystems of the World: Analysis of Grasslands and their Uses*, International Biological Programme 18, (ed. R.T. Coupland), Cambridge University Press, Cambridge, pp. 147–153.

Ricou, G. and Duval, E. (1969) Influence of leafhoppers upon some meadow grasses. *Zeitschrift für Angewandte Entomologie*, **63**, 163–73.

Ridsdill-Smith, T.J. (1975) Selection of living grassroots in the soil by larvae of *Sericesthis nigrolineata* (Coleoptera: Scarabaeidae). *Entomologia Experimentalis et Applicata*, **18**, 75–89.

Ridsdill-Smith, T.J. (1977) Effects of root-feeding by scarabaeid larvae on growth of perennial ryegrass plants. *Journal of Applied Ecology*, **14**, 73–80.

Ridsdill-Smith, T.J. and Matthiessen, J.N. (1988) Bush fly, *Musca vetustissima* Walker (Diptera: Muscidae), control in relation to seasonal abundance of scarabaeid dung beetles (Coleoptera: Scarabaeidae) in south-western Australia. *Bulletin of Entomological Research*, **78**, 633–9.

Riechert, S.E. (1974) Thoughts on the ecological significance of spiders. *Bioscience*, **24**, 352–6.

Riechert, S.E. and Lockley, T. (1984) Spiders as biological control agents. *Annual Review of Entomology*, **29**, 299–320.

Roberts, R.J. (1979a) Insect damage assessment and survey techniques, in *Proceedings of the 2nd Australasian Conference on Grassland Invertebrate Ecology*, (eds T.K. Crosby and R.P. Pottinger), Government Printer, Wellington, pp. 104–7.

Roberts, R.J. (1979b) The theory and use of alternating stocking rates to control pasture pests, in *Proceedings of the 2nd Australasian Conference on Grassland Invertebrate Ecology*, (eds T.K. Crosby and R.P. Pottinger), Government Printer, Wellington, pp. 179–80.

Roberts, R.J. and Morton, R. (1985) Biomass of larval Scarabaeidae (Coleoptera) in relation to grazing pressures in temperate, sown pastures. *Journal of Applied Ecology*, **22**, 863–74.

Roberts, R.J. and Ridsdill-Smith, T.J. (1979) Assessing pasture damage and losses in animal production caused by pasture insects, in *Proceedings of the 2nd Australasian Conference on Grassland Invertebrate Ecology*, Government Printer, Wellington, pp. 124–5.

Robertson, L.N. (1979) Chemical control of *Inopus rubriceps*: a review, in *Proceedings of the 2nd Australasian Conference on Grassland Invertebrate Ecology*, (eds T.K. Crosby and R.P. Pottinger), Government Printer, Wellington, pp. 167–70.

Robertson, L.N., Pottinger, R.P., Gerard, P.J. and Dixon, G.M. (1979) *Inopus rubriceps* control in pasture by grazing management (Diptera: Stratiomyidae), in *Proceedings of the 2nd Australasian Conference on Grassland Invertebrate Ecology*, (eds T.K. Crosby and R.K. Pottinger), Government Printer, Wellington, pp. 185–8.

Rodell, C.F. (1977) A grasshopper model for a grassland ecosystem. *Ecology*, **58**, 227–45.

Rodell, C.F. (1978) Simulation of a grasshopper population in a grassland ecosystem, in *Grassland Simulation Models*, (ed. G.S. Innis), Springer-Verlag, New York, pp. 127–53.

Rogäar, H. and Boswinkel, J.A. (1978) Some soil morphological effects of earthworm activity; field data and x-ray radiography. *Netherlands Journal of Agricultural Science*, **26**, 145–60.

Rogers, L.E. (1972) *The ecological effects of the western harvester ant* (Pogonomyrmex occidentalis) *in the shortgrass plains ecosystem*, US/IBP Grassland Biome Technical Report, No. 206. Colorado State University, Fort Collins, Colorado.

Rogers, L.E. (1974) Foraging activity of the western harvester ant in the shortgrass plains ecosystem. *Environmental Entomology*, **3**, 420–4.

Rogers, L., Lavigne, R. and Millers, J.L. (1972) Bioenergetics of the western harvester ant in the shortgrass prairie ecosystem. *Environmental Entomology*, **1**, 763–8.

Root, R.B. (1973) Organisation of a plant–arthropod association in simple and diverse habitats: the fauna of collards (*Brassica oleracea*). *Ecological Monographs*, **43**, 95–124.

Rorison, I.H. (1971) The use of nutrients in the control of the floristic composition of grassland, in *The Scientific Management of Animal and Plant Communities for Conservation*, (eds E. Duffey and A.S. Watt), Blackwell, Oxford, pp. 65–77.

Rosswall, T. and Paustian, K. (1984) Cycling of nitrogen in modern agricultural systems. *Plant and Soil*, **76**, 3–21.

Rundgren, S. (1975) Vertical distribution of lumbricids in southern Sweden. *Oikos*, **26**, 299–306.

Rundgren, S. (1977) Seasonality of emergence in lumbricids in southern Sweden. *Oikos*, **28**, 49–55.

Rushton, S.P. and Hassall, M. (1983a) The effects of food quality on the life history parameters of the terrestrial isopod (*Armadillidium vulgare*) (Latreille). *Oecologia*, **57**, 257–61.

Rushton, S.P. and Hassall, M. (1983b) Food and feeding rates of the terrestrial isopod *Armadillidium vulgare* (Latreille). *Oecologia*, **57**, 415–19.

Rushton, S.P. and Hassall, M. (1987) Effects of food quality on isopod population dynamics. *Functional Ecology*, **1**, 359–67.

Rushton, S.P., Luff, M.L. and Eyre, M.D. (1989) Effects of pasture improvement and management on the ground beetles and spider communities of upland grasslands. *Journal of Applied Ecology*, **26**, 489–503.

Russell, G.B., Sutherland, O.R.W., Lane, G.A. and Biggs, D.R. (1979) Is there a common factor for insect and disease resistance in pasture legumes?, in Proceedings of the 2nd *Australasian Conference on Grassland Invertebrate Ecology*, (eds T.K. Crosby and R.P. Pottinger), Government Printer, Wellington, pp. 95–7.

Ryke, P.A.J. and Loots, G.C. (1967) The composition of the micro-arthropod fauna in South African soils, in *Progress in Soil Biology*, (eds O. Graff and J.E. Satchell), Vieweg und Sohn, Braunschweig, Germany, pp. 538–45.

Ryl, B. (1980) Enchytraeid (Enchytraeidae, Oligochaeta) populations of soils of chosen crop-fields in the vicinity of Turew (Poznan Region). *Polish Ecological Studies*, **6**, 277–91.

Ryszkowski, L. (1975) Energy and matter economy of ecosystems, in *Unifying Concepts in Ecology*, (eds W.H. van Dobben and R.H. Lowe-McConnell), W. Junk, Wageningen, The Netherlands, pp. 109–26.

Salt, G. (1952) The arthropod population of the soil in some East African pastures. *Bulletin of Entomological Research*, **43**, 203–20.

Salt, G. and Hollick, F.S.J. (1944) Studies of wireworm populations. 1. A census of wireworms in pasture. *Annals of Applied Biology*, **31**, 52–64.

Salt, G., Hollick, F.S.J., Raw, F. and Brian, M.V. (1948) The arthropod population of pasture soil. *Journal of Animal Ecology*, **17**, 139–50.

Samways, M.J. (1979) Practical pest management in Brazil. *Outlook on Agriculture*, **10**, 78–84.

Sands, W.A. (1961) Foraging behaviour and feeding habits of five species of *Trinervitermes* in West Africa. *Entomologia Experimentalis et Applicata*, **4**, 277–88.

Sands, W.A. (1965a) Termite distribution in man-modified habitats in West Africa, with special reference to species segregation in the genus *Trinervitermes* (Isoptera, Termitidae, Nasutitermitinae). *Journal of Animal Ecology*, **34**, 557–71.

Sands, W.A. (1965b) Mound population movements and fluctuations in *Trinervitermes ebenerianus* Sjostedt (Isoptera, Termitidae, Nasutitermitinae). *Insectes Sociaux*, **12**, 49–58.

Sands, W.A. (1977) The role of termites in tropical agriculture. *Outlook on Agriculture*, **9**, 136–43.

Sardar, M.M.A. (1980) The abundance and trophic habits of the mesostigmata (Acari) of the soil of grazed grassland. Ph.D. thesis, University of Nottingham, Sutton Bonington.

Satchell, J.E. (1955) Some aspects of earthworm ecology, in *Soil Zoology*, (ed. D.K.McE.Kevan), Butterworths, London, pp. 180–201.

Satchell, J.E. (1967) Lumbricidae, in *Soil Biology*, (eds A. Burges and F. Raw), Academic Press, London, pp. 259–322.

Satchell, J.E. (1983) Earthworm microbiology, in *Earthworm Ecology – From Darwin to Vermiculture*, (ed. J.E. Satchell), Chapman & Hall, London, pp. 351–64.

Satchell, J.E. and Lowe, D.G. (1967) Selection of leaf litter by *Lumbricus terrestris*, in *Progress in Soil Zoology*, (eds O. Graff and J.E. Satchell), North Holland, Amsterdam, pp. 102–19.

Sayre, R.M. (1973) *Theratromyxa weberi*, an amoeba predatory on plant-parasitic nematodes. *Journal of Nematology*, **5**, 258–64.

Schmidtmann, E.T. (1985) Arthropod pests of dairy cattle, in *Livestock Entomology*, (eds R.E. Williams, R.D. Hall, A.B. Broce and P.J. Scholl), Wiley, New York, pp. 223–38.

Schnürer, J., Clarholm, M. and Rosswall, T. (1986) Fungi, bacteria and protozoa in soil from four arable cropping systems. *Biology and Fertility of Soils*, **2**, 119–26.

Schonborn, W. (1961) Untersuchungen über die Schichtung im Hypolithion. *Biologisches Zentralblatt*, **80**, 179–97.

Schowalter, T.D. (1981) Insect herbivore relationship to the state of the host plant: biotic regulation of ecosystem nutrient cycling through ecological succession. *Oikos*, **37**, 126–30.

Schumacker, A.M. and Whitford, W.G. (1976) Spatial and temporary variation in Chihuahuan desert ant faunas. *Southwestern Naturalist*, **21**, 1–8.

Schuster, R. (1956) Der anteil der oribatiden an den zersetzungsvorgängen im boden. *Zeitschrift für Morphologie und Ökologie der Tiere*, **45**, 1–33.

Schuster, R. and Schuster, I.J. (1977) Ernahrungsund fortpglanzungsbiolgische Studien an der Milbenfamilie Nanorchestidae (Acari, Trombidiformes). *Zoologischer Anzeiger*, **199**, 89–94.

Scriber, J.M. (1977) Limiting effects of low leaf-water content on the nitrogen utilization, energy budget and larval growth of *Hyalophora cecropia* (Lepidoptera: Saturniidae). *Oecologia*, **28**, 269–87.

Seastedt, T.R. (1984) The role of microarthropods in decomposition and mineralization processes. *Annual Review of Entomology*, **29**, 25–46.

Seastedt, T.R., Ramundo, R.A. and Hayes, D.C. (1988) Maximization of densities of soil animals by foliage herbivory: empirical evidence, graphical and conceptual models. *Oikos*, **51**, 243–8.

Sedlacek, J.D., Barrett, G.W. and Shaw, D.R. (1988) Effects of nutrient enrichment on the Auchenorrhyncha (Homoptera) in contrasting grassland communities. *Journal of Applied Ecology*, **25**, 537–50.

Shannon, C.E. and Weaver, W. (1949) *The Mathematical Theory of Communication*, University of Illinois Press, Urbana, Illinois.

Shantz, H.L. (1954) The place of grasslands in the earth's cover of vegetation. *Ecology*, **35**, 142–5.

Sharga, U.S. (1933) Biology and life history of *Limothrips cerealium* Haliday and *Aptinothrips rufus* Gmelin feeding on Gramineae. *Annals of Applied Biology*, **20**, 308–26.

Sharpley, A.N. and Syers, J.K. (1976) Potential role of earthworm casts for the phosphorous enrichment of run-off waters. *Soil Biology and Biochemistry*, **8**, 341–6.

Sharpley, A.N. and Syers, J.K. (1977) Seasonal variation in casting activity and in the amounts and release to solution of phosphorous forms in earthworm casts. *Soil Biology and Biochemistry*, **9**, 227–31.

Sharpley, A.N., Syers, J.K. and Springett, J.A. (1979) Effect of surface-casting earthworms on the transport of phosphorous and nitrogen in surface run off from pasture. *Soil Biology and Biochemistry*, **11**, 459–62.

Sheals, J.G. (1956) Soil population studies. 1. The effects of cultivation and treatment with insecticides. *Bulletin of Entomological Research*, **47**, 803–22.

Sheals, J.G. (1957) The Collembola and Acarina of uncultivated soil. *Journal of Animal Ecology*, **26**, 125–34.

Shipitalo, M.J. and Protz, R. (1988) Factors influencing the dispersibility of clay in worm casts. *Soil Science Society of America Journal*, **52**, 764–9.

Shires, S.W. (1985) A comparison of the effects of cypermethrin, parathion-methyl and DDT on cereal aphids, predatory beetles, earthworms and litter decomposition in spring wheat. *Crop Protection*, **4**, 177–98.

Siegel, M.R. Latch, G.L.M. and Johnson, M.C. (1987) Fungal endophytes of grasses. *Annual Review of Phytopathology*, **25**, 293–315.

Siepel, H. and van de Bund, C. (1988) The influence of management practices on the microarthropod community of grassland. *Pedobiologia*, **31**, 339–54.

Simberloff, D.S. and Wilson, E.O. (1969) Experimental zoogeography of islands: the colonization of empty islands. *Ecology*, **50**, 278–96.

Sinclair, A.R.E. (1975) The resource limitation of trophic levels in tropical grassland ecosystems. *Journal of Animal Ecology*, **44**, 497–520.

Singh, J.S. and Joshi, M.C. (1979) Primary production (in tropical grasslands), in *Grassland Ecosystems of the World, Analysis of Grasslands and their Uses*. International Biological Programme 19, (ed. R.T. Coupland), Cambridge University Press, Cambridge, pp. 197–218.

Singh, U.R. and Singh, A.K. (1975) A preliminary quantitative report of soil micro-arthropods from the floor of a Himalayan grassland biome. *International Journal of Ecological and Environmental Science*, **1**, 175–7.

Slobodkin, L.B. and Sanders, H.L. (1969) On the contribution of environmental predictability to species diversity. *Brookhaven Symposia in Biology*, **22**, 82–95.

Smith, G.D., Newhall, F. and Robinson, L.H. (1964) *Soil-Temperature Regimes: Their Characteristics and Predictability*, USDA Soil Conservation Service SCS-TP-144, 14pp.

Smith, R.L. (1966) *Ecology and Field Biology*, Harper & Row, New York.

Smolik, J.D. (1974) *Nematode studies at the Cottonwood site*, US/IBP Grassland

Biome Technical Report, No. 251, Colorado State University, Fort Collins, Colorado.

Smolik, J.D. and Lewis, J.K. (1982) Effect of range condition on density and biomass of nematodes in a mixed prairie ecosystem. *Journal of Range Management*, **35**, 657–63.

Sohlenius, B. (1980) Abundance, biomass and contribution to energy flow by soil nematodes in terrestrial ecosystems. *Oikos*, **34**, 186–94.

Sohlenius, B., Boström, S. and Sandor, A. (1987) Long-term dynamics of nematode communities in arable soil under four cropping systems. *Journal of Applied Ecology*, **24**, 131–44.

Sohlenius, B., Boström, S. and Sandor, A. (1988) Carbon and nitrogen budgets of nematodes in arable soil. *Biology and Fertility of Soils*, **6**, 1–8.

Solhöy, T. (1972) Quantitative invertebrate studies in mountain communities at Hardangervidda, South Norway. 1. *Norsk Entomologisk Tidsskrift*, **19**, 99–108.

Sørensen, T. (1948) A method of establishing groups of equivalent amplitude in plant sociology based on the similarity of species content and its application to analysis of the vegetation on Danish commons. *Biologiske Skrifter*, **5**, 1–34.

Sotherton, N.W., Moreby, S.J. and Langley, M.G. (1987) The effects of the foliar fungicide pyrazophos on beneficial arthropods in barley fields. *Annals of Applied Biology*, **111**, 75–87.

South, A. (1965) Biology and ecology of *Agriolimax reticulatus* (Müll) and other slugs: spatial distribution. *Journal of Animal Ecology*, **34**, 403–17.

Southwood, T.R.E. (1977) Habitat, the templet for ecological strategies. *Journal of Animal Ecology*, **46**, 337–65.

Southwood, T.R.E. and Jepson, W.F. (1962) The productivity of grassland in England for *Oscinella frit* and other stem-boring Diptera. *Bulletin of Entomological Research*, **53**, 395–407.

Southwood, T.R.E. and Leston, D. (1959) *Land and Water Bugs of the British Isles*, Frederick Warne, London.

Spaull, A.M., Clements, R.O. and Newton, P.G. (1983) The effect of root-ectoparasitic nematodes on grass establishment and productivity, in *Occasional Symposium* No. 14, British Grassland Society, (ed. A.J. Corrall), pp. 310–11.

Spedding, C.R.W. (1971) *Grassland Ecology*, Clarendon Press, Oxford.

Spence, D.H.N. and Angus, A. (1971) African grassland management – burning and grazing in Murchison Falls National Park, Uganda, in *The Scientific Management of Animal and Plant Communities for Conservation*, Blackwell, Oxford, pp. 311–32.

Splittstoesser, C.M. (1981) The use of pathogens in insect control, in *Handbook of Pest Management in Agriculture*, Vol. II, (ed. D. Pimentel), CRC Press, Cleveland, pp. 285–98.

Springett, J.A. (1983) Effect of five species of earthworm on some soil properties. *Journal of Applied Ecology*, **20**, 865–72.

Springett, J.A. and Syers, J.K. (1979) The effect of earthworm casts on ryegrass seedlings, in *Proceedings of the 2nd Australasian Conference on Grassland Invertebrate Ecology*, (eds T.K. Crosby and R.P. Pottinger), Government Printer, Wellington, pp. 44–7.

Standen, V. (1973) The production and respiration of an enchytraeid population in blanket bog. *Journal of Animal Ecology*, **42**, 219–45.

Standen, V. (1979) Factors affecting the distribution of lumbricids (Oligochaeta) in associations at peat and mineral sites in Northern England. *Oecologia*, **42**, 359–74.

Standen, V. (1982) Associations of Enchytraeidae (Oligochaeta) in experimentally fertilized grasslands. *Journal of Animal Ecology*, **51**, 501–22.

Stanton, N.L., Allen, M. and Campion, M. (1981) The effect of the pesticide carbofuran on soil organisms and root and shoot production in shortgrass prairie. *Journal of Applied Ecology*, **18**, 417–31.

Starling, J.H. (1944) Ecological studies on the Pauropoda of the Duke Forest. *Ecological Monographs*, **14**, 291–310.

Steinberger, Y., Freckman, D.W., Parker, L.W. and Whitford, W.G. (1984) Effects of simulated rainfall and litter quantities on desert soil biota: nematodes and microarthropods. *Pedobiologia*, **26**, 267–74.

Stevens, R.J. and Cornforth, I.S. (1974) The effect of pig slurry applied to a soil surface on the composition of the soil atmosphere. *Journal of Science, Food and Agriculture*, **25**, 1263–72.

Stewart, K.M. (1979) The life cycle of *Costelytra zealandica* (Coleoptera: Scarabaeidae), in *Proceedings of the 2nd Australasian Conference on Grassland Invertebrate Ecology*, (eds. T.K. Crosby and R.P. Pottinger), Government Printer, Wellington, pp. 282–4.

Stewart, K.M. and van Toor, R. (1983) Control of grass grub (*Costelytra zealandica*) by heavy rolling. *New Zealand Journal of Experimental Agriculture*, **11**, 265–70.

Stewart, V.I. and Scullion, J. (1988) Earthworms, soil structure and the rehabilitation of former opencast coal mining land, in *Earthworms in Waste and Environmental Management*, (eds C.A. Edwards and E.F. Neuhauser), SPB Academic Publishing, The Hague, pp. 263–72.

Stiling, P. (1988) Density-dependent processes and key factors in insect populations. *Journal of Animal Ecology*, **57**, 581–93.

Stinson, C.S.A. and Brown, V.K. (1983) Seasonal changes in the architecture of natural plant communities and its relevance to insect herbivores. *Oecologia*, **56**, 67–9.

Stockdill, S.M.J. (1959) Earthworms improve pasture growth. *The New Zealand Journal of Agriculture*, **98**, 227–33.

Stockdill, S.M.J. (1966). The effect of earthworms on pastures. *Proceedings of the New Zealand Ecological Society*, **13**, 68–75.

Stockdill, S.M.J. and Cossens, G.G. (1966) The role of earthworms in pasture production and moisture conservation. *Proceedings of the New Zealand Grassland Association*, **28**, 168–83.

Stockli, A. (1957) Die Metazoen fauna von Wiesenund Ackerböden aus der Umgebung von Zürich. *Landwirtschaftliches Jahrbüch der Schweiz*, **6**, 571–95.

Stout, J.D. (1968) The significance of the protozoan fauna in distinguishing mull and mor of beech (*Fagus silvatica* L.). *Pedobiologia*, **8**, 387–400.

Stout, J.D. (1974) Protozoa, in *Biology of Plant Litter Decomposition*, (eds C.H. Dickinson and G.J.F. Pugh), Academic Press, London, pp. 385–420.

Stout, J. and Heal, O.W. (1967) Protozoa, in *Soil Biology*, (eds A. Burges and F. Raw), Academic Press, London, pp. 149–95.

Stradling, D.J. (1978) Food and feeding habits of ants, in *Production Ecology of Ants and Termites*, International Biological Programme 13, (ed. M. V. Brian), Cambridge University Press, Cambridge, pp. 81–106.

Strahler, A.N. (1970) *Introduction to Physical Geography*, 2nd edn, Wiley, New York.

Strenzke, K. (1952) Untersuchungen über die Tiergemeinschaften des Bodens: Die Oribatiden und ihre Synusien in den Boden Norddeutschlands. *Zoologica*, **37**, 1–172.

Strickland, A.H. (1947) The soil fauna in two contrasted plots of land in Trinidad, British West Indies. *Journal of Animal Ecology*, **16**, 1–10.

Striganova, B.R. (1971) Vozrastnyye izmeneniya aktivnosti pitaniya u kivsyakov (Juloidea). *Zoolohichnyĭ Zhurnal Ukrayinȳ*, **50**, 1472–6.

Stringer, A. and Wright, M.A. (1976) The toxicity of benomyl and some related 2-substituted benzimidazoles to the earthworm *Lumbricus terrestris*. *Pesticide Science*, **7**, 459–64.

Strojan, C.L. (1978a) Forest leaf litter decomposition in the vicinity of a zinc smelter. *Oecologia*, **32**, 203–12.

Strojan, C.L. (1978b) The impact of zinc smelter emissions on forest litter arthropods. *Oikos*, **31**, 41–6.

Strong, D.R. Jr (1974) The insects of British trees: community equilibrium in ecological time. *Annals of the Missouri Botanic Gardens*, **61**, 692–701.

Strong, D.R. Jr, Lawton, J.H. and Southwood, T.R.E. (1984) *Insects on Plants: Community Patterns and Mechanisms*, Blackwell, Oxford.

Sturm, H. (1959) Die Nahrung der Proturen. *Naturwissenschaften*, **46**, 90–1.

Subagja, J. and Snider, R.J. (1981) The side effects of the herbicides atrazine and paraquat upon *Folsomia candida* and *Tullbergia granulata* (Insecta, Collembola). *Pedobiologia*, **22**, 141–52.

Sunderland, K.D. (1975) The diet of some predatory arthropods in cereal crops. *Journal of Applied Ecology*, **12**, 507–15.

Sunderland, K.D., Fraser, A.M. and Dixon, F.G. (1986) Field and laboratory studies on money spiders (Linyphiidae) as predators of cereal aphids. *Journal of Applied Ecology*, **23**, 433–47.

Sunderland, K.D. and Vickerman, G.P. (1980) Aphid feeding by some polyphagous predators in relation to aphid density in cereal fields. *Journal of Applied Ecology*, **17**, 389–96.

Suski, Z.W. (1972) Tarsonemid mites on apple trees in Poland. X. Laboratory studies on the biology of certain mite species of the family Tarsonemidae (Acarina, Heterostigmata). *Zeszyty Problemowe Postepow Nauk Rolniczych*, **129**, 111–37.

Suski, Z.W. (1973) A revision of *Siteroptes cerealium* (Kirchner) complex (Acarina, Heterostigmata, Pyemotidae). *Annales Zoologie* (Warsaw), **30**, 510–35.

Sutton, S.L. (1972) *Woodlice*, Ginn, London.

Svendsen, J.A. (1957a) The distribution of Lumbricidae in an area of Pennine moorland (Moor House Nature Reserve). *Journal of Animal Ecology*, **26**, 411–21.

Svendsen, J.A. (1957b) The behaviour of lumbricids under moorland conditions. *Journal of Animal Ecology*, **26**, 423–39.

Svensson, B.H., Boström, U.-L. and Klemedtson, L. (1986) Potential for higher rates of denitrification in earthworm casts than in the surrounding soil. *Biology and Fertility of Soils*, **2**, 147–9.

Swift, M.J., Heal, O.W. and Anderson, J.M. (1979) *Decomposition in Terrestrial Ecosystems*, Blackwell, London.

Syers, J.K. and Springett, J.A. (1983) Earthworm ecology in grassland soils, in *Earthworm Ecology – From Darwin to Vermiculture*, (ed. J.E. Satchell), Chapman & Hall, London, pp. 67–83.

Syers, J.K. and Springett, J.A. (1984) Earthworms and soil fertility. *Plant and Soil*, **76**, 93–104.

Talbot, M. (1953) Ants of an old field community of the Edwin S. George Reserve, Livingston Country, Michigan. *Contributions from the Laboratory of Vertebrate Biology of the University of Michigan*, **63**, 1–13.

Tanaka, M. (1970) Ecological studies on communities of soil Collembola in Mt Soba, Southwest Japan. *Japanese Journal of Ecology*, **20**, 102–10.

Tansky, V.I. (1961) The formation of the thrips fauna (Thysanoptera) on wheat crops in the new soil of Northern Kazakhstan [in Russian]. *Entomologicheskoe Obozrenie*, **40**, 785–93.

Tansley, A.G. (1939) *The British Islands and their Vegetation*, Cambridge University Press, Cambridge.

Tauber, M.J., Tauber, C.A. and Masaki, S. (1986) *Seasonal Adaptations of Insects*, Oxford University Press, New York.

Taylor, J.W. (1907) *Monograph of the Land and Freshwater Mollusca of the British Isles; Testacellidae, Limacidae, Arionidae*, Taylor Brothers, Leeds.

Temirov, T. and Valiakhmedov, B. (1988) Influence of earthworms on fertility of high altitude desert soil in Tajikistan. *Pedobiologia*, **32**, 293–300.

Ter Braak, C.J.F. (1986) Canonical correspondence analysis: a new eigenvector technique for multivariate direct gradient analysis. *Ecology*, **67**, 1167–79.

Tevis, L. Jr (1958) Interrelations between the harvester ant, *Veromessor pergandei* (Mayr) and some desert ephemerals. *Ecology*, **39**, 695–704.

Thambi, A.V. and Dash, M.C. (1973) Seasonal variation in numbers and biomass of Enchytraeidae (Oligochaeta) populations in tropical grassland soils from India. *Tropical Ecology*, **14**, 228–37.

Thiele, H.U. (1977) *Carabid Beetles in their Environment*, Springer-Verlag, New York.

Thomas, D.C. (1944) Discussion on slugs. II. Field sampling for slugs. *Annals of Applied Biology*, **31**, 163–4.

Thompson, A.R. and Gore, F.L. (1972) Toxicity of twenty-nine insecticides to *Folsomia candida*: laboratory studies. *Journal of Economic Entomology*, **65**, 1255–60.

Tischler, W. (1955) Effect of agricultural practice on the soil fauna, in *Soil Zoology*, (ed. D.K.McE. Kevan), Butterworths, London, pp. 215–30.

Tischler, W. (1965) *Agrarökologie*, Fischer Verlag, Jena.

Törmälä, T. (1979) Numbers and biomass of soil invertebrates in a reserved field in Central Finland. *Journal of the Scientific Agricultural Society of Finland*, **51**, 172–87.

Törmälä, T. (1982) Structure and dynamics of reserved field ecosystem in Central Finland. *Biological Research Reports of the University of Jyväskylä*, **8**, 1–58.

Törmälä, T. and Raatikainen, N. (1976) Primary production and seasonal dynamics of the flora and fauna of the field stratum in a reserved field in Middle Finland. *Journal of the Scientific Agricultural Society of Finland*, **48**, 363–85.

Traczyk, T., Traczyk, H. and Pasternak, D. (1976) The influence of intensive mineral fertilization on the yield and floral composition of meadows. *Polish Ecological Studies*, **2**, 39–47.

Trofymow, J.A. and Coleman, D.C. (1982) The role of bacterivorous and fungivorous nematodes in cellulose and chitin decomposition, in *Nematodes in Soil Ecosystems*, (ed. D.W. Freckman), University of Texas Press, Austin, pp. 117–37.

Turnbull, A.L. (1966) A population of spiders and their potential prey in an overgrazed pasture in eastern Ontario. *Canadian Journal of Zoology*, **44**, 557–83.

Turner, J.W. and Franzmann, B.A. (1979) Development of management programmes for lucerne aphids (Hemiptera: Aphididae), in *Proceedings of the 2nd Australasian Conference on Grassland Invertebrate Ecology*, (eds T.K. Crosby and R.P. Pottinger), Government Printer, Wellington, pp. 191–2.

Tyler, G. (1975a) Heavy metal pollution and mineralisation of nitrogen in forest soils. *Nature* (London), **255**, 701–2.

Tyler, G. (1975b) Effect of heavy metal pollution on decomposition and mineralisation rates in forest soils. *Proceedings of the International Conference on Heavy Metals in the Environment*, Toronto, pp. 217–26.

Unwin, R.J. and Lewis, S. (1986) The effect upon earthworm populations of very large applications of pig slurry to grassland. *Agricultural Wastes*, **16**, 67–73.

Usher, M.B. (1979) Natural communities of plants and animals in disused quarries. *Journal of Environmental Management*, **8**, 223–36.

Usher, M.B. (1985) Population and community dynamics in the soil ecosystem, in *Ecological Interactions in Soil: Plants, Microbes and Animals*, (eds A.H.Fitter, D. Atkinson, D.J. Read and M.B. Usher), Blackwell, Oxford, pp. 243–65.

Usher, M.B., Davis, P.R., Harris, J.R.W. and Longstaff, B.C. (1979) A profusion of species? Approaches towards understanding the dynamics of the populations of the micro-arthropods in decomposer communities, in *Population Dynamics*, (eds R.M. Anderson, B.D. Turner and L.R. Taylor), Blackwell, Oxford, pp. 359–84.

Usher, M.B. and Edwards, M. (1984) The terrestrial arthropods of the grass sward of Lynch Island, a specially protected area in Antarctica. *Oecologia*, **63**, 143–4.

Uvarov, P.B. (1928) *Locusts and Grasshoppers: A Handbook for their Study and Control*, The Imperial Bureau of Entomology, London.

Valiela, I. (1974) Composition, food webs and population limitation in dung arthropod communities during invasion and succession. *American Midland Naturalist*, **92**, 370–85.

van Amelsvoort, P.A.M., van Dongen, M. and van der Werff, P.A. (1988) The impact of Collembola on humification and mineralization of soil organic matter. *Pedobiologia*, **31**, 103–11.

van der Drift, J. (1962) The soil animals in an oak-wood with different types of humus formation, in *Progress in Soil Zoology*, (ed. P.W. Murphy), Butterworths, London, pp. 343–7.

van der Drift, J.A. (1975) The significance of the millipede *Glomeris marginata* for oak-litter decomposition and an approach of its part in energy flow, in *Progress in Soil Zoology*, (ed. J. Vanek), Academia, Prague, pp. 293–8.

van der Drift, J. and Witkamp, M. (1960) The significance of the breakdown of oak litter by *Enoicyla pusilla*. *Archives Neerlandaisses de Zoologie*, **13**, 489–92.

van der Maarel, E. (1971) Plant species diversity in relation to management, in *The Scientific Management of Animal and Plant Communities for Conservation*, (eds E. Duffey and A.S. Watt), Blackwell, Oxford, pp. 45–64.

Van Dyne, G.M., Brockington, N.R., Szocs, Z., Duek, J. and Ribic, C.A. (1980) Large herbivore subsystem, in *Grasslands, Systems Analysis and Man*, International Biological Programme 19, (eds A.I. Breymeyer and G.M. Van Dyne), Cambridge University Press, Cambridge, pp. 269–537.

van Emden, H.F., Eastop, V.F., Hughes, R.D. and Way, M.J. (1969) The ecology of *Myzus persicae*. *Annual Review of Entomology*, **14**, 197–270.

van Emden, H.F., and Way, M.J. (1973) Host plants in the population dynamics of insects, in *Insect/Plant Relationships*, (ed. H.F. van Emden), Blackwell, Oxford, pp. 181–99.

Van Hook, R.I. Jr (1971) Energy and nutrient dynamics of spider and orthopteran populations in a grassland ecosystem. *Ecological Monographs*, **41**, 1–26.

Vannier, G. (1970) *Reactions des Microarthropodes aux Variations de l'État hydrique du Sol: Techniques relative a l'Extraction des Arthropodes du Sol*. Editions Centre National de la Recherche Scientifique, Paris.

van Rhee, J.A. (1963) Earthworm activities and the breakdown of organic matter in agricultural soils, in *Soil Organisms*, (eds J. Doeksen and J. van der Drift), North Holland, Amsterdam, pp. 54–9.

van Rhee, J.A. (1965) Earthworm activity and plant growth in artificial cultures. *Plant and Soil*, **22**, 45–8.

van Rhee, J.A. (1969a) Inoculation of earthworms in a newly drained polder. *Pedobiologia*, **9**, 128–32.

van Rhee, J.A. (1969b) Development of earthworm populations in polder soils. *Pedobiologia*, **9**, 133–40.

van Rhee, J.A. (1975) Copper contamination effects on earthworms by disposal of pig waste in pastures, in *Progress in Soil Zoology*, (ed. J.A.Vanek), Academia, Prague, 451–7.

van Rhee, J.A. (1977) A study of the effect of earthworms on orchard productivity. *Pedobiologia*, **17**, 107–14.

Varley, G.C. and Gradwell, G.R. (1960) Key factors in population studies. *Journal of Animal Ecology*, **29**, 399–401.

Venter, J.M. and Reinecke, A.J. (1985) Dieldrin and growth and development of the earthworm, *Eisenia fetida* (Oligochaeta). *Bulletin of Environmental Contamination and Toxicology*, **35**, 652–9.

Verhoef, H.A. and Brussaard, L. (1990) Decomposition and nitrogen mineralization in natural and agroecosystems: the contribution of soil animals. *Biogeochemistry*, **11**, 175–211.

Verhoeff, K.W. (1934) Symphyla and Pauropoda, in *Klassen und Ordnungen des Tierreiches*, V/11/111/1–2 Leiferung, (ed. H.G. Bronn),

Vernon, J.D.R., Findlay, D.C. and Lyons, C.H. (1981) Some observations on earthworm populations in grassland soils. *Pedobiologia*, **21**, 446–9.

Vickerman, G.P. (1978) The arthropod fauna of undersown grass and cereal fields. *Scientific Proceedings of the Royal Dublin Society*, **6A**, 273–83.

Vickerman, G.P. (1980) The phenology of *Oscinella* spp. (Diptera: Chloropidae). *Bulletin of Entomological Research*, **70**, 601–20.

Vickerman, G.P. and Wratten, S.D. (1979) The biology and pest status of cereal aphids (Hemiptera: Aphididae) in Europe: a review. *Bulletin of Entomological Research*, **69**, 1–32.

von Oettingen, H. (1942) Die Thysanopteren des Norddeutschen Graslands. *Entomologische Beihefte-Berlin Dahlem*, **9**, 79–141.

Vossbrinck, C.R., Coleman, D.C. and Woolley, T.A. (1979) Abiotic and biotic factors in litter decomposition in a semiarid grassland. *Ecology*, **60**, 265–71.

Wall, R. and Strong, L. (1987) Environmental consequences of treating cattle with the antiparasitic drug ivermectin. *Nature* (London), **327**, 418–21.

Wallace, H.R. (1963) *The Biology of Plant Parasitic Nematodes*, Edward Arnold, London.

Wallace, M.M.H. (1954) The effect of DDT and BHC on the population of the lucerne flea *Sminthurus viridis* (L.) (Collembola) and its control by predatory mites, *Biscirus* spp. (Bdellidae). *Australian Journal of Agricultural Research*, **5**, 148–55.

Wallace, M.M.H. (1967) The ecology of *Sminthurus viridis* (L.) (Collembola). 1. Processes influencing numbers in pastures in Western Australia. *Australian Journal of Zoology*, **15**, 1173–206.

Wallace, M.M.H. (1974) An attempt to extend the biological control of *Sminthurus viridis* (Collembola) to new areas in Australia by introducing a predatory mite, *Neomolgus capillatus* (Bdellidae). *Australian Journal of Zoology*, **22**, 519–29.

Wallace, M.M.H. and Walters, M.C. (1974) The introduction of *Bdellodes lapidaria* (Acari: Bdellidae) from Australia into South Africa for the biological control of *Sminthurus viridis* (Collembola). *Australian Journal of Zoology*, **22**, 505–17.

Wallwork, J.A. (1967) Acari, in *Soil Biology*, (eds A. Burges and F. Raw), Academic Press, London, pp. 363–95.

Wallwork, J.A. (1970) *Ecology of Soil Animals*, McGraw–Hill, London.

Wallwork, J.A. and Rodriguez, J.G. (1961) Ecological studies on oribatid mites with particular reference to their role as intermediate hosts of anoplocephalid cestodes. *Journal of Economic Entomology*, **54**, 701–5.

Waloff, N. (1975) The parasitoids of the nymphal and adult stage of leafhoppers (Auchenorrhyncha: Homoptera) of acid grassland. *Transactions of the Royal Entomological Society of London*, **126**, 637–86.

Waloff, N. (1980) Studies on grassland leafhoppers (Auchenorrhyncha, Homoptera) and their natural enemies. *Advances in Ecological Research*, **11**, 81–215.

Waloff, N. and Blackith, R.E. (1962) The growth and distribution of the mounds of *Lasius flavus* (F.) (Hymenoptera: Formicidae) in Silwood Park, Berkshire. *Journal of Animal Ecology*, **31**, 421–37.

Waloff, N. and Solomon, M.G. (1973) Leafhoppers of acidic grassland. *Journal of Applied Ecology*, **10**, 189–212.

Waloff, N. and Thompson, P. (1980) Census data of populations of some leafhoppers (Auchenorrhyncha, Homoptera) of acidic grassland. *Journal of Animal Ecology*, **49**, 395–416.

Walter, D.E. (1987a) Trophic behaviour of 'mycophagous' microarthropods. *Ecology*, **68**, 226–9.

Walter, D.E. (1987b) Belowground arthropods of semiarid grasslands, in *Integrated Pest Management on Rangeland*, (ed. J.L. Capinera), Westview Press, Boulder, Colorado, pp. 271–90.

Walter, D.E., Hudgens, R.A. and Freckman, D.W. (1986) Consumption of nematodes by fungivorous mites, *Tyrophagus* spp. (Acarina: Astigmata: Acaridae). *Oecologia*, **70**, 357–61.

Walter, D.E., Hunt, H.W. and Elliott, E.T. (1987) The influence of prey type on the development and reproduction of some predatory soil mites. *Pedobiologia*, **30**, 419–24.

Walter, D.E. and Kaplan, D.T. (1990) Feeding observations on two astigmatic mites, *Schwiebia rocketti* (Acaridae) and *Histiostoma bakeri* (Histiostomatidae) associated with *Citrus* feeder roots. *Pedobiologia*, **34**, 281–6.

Walter, D.E., Moore, J.C. and Loring, S.J. (1989) *Symphylella* sp. (Symphyla: Scolopendrellidae) predators of arthropods and nematodes in grassland soils. *Pedobiologia*, **33**, 113–16.

Walters, M.C. (1964) A study of *Sminthurus viridis* (L.) (Collembola) in the Western Cape Province. *Department of Agriculture Technical Services Entomology Memoirs*, No.16, Pretoria, pp. 1–99.

Wasilewska, L. (1974) Analysis of a sheep pasture ecosystem in the Pieniny mountains (the Carpathians). XIII. Quantitative distribution, respiratory metabolism and some suggestions on predators of nematodes. *Ekologia Polska*, **22**, 651–8.

Wasilewska, L. (1976) The role of nematodes in the ecosystem of a meadow in Warsaw environs. *Polish Ecological Studies*, **2**, 137–56.

Wasilewska, L. and Paplińska, E. (1975) Energy flow through the nematode community in a rye-crop in the region of Poznań. *Polish Ecological Studies*, **1**, 75–82.

Waterhouse, D.F. (1974) The biological control of dung. *Scientific American*, **230** (4), 100–9.

Waterhouse, D.F. (1979) Pasture pests and biological control in Australia, in *Proceedings of the 2nd Australasian Conference on Grassland Invertebrate Ecology*, (eds T.K. Crosby and R.P. Pottinger), Government Printer, Wellington, pp. 12–16.

Waterhouse, F.L. (1955) Microclimatological profiles in grass cover in relation to biological problems. *Quarterly Journal of the Royal Meteorological Society*, **81**, 63–71.

Waters, R.A.S. (1951) Earthworms and the fertility of pasture. *Proceedings of the New Zealand Grassland Association*, **13**, 168–75.

Waters, R.A.S. (1955) Numbers and weights of earthworms under a highly productive pasture. *New Zealand Journal of Science and Technology*, **36**, 516–25.

Watson, A.P., van Hook, R.I., Jackson, D.R. and Reichle, D.E. (1976) *Impact of the Lead Mining-Smelting Complex on the Forest Floor Litter Arthropod Fauna in the New Lead Belt Region of Southwest Missouri*, ORNL/NSF/EATC-30, Oak Ridge National Laboratory, Tennessee.

Watson, R.N. (1979) Dispersal and distribution of *Heteronychus arator* in New Zealand (Coleoptera: Scarabaeidae), in *Proceedings of the 2nd Australasian Conference on Grassland Invertebrate Ecology*, (eds T.K. Crosby and R.P. Pottinger), Government Printer, Wellington, pp. 149–52.

Watson, R.N. and Wrenn, N.R. (1980) An association between *Paspalum dilatatum* and black beetle in pasture. *Proceedings of the New Zealand Grassland Association*, **41**, 96–104.

Weaver, C.R. and Hibbs, J.W. (1952) Effect of spittle bug infestation on nutritive value of alfalfa and red clover. *Journal of Economic Entomology*, **45**, 626–8.

Weber, N.A. (1966) Fungus growing ants. *Science*, **153**, 587–604.

Weis-Fogh, T. (1948) Ecological investigations of mites and collemboles in soil. *Natura Jutlandica*, **1**, 135–270.

Wells, T.C.E. (1971) A comparison of the effects of sheep grazing and mechanical cutting on the structure and botanical composition of chalk grassland, in *The Scientific Management of Animal and Plant Communities for Conservation*, (eds E. Duffey and A.S. Watt), Blackwell, Oxford, pp. 497–515.

Went, J.A. (1963) Influence of earthworms on the number of bacteria in the soil, in *Soil Organisms*, (eds J. Doeksen and J. van der Drift), North Holland, Amsterdam, pp. 260–5.

Werner, F.G. (1973) Foraging activity of the leaf-cutter ant *Acromyrmex versicolor*, in relation to season, weather and colony condition. *US/IBP Desert Biome Research Memorandum RM 73–28, Report 3*.

Wetzel, T. (1964) Untersuchungen zum Auftreten, zur Schadwirkung und zur Bekämpfung von Thysanopteren in Grassamenbeständen. *Beiträge zur Entomologie*, **14**, 427–500.

Wheeler, A.G. (1973) Studies on the arthropod fauna of alfalfa. IV. Species associated with the crown. *Canadian Entomologist*, **105**, 353–66.

Whelan, J. (1976) A comparative study of the acarine fauna of permanent pasture and new leys on cutaway raised bog at Lullymore, Co. Kildare. Ph.D. thesis, National University of Ireland, Dublin.

Whelan, J. (1978) Acarine succession in grassland on cutaway raised bog. *Scientific Proceedings of the Royal Dublin Society*, **6A**, 175–83.

Whelan, J. (1985) Seasonal fluctuations and vertical distribution of the acarine fauna of three grassland sites. *Pedobiologia*, **28**, 191–201.

White, E.G. (1974) Grazing pressures of grasshoppers in alpine tussock grassland. *New Zealand Journal of Agricultural Research*, **17**, 357–72.

White, E.G. (1979a) Energy flow efficiencies in New Zealand grasshoppers (Orthoptera), in *Proceedings of the 2nd Australasian Conference on Grassland Invertebrate Ecology*, (eds T.K. Crosby and R.P. Pottinger), Government Printer, Wellington, pp. 38–40.

White, E.G. (1979b) Modelling the interactive effects of insects and stock animals on herbage production, in *Proceedings of the 2nd Australasian Conference on Grassland Invertebrate Ecology*, (eds T.K. Crosby and R.P. Pottinger), Government Printer, Wellington, pp. 240–2.

White, T.C.R. (1976) Weather, food and plagues of locusts. *Oecologia*, **22**, 119–34.

White, T.C.R. (1978) The importance of a relative shortage of food in animal ecology. *Oecologia*, **33**, 71–86.

Whitford, W.G. (1978) Foraging by seed-harvesting ants, in *Production Ecology of Ants and Termites*, International Biological Programme 13, (ed. M.V. Brian), Cambridge University Press, Cambridge, pp. 107–10.

Whitham, T.G. (1983) Host manipulation of parasites: within-plant variation as a defence against rapidly evolving pests, in *Variable Plants and Herbivores in Natural and Managed Systems*, (eds R.F. Denno and M.S. McClue), Academic Press, New York, pp. 15–41.

Whittaker, J.B. (1965) The distribution and population dynamics of *Neophilaenus lineatus* (L.) and *N. exclamationis* (Thun.) (Homoptera: Cercopidae) on Pennine moorland. *Journal of Animal Ecology*, **34**, 277–97.

Whittaker, R.H. (1972) Evolution and measurement of species diversity. *Taxon*, **21**, 213–51.

Whittaker, R.H. (1975) *Communities and Ecosystems*, 2nd edn, Macmillan, London.

Whittaker, R.H. (1978) Approaches to classifying vegetation, in *Classification of Plant Communities*, (ed. R.H. Whittaker), Junk, The Hague, pp. 1–31.

Whittaker, R.H. and Likens, G.E. (1975) The biosphere and man, in *Primary Productivity of the Biosphere*, (eds H. Lieth and R.H. Whittaker), Springer-Verlag, New York, pp. 305–28.

Wiegert, R.G. (1964) Population energetics of meadow spittlebugs (*Philaenus spumarius* L.) as affected by migration and habitat. *Ecological Monographs*, **34**, 217–41.

Wiegert, R.G. (1965) Energy dynamics of the grasshopper populations in old field and alfalfa field ecosystems. *Oikos*, **16**, 161–76.

Wiegert, R.G. and Evans, F.C. (1967) Investigations of secondary productivity in grasslands, in *Secondary Productivity of Terrestrial Ecosystems*, Vol. 2, (ed. K. Petrusewicz), Polish Academy of Sciences, Warsaw, pp. 499–518.

Wiegert, R.G. and Petersen, C.E. (1983) Energy transfer in insects. *Annual Review of Entomology*, **28**, 455–86.

Wieser, W. (1968) Aspects of nutrition and the metabolism of copper in isopods. *American Zoologist*, **8**, 495–506.

Wieser, W. (1978) Consumer strategies of terrestrial gastropods and isopods. *Oecologia*, **36**, 191–201.

Wightman, J.A. (1979) *Costelytra zealandica* and its environment (Coleoptera: Scarabaeidae), in *Proceedings of the 2nd Australasian Conference on Grassland Invertebrate Ecology*, (eds T.K. Crosby and R.P. Pottinger), Government Printer, Wellington, pp. 70–4.

Wightman, J.A. and Whitford, D.N.J. (1979) Energetics of *Costelytra zealandica* (Coleoptera: Scarabaeidae), in *Proceedings of the 2nd Australasian Conference on Grassland Invertebrate Ecology*, (eds T.K. Crosby and R.P. Pottinger), Government Printer, Wellington, pp. 32–4.

Wigley, P. and Chilcott, C. (1990) *Bacillus thuringiensis* isolates active against the New Zealand pasture pest, *Costelytra zealandica*, in *Proceedings of the Vth International Colloquium on Invertebrate Pathology and Microbial Control*, Adelaide, Australia (ed. D.E. Pinnock), p. 344.

Wilkinson, W. (1977) Effects of direct drilling on soil microarthropods. *Annals of Applied Biology*, **87**, 520.

Willard, J.R. (1974) Soil invertebrates. VIII. A summary of populations and biomass. *Matador Project Technical Report, No 56*, Saskatoon, Saskachewan.

Williams, C.B. (1964) *Patterns in the Balance of Nature and Related Problems in Quantitative Ecology*, Academic Press, London.

Williamson, P. and Evans, P.R. (1973) A preliminary study of the effects of high levels of inorganic lead on soil fauna. *Pedobiologia*, **13**, 16–21.

Williamson, P. and Cameron, R.A.D. (1976) Natural diet of the landsnail *Cepaea nemoralis. Oikos*, **27**, 493–500.

Wint, G.R.W. (1983) The effect of foliar nutrients upon the growth and feeding of a lepidopteran larva, in *Nitrogen as an Ecological Factor*, (eds J.A. Lee, S. McNeill and I.H. Rorison), Blackwell, Oxford, pp. 301–20.

Wood, T.G. (1967a) Acari and Collembola of moorland soils from Yorkshire, England. I. Description of the sites and their populations. *Oikos*, **18**, 102–17.

Wood, T.G. (1967b) Acari and Collembola of moorland soils from Yorkshire, England. II. Vertical distribution in four grassland soils. *Oikos*, **18**, 137–40.

Wood, T.G. (1967c) Acari and Collembola of moorland soils from Yorkshire, England. III. The microarthropod communities. *Oikos*, **18**, 277–92.

Wood, T.G. (1978) Food and feeding habits of termites, in *Production Ecology of Ants and Termites*, International Biological Programme 13, (ed. M.V. Brian), Cambridge University Press, Cambridge, pp. 55–80.

Wood, T.G., Johnson, R.A. and Anderson, J.M. (1983) Modification of soils in Nigerian savanna by soil-feeding *Cubitermes* (Isoptera, Termitidae). *Soil Biology and Biochemistry*, **15**, 575–9.

Wood, T.G. and Sands, W.A. (1978) The role of termites in ecosystems, in *Production Ecology of Ants and Termites*, International Biological Programme 13, (ed. M.V. Brian), Cambridge University Press, Cambridge, pp. 245–92.

Woods, L.E., Cole, C.V., Elliott, E.T. *et al.* Nitrogen transformations in soil as affected by bacterial–microfaunal interactions. *Soil Biology and Biochemistry*, **14**, 93–8.

Woodmansee, R.G. and Wallach, L.S. (1981) Effects of fire regimes on biogeochemical cycles, in *Terrestrial Nitrogen Cycles*, (eds F.E. Clark and T. Rosswall), *Ecological Bulletins* (Stockholm), **33**, 649–69.

Wright, M.A. (1972) Factors governing ingestion by the earthworm *Lumbricus terrestris* (L.) with special reference to apple leaves. *Annals of Applied Biology*, **70**, 175–88.

Wright, M.A. (1977) Effects of benomyl and some other systemic fungicides on earthworms. *Annals of Applied Biology*, **87**, 520–4.

Wright, R.E. (1985) Arthropod pests of beef cattle on pasture or range land, in *Livestock Entomology*, (eds R.E. Williams, R.D. Hall, A.B. Broce and P.J. School), Wiley, New York, pp. 191–206.

Wright, S.J.L., Redhead, K. and Maudsley, H. (1981) *Acanthamoeba castellanii*, a predator of cyanobacteria. *Journal of General Microbiology*, **125**, 293–300.

Yeates, G.W. (1974) Studies on a climosequence of soils in tussock-grassland. 2. Nematodes. *New Zealand Journal of Zoology*, **1**, 171–7.

Yeates, G.W. (1979) Soil nematodes in terrestrial ecosystems. *Journal of Nematology*, **11**, 213–29.

Zajonc, I. (1975) Variations in meadow associations of earthworms caused by the influence of nitrogen fertilizers and liquid-manure irrigation, in *Progress in Soil Zoology*, (ed. J. Vanek), W. Junk, The Hague, pp. 497–503.

Zajonc, I. (1982) Communities of earthworms (Lumbricidae: Oligochaeta) in meadows of the Slovakian Carpathians. *Pedobiologia*, **23**, 209–16.

Zicsi, A. (1969) Über die Auswirkung der Nachfrucht und Bodenbearbeitung auf die Aktivität der Regenwurmer. *Pedobiologia*, **9**, 141–5.

Zlotin, R.I. (1970) *Structure and Productivity of Biogeocoenoses of the Tien Shan Mountains*, Nauka, Moscow [in Russian].

Zlotin, R.I. (1971) Invertebrate animals as a factor of the biological turnover, in *IV Colloque International de la Fauna du Sol*, Dijon 1970, Institut National de la Recherche Agronomique, Paris, pp. 455–62.

Zlotin, R.I. and Khodashova, K.S. (1974) *The Role of Animals in Biological Turnover in Forest-Steppe Ecosystems*, Nauka, Moscow. English translation (ed. N.R. French), 1980, Dowden, Hutchinson and Ross Inc., Stroudsburg, PA.

Zyromska-Rudzka, H. (1976) The effect of mineral fertilization of a meadow on the oribatid mites and other soil mezofauna. *Polish Ecological Studies*, **2**, 157–82.

Index

Page numbers in **bold** type refer to figures, and those in *italics* refer to tables.